From Myth to the Modern Mind:
A Study of the Origins and Growth of Scientific Thought

American University Studies

Series V
Philosophy

Vol. 12

PETER LANG
New York · Berne · Frankfurt am Main

Richard H. Schlagel

From Myth to the Modern Mind

A Study of the Origins and Growth of Scientific Thought

Volume I
Animism to Archimedes

PETER LANG
New York · Berne · Frankfurt am Main

Library of Congress Catologing in Publication Data

Schlagel, Richard H., 1925–
  From Myth to the Modern Mind.

  (American University Studies.   Series V, Philosophy;
vol. 12-   )
  Includes bibliographies and index.
  Contents: v. 1. Animism to Archimedes.
  1. Science, Ancient.   2. Science – Philosophy –
History.   I. Title.   II. Series: American University
Studies.   Series V, Philosophy; v. 12, etc.
Q124.95.S35   1985   509'.3   84-23361
ISBN 0-8204-0219-2 (v. 1)
ISSN 0739-6392

CIP-Kurztitelaufnahme der Deutschen Bibliothek

Schlagel, Richard H.:
From Myth to the Modern Mind: A Study of the Origins
and Growth of Scientific Thought / Richard H.
Schlagel. – New York; Berne; Frankfurt am Main:
Lang.
  (American University Studies: Ser. 5,
  Philosophy; Vol. 12)

NE: American University Studies / 05

Vol. 1. Animism to Archimedes. – 1985.
  ISBN Q-8204-0219-2

© Peter Lang Publishing, Inc., New York 1985

Printed by Lang Druck, Inc., Liebefeld/Berne (Switzerland)

TABLE OF CONTENTS

PREFACE

There is no quest as fundamental nor more fascinating than the attempt to ascertain how man has come to think about the world as he has — once one discards the erroneous supposition that everyone has always conceptualized the world in the same way. How from primitive myths, anthropocentric concepts and theogonic descriptions has mankind arrived at its present mode of scientific inquiry and manner of explanation, utilizing experimentation, mathematical extrapolation, and theory construction? How and why did a scientific way of thinking about the universe first arise and then develop into its current form of methodology and web of beliefs? Are these techniques themselves necessary or final?

Such are the questions this study tries to answer, the present volume describing the origins of scientific thought among the Presocratic philosophers in 6th century Greece and tracing its development to the famous scientists and mathematicians of the School of Alexandria in the Hellenistic period. Although this attempt to evaluate the contribution of these early thinkers to the formation of scientific rationalism, and particularly to discover the nature of the underlying cognitive processes that produced this contribution, will be considerably more conjectural and controversial than the usual philological or classicist study, it should prove to be of more interest to historians and philosophers of science, as well as to anyone curious about man's endeavors to comprehend the universe.

Research on the manuscript was begun during my first sabbatical leave spent in Paris in the academic year 1962-63, but actual writing did not commence until my second sabbatical leave, also passed in Paris, in 1969-70. A first draft of the manuscript was completed in the summer of 1975, the present version, extensively revised and considerably reduced in size, was completed in 1980, which explains why there are no references to publications beyond that date at which time it was accepted by the publishers. The fact that it appears with such inordinate delay is due entirely to internal problems on their part.

I want to thank The George Washington University for two full-year sabbatical leaves, funds for secretarial help, a summer course reduction, and supporting grants from the Columbian College of Arts and Sciences and from The Graduate School of Arts and Sciences without which the book would not have been published. In addition, I owe a special debt of gratitude to Dr. Gerald Donaldson, a scholar in Greek philosophy and a former colleague in our Department, for allowing me to participate in his Seminar on Aristotle and for his helpful comments on several chapters of the book.

Finally, I wish to express my gratitude to my wife, who typed the final manuscript, helped with the proofreading, and whose editorial and stylistic suggestions were invaluable. Also, at times of discouragement, I was sustained by her expressed conviction that, notwithstanding the possible contribution of the work, the real value of such an undertaking lies essentially in the creative endeavor itself.

The George Washington University   RICHARD H. SCHLAGEL
April 18, 1984

INTRODUCTION

In the Foreword to the new translation by Stillman Drake of Galileo's earlier *Dialogue*, Einstein says of Galileo that "he enabled the educated man of his age to overcome the anthropocentric and mythical thinking of his contemporaries," thereby leading them back "to an objective and causal attitude toward the cosmos, an attitude which had been lost to humanity with the decline of Greek culture."[1] This statement expresses a universal belief about the origins of scientific thought, namely, that it emerged from an earlier stage of "anthropocentric and mythical thinking" by attaining a more "objective" attitude toward the universe, a development initiated by the Milesian Greeks in the 6th century B.C. and waning after the Hellenistic Period. Seen from this point of view, the growth of scientific knowledge represents a gradual, if at times interrupted and irregular, progression toward a more objective and comprehensive understanding of the universe. But while this position is commonly accepted, little effort has been made to justify it in terms of an actual study of the growth of a higher level of abstract thought, that of scientific rationalism, from an earlier form of concrete subjectivity.

As a result of extensive research during this century by anthropologists, classicists, philologists, and historians of science a much more exact and complete knowledge of the successive stages of man's intellectual history is now available. Having discarded the naive evolutionary presuppositions of their 19th century predecessors, anthropologists currently present what appears to be a truer characterization of primitive mentality; classicists and philologists have devised better criteria for distinguishing original fragments from spurious attributions and conjectural interpretations to render a more exact picture of Presocratic philosophy; and historians of science have provided a much fuller account of the history of science. As a result, we now possess a better understanding of the successive segments comprising man's intellectual development.

What is still lacking, however, is an interpretation of these various segments from the perspective of an overall continuity, viewing them as a gradual process of intellectual growth (including normal progressions, lapses, crises and revolutions[2]), rather than as separate, independent stages. But even more important is the need for an understanding of the kinds of cognitive processes that brought about this development. While historians of science have concentrated on what Piaget refers to as "an anecdotal history of discoveries," and philosophers of science (e.g., Carnap, Hempel, Nagel, and Popper) have been concerned mainly with the methodology of science and with the confirmation (or "falsification") of scientific theories, as well as with such formal problems as the hypothetico-deductive nature of scientific explanation, the structure of scientific theories, the justification of induction, probability theory, etc., the "history of scientific thought itself"[3] has been largely neglected. Until recently, most philosophers approached science more as a *completed* system of laws, theories, and mathematical formalisms than as a *developmental, creative* process of thought.

Reichenbach, for example, in a highly influential statement maintained that philosophers should be concerned primarily with the confirmation of "completed theories" of science rather than with the creative processes underlying their discovery, drawing the well-known distinction between the "context of discovery" and the "context of justification."

> The philosopher of science is not much interested in the thought processes which lead to scientific discoveries; he looks for a logical analysis of the completed theory, including the relationships establishing its validity. That is, he is not interested in the context of discovery, but in the context of justification.[4]

Now, however, owing to the failure of the research program of the positivists and to the impact of such philosophers, scientists, and historians of science as Hanson, Polanyi, Kuhn, Toulmin, Holton, and Feyerabend, those concerned with understanding science are no longer focusing just on methodological and formal problems but also on the epistemological questions relating to the nature and development of scientific thought itself. Thus in this post-positivistic era we are experiencing a "problemshift" (Lakatos' term) away from the critique of completed scientific theories to the historical analyses of actual scientific developments to gain an understanding as to how the scientific mind functions. For example, in case histories documenting the intellectual development of outstanding scientists such as Kepler, Einstein, and Bohr, Holton has attempted to show "in what respects the traditional view of the way the scientific mind works has to be changed and supplemented."[5] And Feyerabend, aware of the failure of the positivists and of Popper to abstract from the past achievements of science a methodological procedure that would account for all past advances and ensure future progress, now concludes that "anything goes" as far as inquiry is concerned.[6] But a longer historical perspective might show that such an extreme, pessimistic conclusion is not warranted: i.e., that a pattern can be detected in the historical development of science.

This problemshift toward the cognitive, developmental, and historical aspects of science defines the orientation of the present study, the purpose of which is twofold: (1) to describe from a total historical perspective the salient features in the growth of scientific rationalism from its origins in the earlier mythopoetic, animistic consciousness to its present level, and (2) to analyze and describe the thought processes constituting this intellectual development. Starting with various anthropological conceptions of primitive mentality and psychological studies of different mental levels and their developments, this information is used to describe the kinds of cognitive changes manifested in man's continuous effort to achieve a more objective, systematic, and comprehensive understanding of the universe.

In contrast to Kant who redefined metaphysics as the science of discovering the *a priori* conditions of knowledge (i.e., the innate, unalterable, universal and necessary cognitive structures presupposed by any experience and all knowledge), the present study attempts an historical analysis of the progressive differentiation of the contents of experience, the development of interpretational concepts, and the transformations in presuppositions and modes of reasoning constituting the growth of knowledge. Man's cognitive makeup is not looked upon as fixed and final owing to the pre-programmed structure of our minds (presumably by a Divine Cyberneticist), but as a continuously developing system involving changes both in the differentiation of content as well as in the conceptual interpretation — in Kant's terms, as regards both content and form.

Empirical studies of cognitive development proved particularly helpful in this investigation, especially the research of the comparative genetic psychologist, Heinz Werner, and the work of Piaget. For example, Piaget's analysis of the origins of intelligence in the child and his description of the development of the child's conception of reality, including the role played in this development by cognitive structures such as causality,

substance, space, time, logical inference, etc., were helpful in interpreting the transformation of these concepts in the thought of the Presocratic philosophers. In addition,
Piaget's claim that certain levels of mental functioning in the child can be achieved only
after fundamental transitions of decentering, when an "egocentric" point of view has
been overcome, can also be applied to revolutionary developments in science, as he
himself pointed out.

> If we may be permitted to make a somewhat daring comparison, the completion of the objective
> practical universe [by the child] resembles Newton's achievements as compared to the egocentrism
> of Aristotelian physics, but the absolute Newtonian time and space themselves remain egocentric
> from the point of view of Einstein's relativity because they envisage only one perspective on the
> universe among many other perspectives which are equally possible and real.[7] (Brackets added)

Individuals and peoples who have not progressed in their cognitive development
beyond a certain level of egocentrism will be limited in their intellectual outlook, as
before the Copernican era. The overcoming of such limitations, both individually and
culturally, opens up vast cognitive dimensions incorporating radical shifts in perspective
and fundamental revisions in presuppositions and concepts. Recall, for example, the
relatively insignificant transposition of the sun and the earth, with a threefold motion attributed to the latter by Copernicus, and the subsequent expansion of the finite universe
into infinite space.

Or, following Piaget's other suggestion, consider the contrast between relativity
theory and Newtonian mechanics. Just as the geocentric view of the universe was based
on everyday observations so, as far as our usual terrestrial experiences are concerned,
where ordinary distances and velocities are negligible as compared to the velocity of
light, Newtonian science seems to be a close approximation to this domain, though at
one degree removed. That is, while Newtonian science reduces physical objects such as
the planets to point masses and the qualitative experience of distance, duration, and
motion to metrical space, time, and velocity, the simultaneous juxtaposition of point
masses in absolute space does seem to be an abstract representation of the world of
ordinary experience. Change can then be explained as due to the rearrangement of the
indivisible solid particles (the physical interpretation of point masses) as a result of
motion, impact, and the influence of such forces as inertia and gravity. This theoretical
framework seems to confirm our ordinary intuitions regarding the independence and
invariance of space, time, velocity, mass, etc., as well as provide a theoretical justification
of such concepts as absolute space and time, simultaneity, force, and so on.

Then came Einstein. Reflecting on the "asymmetries" in Maxwell's electrodynamics,[8] he systematically rejected most of the intuitive basis of Newtonian mechanics.
More precisely, he showed that rather than Newtonian mechanics being sufficient for an
explanation of all physical motions, it applies only to such "limiting conditions" wherein
the magnitude of distances and velocities are insignificant compared to the velocity of
light. As expressed in terms of the mathematical formalism, the Galilean-Newtonian
transformations can be seen as a "limiting case" of the newer Lorentz-Einsteinian transformations when the velocity of light is considered infinite or instantaneous — as it is
for all practical purposes within the limited dimensions of the earth.

When one attempts to take into account stellar distances and velocities approaching
that of light, however, basic concepts of Newtonian science prove inadequate: space and
time cease to be independent; space, time, and mass no longer are invariant but depend
upon the velocity of the system (time dilating, space contracting, and mass increasing
as the system's velocity approaches that of light); a limiting velocity is introduced, the

constant velocity of light; inertial and gravitational forces are found to be equivalent; mass and energy are convertible; there is no privileged system such as mass at rest in the ether or in absolute space, but all systems are equivalent; hence the meaning of 'simultaneity' is denied or relativized, but the laws of nature become invariant. At first all of this sounds very strange, if not paradoxical, and yet the mind soon becomes conditioned to the use of these newer concepts. What is important for this study is the attempt to understand the reasons for and the nature of the conceptual processes that underlie these changes, both past and present.

Even more startling than the innovations of relativity theory have been the recent developments in quantum mechanics — more startling because they represent an even greater challenge to traditional scientific assumptions and principles, as well as to common sense intuitions. For example, the following presuppositions were at least implicit conditions of scientific knowledge since the emergence of modern science some four hundred years ago: (1) that all events have precise, determinate causes; (2) that assuming a complete knowledge of the relevant antecedent conditions, all future physical occurrences could be precisely predicted; (3) that in all physical changes or transitions from state A to state B there are continuous transitional states or processes; (4) that the creation of new matter or energy is impossible; (5) that the minimum properties for anything to be a particle are both precise localization in space and possession of a definite velocity; (6) that physical processes can be defined in terms of a consistent set of concepts (not as a wave under certain conditions and a particle under others, as is true in quantum mechanics for light radiation and for de Broglie's "matter waves"); (7) that objective knowledge presupposes that one can ascertain the independently existing properties of physical phenomena (not that the object, apparatus, and observer must be considered an indissoluble unity, as in the "Copenhagen Interpretation" of quantum mechanics); (8) that the use of statistics represents a limitation of knowledge, not an objective indeterminacy in nature; and (9) that the mathematical formalism must be representative at least of the structure of nature, not that nature is defined by the mathematical formalism, as Heisenberg claimed. While there are basic differences of opinion regarding some of these issues (e.g., as to whether "God plays dice" or whether there are hidden parameters in terms of which probability functions eventually can be reduced to deterministic explanations, or whether quantum mechanics should be considered "complete" or "incomplete"), yet it would seem that the above assumptions, which were the foundations of Newtonian mechanics (and which also apply to the macroscopic world), do not hold for micronature.

Along with the studies of Piaget, Heinz Werner's comparative investigations of the mental development of different types of individuals, such as the child and the primitive, proved useful in interpreting the development of ancient Greek thought. More specifically, his description of different modes of cognitive functioning, such as "physiognomic perception," the "labile and diffuse character of primitive mental organization," "concrete and syncretic thought," etc., provided concepts for interpreting the philosophical thought of the Presocratics and of Plato and Aristotle. According to Werner, for example, the prevalence of animism and anthropomorphism in the thought of the primitive is not due, as commonly supposed, to a "projection" of subjective images and feelings onto an objective world; instead, they are the result of a natural form of primitive "physiognomic perception" which antedates later cognitive distinctions between the subjective and the objective contents of experience. Just as physiognomic characteristics, such as facial expressions, are instinctively interpreted by the child as manifestations of inner conscious states, so the primitive naturally perceives physical events as manifestations of underlying psychic forces, such as *mana, Wakenda*, daemons, souls, etc. Such an interpretation not only provides a psychological explanation of primitive animism but also makes it possible to identify animistic aspects of the thought of the Presocratics.

For example, what makes the thought of the 5th century Pythagoreans and Parmenideans so puzzling and yet so fascinating is the fact that they were unable to make the conceptual distinction, so clear to us, between the mode of existence of abstract entities, such as numbers and logical principles, and that of spatial entities — every form of being had to be spatial. Since the conception of a purely non-spatial form of existence was not conceived until Plato, the Presocratics necessarily conflated the formal properties of numbers and logical relations with the property of spatial extension. While this had its advantage for the Pythagoreans in that it enabled them to conceive of the generation of the three-dimensional physical universe from numbers, as a mode of thinking it was diffuse and syncretic — although their mathematical speculations were prescient and had far-reaching consequences. The same diffuse concreteness contributed to Parmenides' confusion of logic with ontology.

These examples illustrate the major presupposition and purpose of this monograph, that an interdisciplinary approach to the development of Western thought, one including anthropological and psychological, as well as historical data, can illuminate this development in a way in which purely historical and philological investigations cannot. Such an approach will be more conjectural than those of the latter, but this will be justified if it can provide additional insight into the development of thought to complement the more traditional studies. Thus the purpose of the present book is to describe the growth of scientific rationalism from its origin in the earlier animistic, mythopoetic tradition to the thought of Archimedes, the culminating figure of the Hellenistic period.

NOTES

INTRODUCTION

[1] Galileo Galilei, *Dialogue Concerning the Two Chief World Systems*, 2nd ed. , trans. by Stillman Drake (Berkeley: University of California Press, 1970), p. vii.

[2] Cf. Thomas Kuhn, *The Structure of Scientific Revolutions*, 2nd ed. rev. (Chicago: Univ. of Chicago Press, 1970).

[3] Jean Piaget, *Psychology and Epistemology*, trans. by Arnold Rosen (New York: The Viking Press, 1972), p. 102.

[4] Hans Reichenbach, "The Philosophical Significance of Relativity Theory," in *Albert Einstein: Philosopher-Scientist,* ed. by Paul Arthur Schilpp (Evanston: The Library of Living Philosophers, Inc., 1949), p. 292. It should be noted, however, that Reichenbach's main concern in this article is to show that relativity theory obviates Kant's interpretation of physical concepts such as space, time, causality, etc. as *a priori* cognitive structures, a position similar to the one presented in this book.

[5] Gerald Holton, *Thematic Origins of Scientific Thought* (Cambridge: Harvard Univ. Press, 1973), p. 11.

[6] Paul Feyerabend, *Against Method* (London: Verso Edition, 1978), Ch. 1.

[7] Jean Piaget, *The Construction of Reality in the Child*, trans. by Margaret Cook (New York: Basic Books, Inc., 1954), p. 367.

[8] Cf. Albert Einstein, "On the Electrodynamics of Moving Bodies," reprinted in *The Principle of Relativity*, H. A. Lorenz, A. Einstein, H. Minkowski, and H. Weyl, trans. by W. Perrett and G. B. Jeffery (New York: Dover Pub., Inc., 1923), pp. 37-65.

CHAPTER I

THE ORIGIN AND NATURE OF KNOWLEDGE

It is through wonder that men begin to philosophize [. . .] [1]
Aristotle

Considering the vast expanse and infinite complexity of the universe, particularly in contrast to the finiteness of his own existence, it is hardly surprising that from the very beginning man has been beset by awe and wonder when confronting and coping with the world. Was it the configuration of the stars and the wandering of the planets, giving rise to names and tales which fixed for the ancient mariner the positions of the "celestial bodies" as signposts on his nocturnal voyages, that first stirred man's speculative curiosity, as has been suggested by some scholars? We who live in modern cities whose streets are narrow canyons flanked by walled surfaces of glass and steel, and whose view of the night sky is dimmed by the jeweled surfaces of these glowing facades, easily forget how our early ancestors felt themselves enveloped by a semi-circumferencial, space-filled globe, domed with stars. It is only occasionally when we are transported to the sea or mountains that we sense once again the tremendous expanse of nature.

Or rather than the dramatic dominance of the night sky, was it the cyclical succession of birth and death, of growth and decay, that first aroused man's questioning and quest for knowledge? Almost all cosmologies begin with a begetting. For primitive man his own birth was as much a mystery as the legendary origin of the tribe, for birth, as death, was never conceived of as a strictly natural sequence, but believed to be caused ultimately by the intervention of some ultramundane force or mysterious power.

However, neither of these views proved to be an adequate "Ariadne thread" guiding one through the labyrinth of primitive speculation and leading to a unitary solution to the puzzle of the origin of man's quest for knowledge. Considering again the awesome complexity of the universe, one begins to sense the overwhelming challenge posed by nature, as very few events, if any, display their own causes or rationale. We observe that such and such exists, or that this or that happens, but why things exist and how they occur is not at all apparent — only familiarity dulls our sense of wonder and curiosity. Still, then, as now, certain recurring patterns and glittering regularities stood out as foreground configurations against a more remote, diffuse background. Thus the successive alternation of day and night correlated with the appearance and disappearance of the sun, as well as the daily and seasonal variations in temperature apparently related to the sun's movements as it rises and sets in the sky, are among the most visible changes in nature. So too are the nightly reappearance and morning disappearance of the stars and the periodic phases of the moon, the latter providing the oldest of all calendric units.

As the centuries passed certain features began to emerge in a solar gestalt against a stellar background: that the earth appears to be stationary in the center of a spherical universe; that among the celestial bodies, the planets appear larger and closer than the remotely sprinkled stars; that diurnal and seasonal variations are correlated with the various positions of the sun; that the planets move in individual orbits in a great void between the earth and the stars; that in contrast to the planets, the stars do not change their position with respect to each other, but remain "fixed" in position while revolving around the earth diurnally. We are so accustomed now to "seeing" these astronomical phenomena within an accepted conceptual framework that it is only with difficulty that we recognize that the data are mere successive appearances which, in antiquity, required millennia of observations and interpretation before the appearances began to fit together into the theoretical model of the ancient two-sphere universe: i.e., the solid spherical earth in the center of a larger space-filled globe on the inner surface of which were located the fixed stars and within which the planets wandered.

By peeling off the interpretational layers one begins perhaps to approach more and more the elemental, observational standpoint of early man. One observes that the earth looks flat except for such indirect evidence as the curvature of the horizon line, the fact that the masts of receding sailing ships remain visible after their hulls dip below the horizon, that at different latitudes the visibility of celestial bodies vary in a way that would be inconsistent with successive observations along a plane surface, and that during a lunar eclipse the shadow of the earth on the moon has a circular outline. One realizes that except for the sun and the moon, the planets are barely distinguishable from the stars, and that their various positions are merely different individual appearances unless interpreted as the continuous motion, too slow to be registered as such by our sight, of the same entities. Moreover, what becomes of these celestial bodies when they are no longer visible? After crossing the sky in a chariot, does the sun, as in a popular Greek myth, sail in a golden bowl round the stream of Okeanos — or is it rekindled daily, as believed by the Presocratic philosopher Heraclitus? How does one explain eclipses, variations in brightness, and the apparent retrograde motions of the planets? Of what stuff are the planets and stars made and what causes their motions? Have they always existed or were they once created, as by the Divine Craftsman of Plato?

Nor are the phenomena in our more immediate terrestrial domain any less puzzling when one lifts the veil of familiarity. Why does smoke rise and physical objects fall? When the propelling agent ceases, what impels a projectile to continue to move? What causes thunder and lightning, floods and famine, fever and fire, draught and epidemics? What happens when things undergo change? What is the significance of dreams and hallucinations, the interpretation of the former and the cultivation of the latter often playing a significant role in early religions? Why do some men possess an unusual degree of physical strength and psychic power or charismatic force? What is the significance of kinship relationships, tribal laws, vested authority, and taboos? Why do certain individuals and societies prosper at certain times and then suffer a demise? What fates control the destinies of individuals and kingdoms? Such questions as these, asked by every man at certain times, generated our myths, religions, cosmologies, and sciences.

THE CONDITIONS OF KNOWLEDGE

Thus the world as confronted in experience is not self-explanatory, but problematical; moreover, human experience itself is limited, fragmentary, and seemingly deceptive. All knowledge of the world is ultimately based upon sensory experience, however theoretical and abstract such knowledge may become, with sight the primary source of empirical data. Yet, as we know, our eyes are sensitive only to a very small fraction (from .00004 cm. to .00007 cm.) of the wave-lengths constituting the spectrum: we cannot 'see' radio waves, X-rays, ultraviolet waves, though our skin is sensitive to the latter. We cannot perceive the universe in terms of its submicroscopic or cosmic dimensions, and thus are limited to the macroscopic domain.

One of the reasons Renaissance astronomers, most of them Aristotelians, refused to acknowledge the novel telescopic observations of Galileo was their belief that our senses disclose the world as it truly is, and hence there could be no new astronomical phenomena to be discovered. It is difficult today to appreciate the sense of wonder Galileo himself must have felt when he first realized that his telescope disclosed innumerable new stars wherever he looked, hinting at a tremendously enlarged universe. Furthermore, accepting Galileo's arguments in defense of the Copernican theory required a radical shift in perspective, imagining the solar system not as it naturally appears to an observer from the earth, but as it would appear from any other planet. While this altered perspective seems simple enough to us after more than three centuries of conceptual conditioning, at the time it seemed to throw out of joint the most basic experiences and fundamental convictions acquired from antiquity by mankind. Developments in 20th century physics in relativity theory and quantum mechanics also were revolutionary because they showed that what formerly were assumed to be irreducible, independent physical properties of the universe are characteristics merely of the "limiting conditions" of our terrestrial or macroscopic domain. As Piaget states:

> In the field of thinking, the whole history of science from geocentrism to the Copernican revolution, from the false absolutes of Aristotle's physics to the relativity of Galileo's principle of inertia and to Einstein's theory of relativity, shows that it has taken centuries to liberate us from the systematic errors, from the illusions caused by the immediate point of view as opposed to "decentered" systematic thinking. [2]

As scientific knowledge has developed it has become less and less representative of common sense experience. One recent interpreter, Milič Čapek, has even questioned whether the tactual-visual experiential source of our traditional scientific imagination and concepts is adequate today as a basis for representing current scientific theories, suggesting that perhaps other modes of sensory experience, such as melodic variations in music, might prove more fruitful.

> Today we are in the midst of a far more radical transformation of our view of nature. The most revolutionary aspect of this transformation consists in the fact that the words "picture" and "view" lose entirely their etymological meaning. As the so-called primary qualities of matter now join the secondary qualities in their exit from the objective physical world, it is clear that the future conception of matter ought to be devoid of *all* sensory qualities, including even those which are subtly and implicitly present in seemingly abstract mathematical notions [. . .]. The positive significance of the auditory models is in the discovery of *imageless dynamical patterns* structurally similar to those which, according to growing empirical evidence, constitute the nature of physical reality. [3]

That our ordinary experience, language, and cognitive picture of the world have been conditioned by the number and nature of our sense organs, adapted to the macroscopic dimensions of nature, can hardly be doubted. Thus it is not surprising that our conceptual imagination based on this limited experience finds it difficult to conceive a universe radically different from that of ordinary experience. Perception seems to disclose manifest qualities and processes the underlying conditions of which remain hidden from our senses, though somewhat amenable to experimental discovery and conceptual interpretation. As Hume pointed out, there are no "necessary connections" linking our sensory impressions. Only the repetitive transition from one perception to another giving rise to a conditioned expectation is the source of the illusory belief that *direct observation* discloses necessary connections among phenomena. Thus Hume maintained that we have neither "intuitive" nor "demonstrative" evidence of any necessary transition from cause to effect: our knowledge of such connections must be derived from observation and hence is always contingent.

Nonetheless, experience does disclose similarities, regular occurrences, and repetitive patterns which are the basis of all knowledge. Things do exhibit similar properties and functions on which classifications are based — and without classifications everything would be unfamiliar and disorganized. Nature does manifest regular occurrences and uniformities which are the grounds of our common sense inferences and inductive generalizations; structures and forms do reappear, and like causes normally produce like effects under similar conditions. Thus mankind has built up a lore of common sense truths in the form of inductive generalizations, causal inferences, and natural laws. With varying degrees of exactness scientists have gradually uncovered the causes of the tides, combustion, lightning, radiation, genetic traits, and are now on the threshold of explaining cancer and the origin of life on this planet.

Even when the search for more ultimate causes or explanations prove illusive, functional correlations among phenomena are discoverable which can be quantified and expressed as scientific laws in mathematical equations. In fact, modern science began when "natural philosophers" gave up the search for "occult" causes, seeking instead uniformities expressible in mathematical terms. For example, in his *Dialogues Concerning Two New Sciences*, when the question of the cause of the acceleration of naturally falling objects is raised, Galileo has Salviati (who speaks for Galileo) reply:

> The present does not seem to be the proper time to investigate the cause of the acceleration of natural motion concerning which various opinions have been expressed by various philosophers [. . .]. At present it is the purpose of our Author merely to investigate and to demonstrate some of the properties of accelerated motion (whatever the cause of this acceleration may be) [. . .].[4]

Like Galileo, Newton derided the search for "occult qualities" (which the Scholastics invoked as explanatory agencies for certain obscure occurrences) and the use of "feigned hypotheses" or *non-empirical* suppositions to explain natural events, stating that the essential function of science is to "deduce" laws and explanatory principles from the phenomena. But Newton also postulated various "forces" to account for the "production" of motion, even though he could not explain the origin of these forces. As he states in the "Preface" to the first edition of the *Principia*:

> [. . .] rational mechanics will be the science of motions resulting from any forces whatsoever and of the forces required to produce any motions, accurately proposed and demonstrated [. . .] and therefore I offer this work as the mathematical principles of philosophy, for the whole burden of philosophy seems to consist of this: from the phenomena of motions to investigate the forces of nature and then from these forces to demonstrate the other phenomena [. . .].[5]

Yet the discovery of scientific laws, on which theories are based and which they attempt to explain, are prerequisites for, and milestones in, the advance of science: e.g., Kepler's three laws of planetary motion, Galileo's law of gravitational acceleration, Newton's law of universal gravitation, Maxwell's equations describing the structure of electromagnetic fields, Planck's constant of radiation, and Einstein's equation describing the equivalence of mass-energy.

Nonetheless, the sensory foreground of experience is ultimately inexplicable except in relation to a background context of substructures and processes on which the foreground phenomena depend. The purpose of experimentation is to uncover these underlying causes, while theories attempt to account for the unobservable occurrences in terms of inferred entities and postulated processes.

As beings who exist within the world, and who themselves comprise part of the conditions which give rise to ordinary phenomena, we do not experience the occurrence of these conditions, but the resultant phenomena. For example, our eyes and ears are necessary conditions for seeing colored objects and hearing sounds, but we do not perceive our eyes and ears functioning in this role. Similarly, we know that to perceive objects light must be reflected from the object to our eyes, but we do not see this radiation any more than we see the transmission of nerve discharges along the optic nerves to the brain. All of experience is the outcome of tremendously complex processes, physical, chemical, as well as physiological, which are not directly observed. The purpose of experimentation, once again, is to induce nature to reveal more of these underlying states, processes, and conditions.

But experimentation itself is insufficient for an explanation of natural events. The progress of science also has depended upon the "free creation" (Einstein's term) of theoretical constructs, hypothetical entities, and inferred processes (e.g., atoms and molecules, light and sound waves, electromagnetic fields, etc.) representing the background conditions and causes of foreground phenomena. Thus the physical cause of sounds is described as a transmission of waves having a certain frequency, length and amplitude correlated with pitch and volume and requiring a medium for their transmission (which is partially confirmed by the fact that sound waves are not transmitted in a vacuum). We imagine the increased pressure of gases with added temperature (volume remaining constant) as due to the increase in average kinetic energy of the gas molecules striking the inner surface of the containing vessel. We conceive of radiation as a process of "freeing" electrons or quanta of energy and explain the shiny metallic surface of metals when homogeneous light is reflected on them as due to the "ejection" of electrons by light quanta.

Pierre Duhem criticized British scientists for relying on imaginative models as aids to scientific explanation, praising Continental scientists for dispensing with such visual aids and for preferring deductive mathematical theories.[6] To a certain extent his judgment has been vindicated by developments in mathematical physics in the 20th century which seem to defy pictorial representation, although research in atomic physics and molecular biology tend to refute his extreme criticism of the atomic model and premature prediction of its uselessness. Increasingly, the aim in physics has shifted from giving a pictorial representation of the causes of phenomena (traditionally relying on mechanical models), to reducing the multiplicity of phenomena and independence of explanations to a more unified set of explanatory concepts, principles, or equations from which the disparate phenomena can be deduced and/or predicted, and thus 'explained.'

Progress in theoretical physics, accordingly, is partially measured by the gradual assimilation of diverse phenomena and independent explanations into more unified theoretical frameworks. One of the great achievements of 19th century physics was the unification of electricity and magnetism into electromagnetic fields, with the discovery that light too was a form of electromagnetic phenomena. Even more dramatically, in the 20th century Einstein dissolved the dichotomies between space and time, inertial and gravitational fields, mass and energy, and materiality and the metrical structure of space-time. However, his further attempt to integrate gravitational and electromagnetic fields under a unified set of equations proved illusive, along with the attempt to relate relativity mechanics (which applies essentially to cosmic structures) to quantum mechanics (which deals primarily with sub-microscopic structures). [7]

While the equations used in such powerful theories usually involve more complex mathematics and thus are more difficult to comprehend, they do result in greater theoretical simplicity and elegance, the goal of theorists from Plato to Einstein. And yet, unless one is willing to accept a sophisticated form of neo-Pythagoreanism, admitting only mathematical correlations as the very nature of things, one will continue to try to represent the kinds of physical entities and structures abstractly symbolized in scientific equations in a more concrete manner. But if, as research in quantum mechanics seems to indicate, this is no longer possible, then it becomes necessary to revise our conception of knowledge in accordance with the new experimental findings in particle physics and their theoretical implications.

CONCEPTUAL-LINGUISTIC FRAMEWORKS

If the ultimate aim of thought is to catch as much of the diversity of phenomena as possible within a unified theoretical network, whether representational or mathematical, this raises the question as to the origin of the framework. As already indicated, ordinary experience at least exhibits similarities and regularities which serve as a basis of knowledge — but how does this experience become transposed or transformed into knowledge? The traditional way of answering this question has been to analyze knowledge into two essential components: (1) the empirical or sensory content derived from the complex neuronal processing of our sensory systems, and (2) the various interpretational, conceptual-linguistic frameworks contributed by our cerebral-conscious processes. Although neither of these aspects can be completely isolated, they can be distinguished and described. The sensory content has been variously designated as "sensory impressions," "sense data," or "the given," indicating that content of knowledge resulting from the effects or "impressions" of the physical world on our sense organs as processed by our brains, and hence "given" independently of the will.

In contrast to the imposed "givenness" of the sensory content, the conceptual-linguistic framework represents the "creative" interpretation, as expressed in various symbolic frameworks, contributed by the individual himself. This interpretational aspect is manifested in such cognitive processes as sign recognition, conceptualization, verbal expression, theory construction, etc. Knowledge, which consists of conceptual interpretations or judgments as expressed in true statements (since it is somewhat contradictory to refer to "false knowledge," although we do countenance "approximate knowledge"), is inseparable from a conceptual-linguistic framework. While there might be a sense in which it could be said that animal behavior sometimes involves mistaken judgments, the designations 'true' or 'false' presuppose some assertion or claim to knowledge, which in turn involves a symbolic expression.

Although different schools of thought have placed different emphases on either of these aspects of knowledge, rationalists basing knowledge on self-evident principles exhibited in the conceptual framework, while empiricists derive all knowledge from the original sensory contents, the predominant tendency in the 20th century has been to assume that an indubitable sensory content, clearly distinguishable from any conceptual interpretation, is necessary as a foundation of knowledge. This presupposition was held by philosophers as diverse as Bertrand Russell, G. E. Moore, A. J. Ayer, C. I. Lewis, H. H. Price, H. Feigl, and R. Carnap. The presuppositional framework of the positivists especially,

22

which originally included such tenets as the verifiability criterion of meaning, indubitable protocol statements, an absolute demarcation between analytic and synthetic statements, and the reduction of theoretical constructs to observables, depended upon this distinction.

Recently, however, the sharp separation of the sensory content from the theoretical interpretation, as the ultimate ground of all knowledge, has been criticized from two points of view. First, philosophers of science such as Norwood Hanson and Paul Feyerabend have denied that sensory observation or evidence can occur apart from a theoretical interpretation, Hanson claiming that all observations are "theory laden." [8] Feyerabend, in turn, has argued that since "empirical facts" are dependent upon theoretical interpretations, the discovery of new facts requires developing alternative theories to those commonly accepted. Moreover, when there is a conflict between theory and facts, one always has the option of rejecting the so-called facts.

> Facts and theories are much more intimately connected than is admitted by the [sense] autonomy principle. Not only is the description of every single fact dependent on *some* theory [. . .] but there also exist facts which cannot be unearthed except with the help of alternatives to the theory to be tested, and which become unavailable as soon as such alternatives are excluded.[9] (Brackets added)

Secondly, recent investigations of the visual performance of baby chicks, primates, and human infants have indicated that rather than such performances being dependent merely on experience and conditioning, such capabilities are based on innate hereditary structures. That is, due to evolutionary developments the various species have acquired an innate, preprogrammed neural system that is the basis of their initial responses to the world, and on which subsequent learning and development depend. The empiricists' notion of the mind as a "blank tablet" subsequently filled with unformed, unprocessed, or uninterpreted sensory data has little support in recent empirical investigations. [10]

Moreover, the thesis that all 'facts' are theory laden or that all knowledge is framework dependent is supported by investigations in linguistics. The American linguist, Benjamin Whorf, particularly has brought out the fact that language is not merely a vehicle for reproducing thought, but itself shapes or structures thought. We do not confront nature as it is, but our descriptions of nature always depend upon a background linguistic system.

> We dissect nature along lines laid down by our native languages. The categories and types that we isolate from the world of phenomena we do not find there because they stare every observer in the face; on the contrary, the world is presented in a kaleidoscopic flux of impressions which has to be organized by our minds — and this means largely by the linguistic systems in our minds. We cut nature up, organize it into concepts, and ascribe significances as we do, largely because we are parties to an agreement to organize it in this way — an agreement that holds throughout our speech community and is codified in the patterns of our language. [11]

That thought processes and world views vary with linguistic frameworks can be illustrated not only in connection with contrasting grammatical structures, but also in relation to historical developments. What is particularly fascinating about *intellectual* history is the continual transformation and elaboration, usually gradual but at times revolutionary, of the extant conceptual-linguistic framework. For example, during the Hellenic period (the time-span mainly covered in this volume) the philosopher-scientists had to redefine older terms as well as forge new concepts and words (e.g., 'chaos,' 'arche,' 'physis,' 'apeiron,' 'nous,' 'atmos,' 'eidia,' 'enteleche,' etc.) to express their evolving philosophical and cosmological theories. Aristotle is a particularly apt example because he invariably begins his consideration of a theoretical problem with an examination of the various senses of the key words used in stating the problem. And during the medieval period the various attempts to synthesize Greek philosophy with the Eastern mystical heritage of the Hebraic-Christian religion effected another radical change in conceptual orientation and terminology, so that studying medieval philosophy may be compared to learning a foreign language.

Then in the 17th and 18th centuries the founders of modern classical science were confronted not only with discovering new laws and forging a new method of inquiry, but with replacing an older conceptual framework, the organismic cosmology of Aristotle, with a new schema of interpretation based on mechanical concepts and the corpuscular-kinetic theory of the universe. As their frequent criticism of such Aristotelian-Scholastic terms as "occult qualities," "substantial forms," "final causes," "natural tendencies," etc. indicate, savants such as Galileo, Newton, Boyle, Gassendi, and Locke were faced with replacing the entrenched academic language of Scholasticism with the newer mechanistic terminology of "insensible particles," "primary qualities," "powers," "mass," "force," etc. And in the 20th century, while Einstein did not replace such basic terms in Newtonian mechanics as "space," "time," "mass," "gravity," and "inertia" with new ones, he did change fundamentally the meanings of these terms within the theoretical framework of relativity theory. Even so-called "ordinary language" is subject to the infiltration of technical terms from current specialized terminologies: e.g., "quantum jumps" from quantum mechanics, "negative feedback" from cybernetics, "cloning" from biogenetics — not to mention such general terms as "counterproductive," "ecology," "biodegradable," etc.

And it is not only individual words and their meanings that change but the underlying fabric of conceptual implications and interconnected meanings also change. When this happens previously accepted modes of implication and patterns of reasoning along with assumed basic presuppositions or self-evident principles are called into question. At such times it often appears as if there were no solid foundation to knowledge at all. For example, in the early 17th century, following the Copernican Revolution and during the time that Kepler, Descartes, and Galileo were creating the foundations of the new science of mechanics, the English poet, John Donne, wrote that the "new Philosophy calls all in doubt [. . .] 'tis all in pieces, all coherence gone." And so it might have seemed.

The theory of evolution, the construction of non-Euclidean geometries, and especially relativity theory have had similar dislocating, disconcerting effects on traditional ideas. But even relativity theory did not undermine fundamental principles of thought as radically as has quantum mechanics. For nearly 2500 years two of the basic principles underlying all thought have been Parmenides' logical maxim that "being cannot come from non-being nor cease to be," and Aristotle's logical law that "something either is what it is or is something else." Yet the encounter in quantum mechanics with "spontaneous quantum jumps," "indeterminate physical states," and "matter-waves" has led some logicians to conjecture whether our traditional laws of logic (i.e., our two-valued logic based on Aristotle's law of non-contradiction and excluded middle) are still applicable to micronature. As Quine has stated:

Revision even of the logical law of the excluded middle has been proposed as a means of simplifying quantum mechanics; and what difference is there in principle between such a shift and the shift whereby Kepler superseded Ptolemy, or Einstein Newton, or Darwin Aristotle? [12]

It could be argued, however, that there is a difference in "principle" between altering the laws of logic and revising scientific theories, at least in the sense that the laws of logic are so deeply embedded in our whole cognitive-linguistic framework (as Quine himself pointed out) that one does not know how he could alter the former without undermining the latter. Thus while one has grown accustomed to theoretical shifts in science, one is still reluctant to contravene or abridge the laws of logic — and yet this may be necessary as Quine and others have suggested.

Such conjectures as these raise even more fundamental questions. If theoretical developments in physics were to require a revision in the laws of logic, then what status would these laws have? Would they be thought of as similar to scientific laws and thus subject to empirical verification? Or would they be elevated to the status of mathematical axioms which, as formal postulates, are immune from falsification? Or would such laws merely have a limited application, rather than the universal and necessary status they occupied in the past? That is, would it mean that while the laws of traditional logic apply to man's thinking about the world of macroscopic objects they would not apply to the world of micronature, so that one would adjust his logical principles, as one does optical resolution, to different dimensions of experience? Rather than the laws of logic being "laws of reality" or "laws of thought," or even the "logical form" of language, previous interpretations, they would be principles conditioned by and hence relative to man's cognitive response to the macroscopic dimension of experience.

Because of the revolutionary impact of developments such as relativity theory and quantum mechanics, as well as research in linguistics, analytic philosophy, and the philosophy of science there is a growing tendency to describe all knowledge as framework dependent. On this view, although particular knowledge claims may be dealt with separately, as if they had an independent status, ultimately they derive their meaning and truth-value from the total theoretical framework within which they occur. Thus epistemological questions regarding the meaningfulness and truth of statements, and ontological questions pertaining to the existence or non-existence of entities, cannot be separated from broader questions relating to the entire framework and its foundations. [13]

But it should not be overlooked, as it often has been by ordinary language philosophers, that language is not a self-contained system. Although all knowledge claims are made within the context of some conceptual-linguistic framework, *the framework itself does not establish the truth of such claims.* All knowledge is knowledge about something; all factual truth-claims refer beyond themselves to an independent, external 'reality.' If it is true that we can interpret phenomena only within the context of a theoretical framework, it is also the case that such interpretations lead to further observations or the discovery of new phenomena or laws which in turn result in the continual revision of the framework. Thus while a framework focuses our view of the world in a particular way, it does not completely determine what it is we see, or preclude looking at things in a different way. The history of science is full of examples confirming this.

The very fact that the two-sphere model of the universe of Aristotle and Ptolemy incorporated the most obvious astronomical phenomena (the revolution of the planets and stars around a centrally located, stationary earth) enabled later astronomers to examine more critically specific data, such as retrograde motion, variations in brightness, and discrepancies in orbital periods, which in turn eventually resulted in discarding the geocentric model in favor of the heliocentric theory. And while the combination of the principle of inertia and the universal law of gravitation enabled Newton to explain the orbital velocities of the planets as described by Kepler's three laws, they did not account for the anomalous displacement of the perihelion of Mercury. However, one of the confirming consequences of the general theory of relativity was that it could explain and predict the discrepancy between the slow rotation of the ellipse of Mercury and the orbital motion predicted on the basis of Newton's law. But without the celestial mechanics of Newton one would have been unaware of the precise irregular rotation of Mercury.

Similarly, Maxwell's development of the theory of electromagnetic waves, which presupposed an ether, eventually led to the Michelson-Moreley experiments and Einstein's interpretation which eliminated the ether. In each of these cases the acceptance of older theories did not preclude the discovery of new facts which in turn led to a revision of the theories. Thus science is the history of the continual application, extension, and modification of preexistent frameworks due to the interaction of the investigator with the world.

ABSTRACTING

Since the purpose of a theoretical framework is to enable one to fixate, refer to, contemplate, describe, etc. some domain of experience, and since no system can represent all of reality, each framework is distinguished by the set of abstractions incorporated into the framework. That is, insofar as each cognitive-linguistic system is unavoidably limited, it will focus upon and abstract out of the seemingly infinite complexity of phenomena those features that it deems most significant and fruitful for interpretation. The grammatical categories of each natural language, for example, in terms of nouns, verbs, adjectives, adverbs, etc. contain for any community the basic set of abstractions in terms of which its users view the world, as Whorf pointed out.

The conscious process of abstraction, which is based on the discriminating capacities of our sense organs and the information processing of our central nervous system, involves three functions: (1) the discrimination of various sensory contents, forms, relations, events, etc., (2) the ability to focus attention upon the discriminated components, and (3) the capacity to fixate these data in one's consciousness so as to recognize them when they recur. Once having acquired a natural language based on such primary abstractions, further interpretations of the world and extensions of knowledge depend upon forming different abstractions on various levels and imaginatively rearranging those abstractions into new concepts, models, and theories.

Because of the concreteness and variety of the abstractions, this is graphically illustrated in the theorizing of the Presocratic philosophers. Not only did they attempt to account for the stuff of the world in terms of elements abstracted from experience, for example water (Thales), air (Anaximenes), and fire (Heraclitus), they also attempted to explain the origin and processes of the world in terms of occurrences abstracted from common experiences, such as "separating-out" (Anaximander), "condensation and rarification" (Anaximenes), "Love and Strife" (Empedocles), "*nous*" or "mind" (Anaxagoras), and mechanical impact (Democritus). The material ontology of Democritus, as well as that of Newtonian science, consisted of "insensible particles" or atoms defined in terms of "primary qualities" such as solidity, shape, size, position, and motion abstracted from the ordinary objects of experience and interpreted within the framework of a mechanical model of interaction. The abstract theoretical terms used in contemporary theories also reveal their origins in more basic experiences; e.g., 'natural *selection*,' 'sound *waves*,' 'chemical *reactions*,' 'force *fields*,' 'electron *spin*,' 'quantum *jumps*,' etc.

Just as various kinds of abstractions can be formed from existing things (abstracting a leg from a chair, then the shape, color, hardness, woodenness, etc. from the leg), so we can form conceptual abstractions from more basic abstractions: e.g., infinite space, complex numbers, unconscious desires, etc. We can isolate the physical surface of a table and from that distinguish its plane surface and rectangular shape, abstracting from the latter the right angles, lines, points, etc. It was from such arithmogeometric abstractions that the earliest Pythagoreans attempted to reconstruct the world. One can abstract aspects such as change and, with Heraclitus, make the continuous flux of things the fundamental category of reality. Or one can even abstract the logical principle of contradiction and base a whole ontology on that, as did Parmenides. We can refer to Abélard and Héloïse, a man and a woman, a pair of lovers, or with Bertrand Russell, the class of paired objects, in terms of successive abstractions.

Much of modern art from Brach to Pollock consists of the visual manipulation of abstractions such as perspectives (presenting front, rear, and side perspectives on one plane or surface), colors, textures, geometrical forms and patterns, etc. Mathematics, that most abstract of disciplines, is the science that deals with the formalized manipulation of the most abstract of abstractions, mathematical symbols.

> The abstractions of mathematics are distinguished by three features. In the first place, they deal above all else with quantitative relations and spatial forms, abstracting them from all other properties of objects. Second, they occur in a sequence of increasing degrees of abstraction, going very much further in this direction than the abstractions of other sciences [. . .] . Finally, and this is obvious, mathematics as such moves almost wholly in the field of abstract concepts and their interrelations. While the natural scientist turns constantly to experiment for proof of his assertions, the mathematician employs only argument and computation. [14]

Situations too can be theoretically altered and manipulated. Einstein repeatedly describes how he arrived at new insights by imagining unusual but concrete situations, *gedanken experiments,* such as a man traveling at the speed of light observing a light wave, or elevators falling in or being hoisted out of gravitational fields. [15] And Heisenberg arrived at his "principle of indeterminacy" by imagining what would be involved theoretically in observing an electron through a gamma ray microscope. [16]

> Creative imagination is a reworking, remodeling, reshaping of abstractions, a kind of mental playing around with the part and aspects that have been abstracted from experience [. . .] . Why is it that the imaginative altering of abstractions, or creative imagination, is so important to man? For one thing, without it he could form no new philosophical or scientific hypotheses, in which case philosophy and science would be impossible. [17]

The discoveries and theories of science are rarely if ever based on direct abstractions from the world, but are arrived at by reflecting on problems and conducting experiments in terms of previously established assumptions, concepts, and theoretical frameworks. Copernicus, for example, was driven to construct a new model of the universe because of the lack of symmetry and coherence in the Ptolemaic system. Kepler, in turn, working with the exact astronomical observations supplied by Tycho Brahe, could not derive accurate predictions of the orbital motions of the planets based on the ancient assumptions of uniform and circular motions. Galileo had already deduced the inadequacy of Aristotle's law that free falling objects accelerate proportional to their weights before he began his incline plane experiments. It can be shown that every aspect of Newtonian mechanics had already been formulated either by his predecessors or by his contemporaries, but that it took the genius of Newton to integrate these aspects into a unified science of mechanics.

Nor are the concepts of science usually based on direct abstractions from experience, but are derived by reflecting on problems in terms of abstractions incorporated into an established theory. Concepts such as inertia, mass, atomic weight, entropy, positron, parity, quarks, etc., although introduced as a result of experimentation and to satisfy the requirements of the mathematical formalism, are usually defined in terms of lower level abstractions already incorporated into the language of the theoretical framework. Experiments in relativity theory and quantum mechanics, for instance, are described in terms of the background concepts of Newtonian mechanics which in turn is discussed within the context of ordinary language. Thus theories seem to be nested within one another. In place of the metaphor of a net often used to characterize a conceptual framework, one would do better to refer to a vertically stratified matrix in which each successive layer becomes more abstract and refined. At the foundation are those concrete concepts abstracted from ordinary sensory experiences, while at the top are those abstract mathematical calculi waiting to be given a physical interpretation and thus become a part of the growing structure of science.

Accordingly, we have now answered the question originally asked as to the origin of the framework: it originates by successive processes of creative abstraction from the original contents of experience. As David Bohm asserts:

> The task of science is, then, to find the right kind of things that should be abstracted from the world for the correct treatment of problems in various contexts and sets of conditions [. . .]. When this does not happen, we must, of course, revise our abstractions until such success is obtained in these efforts. Scientific research thus brings us through an unending series of such revisions in which we are led to conceptual abstractions of things that are relatively autonomous in progressively higher degrees of approximation, wider contexts, and broader sets of conditions.[18]

To a great extent, therefore, the power of human thought is dependent upon the capacity to abstract various qualities, forms, aspects, relations, etc. from the myriad of possible distinctions in experience, to incorporate these abstractions into various frameworks and, finally, to creatively reconstruct the lower order concepts into new theoretical frameworks. Or, looking at the same process from the standpoint of the end results rather than the original activity, all intellectual history can be described in terms of the kinds or levels of abstractions utilized in man's successive efforts to comprehend the universe. In a sense, therefore, the present study can be looked upon as an attempt to attain a certain perspective (and no study can be more than one among an innumerable number of possible perspectives) on Western thought by observing its growth through the lens of abstraction.

NOTES

CHAPTER I

[1]   Aristotle, *Met.*, 982b.

[2]   Jean Piaget, *Comments* (on L. S. Vygotsky,*Thought and Language*, 1962), trans. by Parsons, Hanf-
mann, and Vakar (Cambridge: The M.I.T. Press, 1962), p. 3.

[3]   Milič Capek, *The Philosophical Impact of Contemporary Physics* (Princeton: D. Van Nostrand Co.,
1961), p. 379.

[4]   Galileo Galilei, *Dialogues Concerning Two New Sciences*, trans. by Crew and de Salvio (New York:
McGraw-Hill Co., Inc. [1914] 1963), p. 160.

[5]   "Newton's Preface To The First Edition," *Mathematical Principles*, Andrew Motte's trans., revised
by Florian Cajori (Berkeley: University of California Press, 1962), pp. XVII-XVIII.

[6]   Cf. Pierre Duhem, *The Aim and Structure of Physical Theory*, trans. by Philip Wiener (Princeton:
Princeton Univ. Press, 1954), ch. IV.

[7]   Cf. A. Einstein, "The Fundamentals of Theoretical Physics," in *Ideas And Opinions*, new translations and revisions by Sonja Bargmann (New York: Bonanza Books, 1954), p. 323.

[8]   Cf. Norwood Hanson, *Patterns of Discovery* (Cambridge: At The University Press, 1961), p. 54ff.

[9]   Paul Feyerabend, *Against Method* (London: Verso Edition, 1978), p. 39.

[10] Cf. *Scientific American, Perception: Mechanisms and Models* (San Francisco: W. H. Freeman and
Co., [1950] 1972), sec. VI.

[11] Benjamin Whorf, *Language, Thought, and Reality*, ed. by John Carroll (Cambridge: The M.I.T. Press
and John Wiley & Sons, Inc., 1956), p. 213.

[12] Willard Van Quine, *From a Logical Point of View* (Cambridge: Harvard Univ. Press, 1963), p. 10.

[13] Cf. Rudolf Carnap, "Empiricism, Semantics, and Ontology," *Semantics and the Philosophy of
Language*, ed. by Leonard Linsky (Urbana: The Univ. of Illinois Press, 1962), ch. 11.

[14] A. D. Aleksandrov, A. N. Kolmogorov, M. A. Lavrent'ev (eds.), *Mathematics: Its Content, Methods,
and Meaning*, Vol. I, second edition, trans. by Gould Bartha (Cambridge: The M.I.T. Press [1956]
1963), p. 2.

[15] Cf. A. Einstein, *Relativity*, authorized trans. by R. W. Lawson (New York: Crown Pub., Inc., 1961),
ch. XX.

[16] Cf. Max Jammer, *The Conceptual Development of Quantum Mechanics* (New York: McGraw-Hill
Book Company, 1966), p. 325ff.

[17] H. G. Alexander, *Language and Thinking* (Princeton: D. Van Nostrand Co., Inc., 1967), p. 191.

[18] Davic Bohm, *Causality and Chance in Modern Physics* (New York: Harper & Brothers, 1957),
p. 146.

CHAPTER II

ANTHROPOLOGICAL CONCEPTIONS OF PRIMITIVE MENTALITY

> [. . .] *among primitive peoples there exists the same
> distribution of temperament and ability as among us.* [1]
> Radin

Because an understanding of any phenomenon is usually enhanced by contrasting it with something else, before tracing out the origin and development of a higher level of objective rational thought, the main purpose of this volume, we shall attempt to delineate a more subjective, irrational form of thinking. By analyzing the contents of myths, reviewing various ethnological studies of "primitive mentality," and aided by comparative studies of mental functions and development, criteria of a lower form of thought processes begin to emerge. It is in terms of this more subjective, more 'primitive' mode of thought that the achievements of the Presocratic philosophers will be compared.

There was a time not long ago when reading the memoirs of missionaries or travelers who visited exotic lands and lived with tribes allowed one to pose as an anthropologist or ethnologist. Now, however, that time has passed. As Lévi-Strauss maintains, now that ethnology has reached the level of a disciplined science it no longer should be considered an open field for amateurs — and anyone who has not had the benefit of a least one field trip, the *rite de passage* of anthropologists, is certainly an amateur, however widely read he may be in the literature. Thus anyone venturing into this field is apt to suffer all the misgivings of a non-specialist intruding into a very restricted area. These apprehensions aside, some consideration of ethnological studies of more primitive peoples seems useful as a background for recognizing more highly developed modes of thought. Bearing in mind the following statement by Paul Radin, this chapter is therefore offered as a selective review of some interpretations of the modes of thinking of primitive peoples proposed by leading ethnologists:

> [. . .] in the present condition of our knowledge any attempt to describe the intellectual view of life of primitive peoples is destined to be tentative, provocative of further investigation and interpretation rather than permanent and final. [2]

As in other disciplines, each generation tends to interpret the mental outlook of primitive peoples according to the assumptions, prejudices, and resources of its own period. Thus Lévi-Strauss criticizes the 19th century English school of anthropologists for evaluating primitive mentality in terms of their own Victorian preconceptions, while he finds a clue for his interpretation in the latest developments in symbolic logic, Boolean algebra, and cybernetics. [3] But one cannot help wondering why a more exact understanding of primitive mentality should have had to depend upon the development of these more recent sophisticated techniques of interpretation, especially as Lévi-Strauss acknowledges that the philosopher Henri Bergson had a better understanding of primitive mentality than many contemporary ethnologists precisely because his personal philosophical outlook tended to approximate that of the primitive. [4]

Yet, even if one's philosophical outlook is quite different from that of Bergson, one cannot but recognize in the written accounts of primitive peoples modes of thinking which are similar to those of an earlier period in his own mental development, as well as characteristic of peoples whose present mental level is relatively undeveloped. There are obvious dangers in attributing an *exact* similarity to the mental outlooks of such diverse individuals, and yet the assumption of *some* similarity would seem to be a necessary pre-supposition for any understanding of 'primitive mentality.' As Werner states, one should recognize that

> the European mentality is, genetically considered, highly variable; that man possesses more than one level of behavior; and that at different moments one and the same man may belong to different genetic levels. In this demonstrable fact that there is a plurality of mental levels lies the solution of the mystery of how the European mind can understand primitive types of mentality.[5]

EARLY ETHNOLOGICAL INTERPRETATIONS

While any classification is somewhat arbitrary, one convenient way of distinguishing the various ethnological interpretations of the mentality of primitive peoples is to classify them into the following three types: (1) The first type is the earliest and includes those two classics of modern ethnology, Tylor's *Primitive Culture*[6] and Frazer's *The Golden Bough.*[7] Both of these works were written from the late 19th century evolutionary point of view according to which human culture, social institutions, and modes of thought were believed to have undergone a continuous evolutionary development from the earliest stages, as illustrated in surviving primitive societies, to the latest and most advanced forms of this evolution, as manifested in contemporary Western civilization. On the assumption that the earlier is the more primitive, one could use a simple chronological yardstick to measure a progressive development from the most primitive, simplest forms of thought to the latest, most advanced. Thus Frazer could remark at the conclusion of his epic work that "the movement of the higher thought, so far as we can trace it, has on the whole been from magic through religion to science."[8] Because this approach has been thoroughly criticized and completely rejected, no further discussion will be given to it.

(2) The second type of interpretation derives from what has been called the "French School" of anthropologists (in contrast to the "English School" mentioned above), the major representatives of which were Durkheim and Lévy-Bruhl. According to Durkheim's well-known study of primitive religions, religion is essentially a social affair whose forms of experience and belief are the product of collective behavior and thought expressing objective social realities.[9] Primitive thought is determined by "collective representations" (i.e., perceptual and ideational contents derived from the social milieu) precluding in-dividual deviations and thereby insuring homogeneity of thought.

Lévy-Bruhl undertook "the study, by means of the collective-representations of primitives, of the mental processes which regulate them."[10] Although Lévy-Bruhl's theory has been severely criticized by later ethnologists (e.g., Radin, Malinowski, Lévi-Strauss, etc.), it is worth considering because in the early decades of the century it was extremely influential and still represents a common interpretation of primitive mentality. Moreover, his views have been misunderstood and therefore unfairly criticized and rejected, whereas there still is something to be learned from his interpretation of the mental processes of the primitive.

(3) The third type of interpretation, to be discussed later, consists of the tendency

of more recent ethnologists to minimize any difference between the logical thought of the primitive and that of modern man.

Contrary to the usual descriptions of Lévy-Bruhl's theory,[11] he did not deny that primitive peoples were capable of a rational form of thought. Instead, he divided the thinking of primitive peoples into two types, claiming that only one type was logically different from our own: (1) the kind of thinking manifested in everyday practical affairs which, in its objectivity and patterns of reasoning, tends to approximate our own; and (2) the mode of thinking determined by the "collective representations" of the group which does not obey the logical laws of our thinking.

> Considered as an individual, the primitive, in so far as he thinks and acts independently of [. . .] collective representation [. . .] will usually feel, argue and act as we should expect him to do. The inferences he draws will be just those which would seem reasonable to us in like circumstances [. .]. But though on occasions [. . .] primitives may reason as we do [. . .] it does not follow that their mental activity is always subject to the same laws as ours. In fact, as far as it is collective, it has laws which are peculiar to itself [. . .].[12]

The mode of primitive thought which is determined by collective representations Lévy-Bruhl called "prelogical." This proved to be an unfortunate designation because it has been interpreted as meaning that primitive peoples exhibit a *non-rational* form of thought *preceding in time* the development of logical thought. But both of these contentions were explicitly denied by Lévy-Bruhl:

> By *prelogical* we do not mean to assert that such a mentality constitutes a kind of antecedent stage, in point of time, to the birth of logical thought [. . .]. It is not *antilogical*; it is not *alogical* either. By designating it "prelogical" I merely wish to state that it does not bind itself down, as our thought does, to avoiding contradiction. It obeys the law of participation first and foremost.[13]

That the primitive's thinking is not as determined by the laws of contradiction as our own, but by the "law of participation," is a primary thesis of Lévy-Bruhl. But what, precisely, does this mean?

According to Lévy-Bruhl, when *we* perceive an object we have a mental representation of the object determined (we assume) primarily by the physical effects of that object on our senses. In contrast to *our* percept, the primitive's conscious representation of the same object is imbued with "mystic qualities:" it is charged with emotions, feelings, and meanings which seem to emanate from it, giving it a different kind of reality. In Lévy-Bruhl's terms, the primitive's representations "participate in" these "mystic qualities" (again an unfortunate term because he means primarily subjective and socially engendered feelings and meanings) and it is this participation which determines primitive thought. "What is really 'representation' to us is found blended with other elements of an emotional or motor character, coloured and imbued by them, and therefore implying a different attitude with regard to the objects represented."[14]

Lévy-Bruhl thus employs the word 'mystic,' not in the religious sense, but in the "strictly defined sense in which 'mystic' implies belief in forces and influences and actions which, though imperceptible to sense, are nevertheless real."[15] It is the subjective influence of these socially engendered 'mystic' qualities that controls much of primitive thought, rather than the law of contradiction, and this is what is meant by saying that primitive "prelogical" thought is determined by "collective representations" and the "law of participation." Rather than secondary causes or empirical conditions, it is these mysterious occult qualities, often transcending spatial and temporal relations and defying Western comprehension, that are believed to be the ultimate determination of events.

While the unfortunate terminology of "mystical" and "prelogical" in Lévy-Bruhl's first book, *How Natives Think*, led to misunderstanding and criticism of his views, his second book, *Primitive Mentality,* is less subject to misinterpretation. As discussed in the first chapter, the world as we experience it is not self-explanatory; i.e., events occur for which there is no evident reason and thus man has been forced to try to find or invent the reasons for or causes of events. Since the rise of modern science we have learned to seek for explanations of events in secondary causes: in perceptible antecedent physical conditions and/or imperceptible underlying physical states and processes. Primitive man also sought to understand the reasons for events, but his conception of these reasons, as his perception of nature, was radically different from ours. He thought of the causes of events in terms of the "mystic powers" which enveloped these events. Accordingly, the primitive has an

> implicit faith in the presence and agency of powers which are invisible and inaccessible to the senses, and this certainly equals, if it does not surpass, that afforded by the senses themselves. To the prelogical mind these elements of reality — much the most important in his eyes — are no less matters of fact than the others. It is these which give the reason for what occurs [. . .] to the primitive the surrounding world is the language of spirits speaking to a spirit.[16]

Thus where we have learned to select and abstract from experience *physical* properties, conditions, and structures as the causes of phenomena, primitive man spontaneously apprehends *mystical* properties (i.e., feelings of fear or dread and a sense of psychic power or intention associated with and emanating from things) as the reason for their occurrence. It is in terms of these *different kinds of abstractions* that Lévy-Bruhl accounts for the prevalence among primitives of explanations in terms of spirits, demons, psychic forces, etc.

As these mystic powers function in a way that is radically different from that of the secondary causes investigated by modern scientists, primitive thinking about the world is necessarily different from ours; i.e., the primitive will have quite a different conception of the nature of these causes, how they operate, how they can be controlled, as well as the kinds of inferences one can draw from them. His world, instead of being composed of unseen atoms and energy, inferred physical fields and forces, will be peopled by spirits, demons, ghosts of dead ancestors, etc. It is not that he is indifferent to the empirical connections among events, but like us, he is aware of the necessity of seeking more basic causes. Secondary empirical causes are merely the means used by the fundamental occult powers to achieve their ends. This explains why divination and magic, which are attempts to 'divine' and influence these occult powers, play such an important role in primitive society.

As such, this view of primitive mentality does not deserve the caricatured description and wholesale rejection often given it by later anthropologists (particularly Malinowski). While his use of materials may have been somewhat uncritical by modern standards, while he lacked the benefit of firsthand experience of primitive peoples, and while he still seemed to be subject to some of the presuppositions of the 19th century evolutionary school, Lévy-Bruhl was an acute analyzer of modes of thought based on anthropological reports of the behavior of primitives. Moreover, his distinction between two domains of experience and thought, the more practical context in which the exigencies of life force the primitive to think much as we do, and the level of institutionalized collective thought, the kind of thought manifested in myth, magic, and religion, will have echoes even in such severe critics as Radin and Malinowski, and will find some support in our own culture where both the scientific and the religious mind continue to coexist to some extent.

MORE RECENT INTERPRETATIONS OF PRIMITIVE MENTALITY

The third type of interpretation of primitive mentality consists of recent investigations in ethnology that reject both the evolutionary position that primitive mentality represents a very early, elementary form of mental development, along with the thesis that primitive thought is so homogeneous that it lacks individuals capable of an objective, abstract, rational outlook. This group includes Boas, Radin, Malinowski, Radcliffe-Brown, Lévi-Strauss, etc. who in spite of their differences, have undertaken to show that primitive peoples exhibit rational, scientific thought. In any cross-section of a society of sufficient size, regardless of its level of cultural development, it is maintained there can be found in various proportions individuals possessing the same diverse mental capacities as found in more advanced societies. For example, in contrast to Durkheim and Lévy-Bruhl, Radin states:

> It must be explicitly recognized that *in temperament and in capacity for logical and symbolical thought, there is no difference between civilized and primitive man.* A difference exists — and one that profoundly colors primitive man's mental and possibly emotional life; but that is to be explained by the nature of the knowledge the primitive man possessed, by the limited distribution of individuals of certain specific temperaments and abilities and all that this implied in cultural elaboration.[17] (Italics added)

In his investigation of Indian societies of North America, particularly the Winnebago tribe, Radin found a range of intellectual interests and abilities similar to those in our society: "I feel quite convinced that the idealist and the materialist, the dreamer and the realist, the introspective and the non-introspective man have always been with us."[18] According to Radin, the reason earlier ethnologists did not find examples of speculative, rational thought among primitives is that they failed to distinguish between the man of action and the thinker, basing their conclusions primarily on what they observed about the man of action. Their neglect of a more sophisticated form of thought would be analogous to a description of contemporary mentality based on an investigation of an "average worker" in Western society — what would one learn about the contribution of an Einstein, Gödel, or Russell?

While Radin sought to show that primitive peoples include a wide range of different mental types, including "speculative philosophers," Malinowski attempted to demonstrate, on the basis of his experience with the Trobrianders, that primitive peoples are capable of a truly scientific attitude toward the world. If one means by science "a body of rules and conceptions, based on experience and derived from it by logical inference [. . .], then there is no doubt that even the lowest savage communities have the beginnings of science, however rudimentary."[19] Moreover, even if one demands a stricter criterion,

> that of the really scientific attitude, the disinterested search for knowledge and for the understanding of causes and reasons, the answer would certainly not be in a direct negative [. . .] there is both the antiquarian mind passionately interested in myths, stories, details of customs, pedigrees, and ancient happenings, and there is also [. . .] the naturalist, patient and painstaking in his observations, capable of generalization and of connecting long chains of events in the life of animals, and in the marine world or the jungle.[20]

Thus Malinowski concludes that "primitive man can observe and think, and that he possesses, embodied in his language, systems of methodological though rudimentary knowledge."[21]

As Radin, Malinowski believes that earlier anthropologists had been misled by an undue emphasis on one facet of primitive behavior or culture, thereby arriving at a distorted understanding of the total mentality. While Radin emphasized the necessity of recognizing two fundamentally different types of individuals when describing the mentality of primitives, the man of action and the speculative philosopher, Malinowski underscores the distinction between the sacred and the profane as essential to an adequate interpretation of the total dimension of primitive activities. Although primitive man recognized that survival required understanding as much as possible the natural conditions on which the life of the tribe depended, and developed reliable procedures for coping with these conditions (e.g., techniques of planting, hunting, fishing, etc.), he was even more aware than we that once empirical knowledge and skills have been utilized, the final outcome of any endeavor is often dependent upon unknown forces not susceptible to empirical control.

> It is most significant that in the lagoon fishing, where man can rely completely upon his knowledge and skill, magic does not exist, while in the open-sea fishing, full of danger and uncertainty, there is extensive magical ritual to secure safety and good results.[22]

This division between the empirical dimension of experience and the trans-empirical reality ultimately controlling the empirical (reminiscent of Lévy-Bruhl) defines the basic distinction between the profane and the sacred in primitive society. Science and technology were developed to deal with the domain of the profane, while magic (and religion) were devoted to the sacred.

> Science is founded on the conviction that experience, effort and reason are valid; magic on the belief that hope cannot fail nor desire deceive. The theories of knowledge are dictated by logic, those of magic by the association of ideas under the influence of desire [. . .]. The one constitutes the domain of the profane; the other, hedged round by observances, mysteries, and taboos, makes up half of the domain of the sacred.[23]

Accounts of primitive mentality based solely on a description of the stranger, more exotic aspects of the culture, such as magic and religion, naturally result in a distorted account of this mentality. Perhaps this explains why H. and H. A. Frankfort, in their otherwise excellent introduction to the study of primitive myths, deny the capacity for "speculative thought" to the peoples of the ancient Near East.

> If we look for 'speculative thought' in the documents of the ancients, we shall be forced to admit that there is very little indeed in our written records which deserves the name of 'thought' in the strict sense of that term. There are very few passages which show the discipline, the cogency of reasoning, which we associate with thinking. The thought of the acient Near East appears wrapped in imagination. We consider it tainted with fantasy. But the ancients would not have admitted that anything could be abstracted from the concrete imaginative forms which they left us.[24]

A description of the ancient myths of Egypt and Mesopotamia is the basis of this judgment. But is it likely that people who later discovered geometrical theorems and algebraic functions, who reputedly were able to predict the eclipse of the sun, and who could perform such technological feats as the construction of the pyramids, were incapable, even in the 2nd millenium B.C., of a disciplined form of "speculative thought? " Perhaps the error — if it is an error — lies in misinterpreting the primary function of ancient myths; that is, in assuming that their main function was to give a *rationally* satisfying account of natural phenomena. But if their primary intent, as suggested by Malinowski, was to give *symbolic* satisfaction to deeper needs of the psyche related to an awareness of the profoundly mysterious powers controlling human destiny and natural phenomena, then the meaning and logic of myths necessarily would be different: they would approximate more the content and structure of dreams, fantasy, and poetic imagination.

> Myth is not a savage speculation about origins of things born out of philosophic interest. Neither is the result of the contemplation of nature — a sort of symbolic representation of its laws [. . .]. The deep connection between myth and cult, the pragmatic function of myth in enforcing belief, has been [. . .] persistently overlooked in favor of the etiological or explanatory theory of myth [. . .].[25]

Accordingly, the primary function of myth is to validate magic — not to provide explanations of natural phenomena or to satisfy curiosity — and to "vouch for" the belief in the effectiveness of the magic and ritual of the tribe. As such, myth is more representative of what Piaget calls "autistic thought," than speculative or "directed thought."

> Directed thought is conscious, i.e., it pursues an aim which is present to the mind of the thinker; it is intelligent, which means that it is adapted to reality and tries to influence it; it admits of being true or false [. . .]. Autistic thought is subconscious, which means that the aims it pursues and the problems it tries to solve are not present in consciousness; it is not adapted to reality, but creates for itself a dream world of imagination; it tends not to establish truths, but to satisfy desires [. . .] it works chiefly by images, and in order to express itself, has recourse to indirect methods, evoking by means of symbols and myths the feeling by which it is led.[26]

In dreams the manifest content and structure are determined primarily by subconscious or unconscious factors, so that the meaning of the dreams must be explained by reference to latent associations having an unconscious origin and meaning. Although the content of myths is more empirical and the structure subject to more objective impersonal influences, it still shows signs of "condensation" and "transference" characteristic of dreams, though "this condensation and transference are not so absurd nor so deeply affective in character as in dreams or autistic imagination."[27]

Because of the "autistic" nature of myth, H. and H. A. Frankfort maintain that only "speculative thought attempts to *underpin* the chaos of experience so that it may reveal the features of a structure-order, coherence, and meaning."[28] But myth too consists of a conceptual framework (even if not primarily rational) developed for the interpretation of certain experiences, and exhibits its own inherent systematic order and import. Yet if myth is related to a different domain of experience, the sacred as opposed to the empirical or the profane, then it would be expected to exhibit a different *kind* of structure and significance. The thought of the ancient Near East may be "wrapped in imagination," but so is Western scientific thought in its dependence upon novel constructs and theoretical interpretations. It is not that science is less dependent upon imagination than myth, but that the imaginative content is abstracted from a different domain of experience.

This distinction between different domains of experience explains why a highly sophisticated form of empirical thought and technology can exist alongside the more subjective, symbolic, and to us, fantastic forms of mythical thought. But even individuals themselves exhibit different levels or patterns of thinking in different situations or under different conditions. According to Werner, "the normal adult, even at our own cultural level, does not always act on the higher levels of behavior. His mental structure is marked by not one but many functional patterns, one lying above the other."[29] This mixture of modes of thought was also stressed by Lévy-Bruhl.

In the mentality of primitive peoples, the logical and prelogical are not arranged in layers and separated from each other like oil and water in a glass. They permeate each other, and the result is a mixture which is a very difficult matter to differentiate.[30]

Those who have had the experience of teaching are aware of the tremendous variation in the thinking of young people in general or the multiple, interrelated levels of thought exhibited by any one student — and this is true, surely, not just of young minds. One recalls Pascal who was both an outstanding mathematician and a mystic, and recent scientists such as Jeans and Eddington who retained a marked religious outlook. Moreover, any one society can produce a variety of creative individuals; e.g., in roughly the same period France produced symbolist poets such as Rimbaud and Verlaine, impressionist painters such as Monet and Cézanne, as well as scientists like Becquerel, the Curies, and Poincaré.

As a representative of recent ethnological points of view, Lévi-Strauss, even more than Radin and Malinowski, has sought to demonstrate that primitive thought exhibits the *same degree of rationality* as that of modern man. In fact, within the bewildering and seemingly irrational complexity of such primitive forms of culture as totemism and myth, Lévi-Strauss finds a highly sophisticated and articulated logical schema. In contrast to Lévy-Bruhl who denied that mental operations could be reduced to a single form, maintaining that though the thought of primitive peoples is not controlled by a different logic, yet it does not conform exclusively to our logic either, Lévi-Strauss finds a common logical structure underlying *all* forms of thought, a structure as rigorous in myth as it is in modern science.

If our interpretation is correct, we are led toward a completely different view — namely, that the kind of logic in mythical thought is as rigourous as that of modern science, and that the difference lies, not in the quality of the intellectual process, but in the nature of the things to which it is applied.[31]

Moreover, not only is the logic of primitive thought as "rigorous" as that of the contemporary thinker, it is *of the same type*: "The savage mind is logical in the same sense and the same fashion as ours, though as our own is only when it is applied to knowledge of a universe in which it recognizes physical and semantic properties simultaneously."[32] That is, when we become aware that scientific knowledge is a function of a conceptual-linguistic framework, as well as observations and experiments, and not just a duplication of nature, then the "logic of our thought" is identical with that of the primitive. If the modern thinker discards the notion that he knows nature as it is, and realizes that he knows it only relative to his intellectual or symbolic framework, then he is in a position to recognize that his thinking is similar to that of the primitive. Rather than physical and semantic properties being *in*dependent, as one assumed in the past, one must recognize that they are *inter*dependent: that all knowledge is framework dependent. Lévi-Strauss seems to be saying that the primitive thinker was unconsciously aware of this while the modern thinker has only recently explicitly recognized this.

Lévi-Strauss was guided in his interpretation by developments in structural linguistics, particularly the science of phonemics. By shifting their attention "from the study of *conscious* linguistic phenomena to the study of their unconscious *infrastructure*," modern linguists were able to discover a universal phonemic structure underlying diverse linguistic systems.[33] Lévi-Strauss urges anthropologists to adopt the structuralist approach of linguists, hoping to find a common pattern pervading the diverse forms of primitive organization. That is, can one discover among kinship relations an underlying universal structure similar to that of phonemics? Lévi-Strauss believes that he has discovered such a structure, one which is further evidence of an unconscious unitary logic unifying all modes of thought.

> If, as we believe to be the case, the unconscious activity of the mind consists in *imposing forms upon content*, and if these forms are fundamentally the same for all minds — ancient and modern, primitive and civilized (as the study of the symbolic function, expressed in language, so strikingly indicates) — it is necessary and sufficient to grasp the unconscious *structure* underlying each institution and each custom, in order to obtain a principle of interpretation valid for other institutions and other customs, provided of course that the analysis is carried far enough.[34] (Italics added)

This conception of the mind imposing "forms upon content" is strikingly reminiscent of Kant and of the neo-Kantian, Ernst Cassirer, neither of whom are referred to by Lévi-Strauss. In fact, Lévi-Strauss' statement is almost a direct paraphrase of an earlier statement by Cassirer:

> In the critical view [i.e., the view derived from Kant] we obtain the unity of nature only by injecting it into the phenomena; we do not deduce the unity of logical form from the particular phenomena, but rather represent and create it through them. And the same is true of the unity of culture and of each of its original forms. It is not enough to demonstrate it empirically through the phenomena; we must explain it through the unity of a specific "structural form" of the spirit.[35] (Brackets added)

There is even a close similarity of terminology between Lévi-Strauss' "structuralism" and Cassirer's notion of "structural form," as seen in the following statement by Lévi-Strauss: "One consequence of modern structuralism" is to show that the mind (or brain) possesses an "elementary logic, which is like the least common denominator of all thought [. . .] an original logic, a direct expression of the structure of the mind [. . .]."[36]

However, although Cassirer and Lévi-Strauss agree that the mind imposes "structural forms" on the content of experience, their views do differ markedly. Following Kant, Cassirer interprets the empirical content of myth as disclosing a "mythical intuition" and a "mythical form" which together are distinct, autonomous, and necessary *modes* or *functions* of the mind in the "Spirit's" effort to free itself from subjectivity and achieve a more objective representation of the world. Myth is as necessary a mode of thinking as science in the *development* of the mind toward greater objectivity in its understanding of the world:

> [. . .] we may inquire into the pure essential character of the mythical function [. . .] and set this pure form in contrast with that of the liguistic, aesthetic, and logical functions [. . .]. We shall no longer seek to explain it as the expression and reflection of a transcendent process or of certain constant psychological forces [. . .] but functionally [. . .] in what myth itself is and achieves, in the manner and form of *objectivization* which it accomplishes. It is objective insofar as it is recognized as one of the determining factors by which consciousness frees itself from passive cativity in sensory impression and creates a world of its own [. . .].[37]

In contrast to Cassirer's notion that myth exhibits a unitary, autonomous *function* in the *mind's development* toward greater objectivity, Lévi-Strauss finds the cultural forms of myth and totemism to be diverse manifestations not of a *special function* of the mind, but of a *general logic*, a logic bound by "binary oppositions." Thus there is no *development* of the mind involved, only different manifestations of *a common underlying logic* utilizing different contents (or different kinds of abstractions). From the earliest stages of primitive thought to the present, there has been no fundamental change or development in the "logical process" of this thought, but merely differences as regards the content to which this logic is applied:

> [. . .] the same logical processes operate in myth as a science and [. . .] man has always been think-ing equally well; the improvement lies, not in an alleged progress in man's mind, but in the discovery of new areas to which it may apply its unchanged and unchanging powers.[38]

Thus Lévy-Strauss goes further even than other ethnologists in denying *any* essential difference between the thought process of the primitive and that of the modern thinker. Yet, as Werner states (without agreeing with the thesis), the attitude of most modern ethnologists tends to the view that

> there is no essential difference between civilized and primitive man. The difference, such as it is, lies only in cultural sets; the one people lives in an advanced culture, the other in a primitive one, but the mental functions of the individual are fundamentally the same.[39]

In contrast to Malinowski's attempt to explain the origin of myth, magic and re-ligion as satisfying emotional, unconscious, and psychical needs in the individual, Lévi-Strauss interprets myth and totemism as satisfying primarily intellectual needs, par-ticularly the primitive's need to organize his world into some classificatory scheme: the need to introduce some rational order into the universe. "Classifying, as opposed to not classifying, has a value of its own, whatever form the classification may take [. .]. The thought we call primitive is founded on this demand for order."[40] As all classi-fication depends upon discriminating, selecting, and abstracting certain qualities, forms, or functions within the panorama of experience, primitive classification too is based on certain abstractions isolated from the flux of experience and rendered relatively stable by being named and incorporated into a general schema. In this way the natural kinds of things (e.g., species of animals and plants) are made less alien by being assimilated into the conscious experience of the primitive.

The striking differences among the species of animals, their traits lending themselves to human characterizations, make them particularly suitable for primitive classificatory schema, as illustrated in the totemic organization of primitive societies. In the process, not only are animals invested with anthropomorphic and social characteristics ("the world of animal life is represented in terms of social relations similar to those of human society"), but personal traits and group distinctions are rendered conceptually objective by being fitted into the formal schema of totemism. The distinctive nature of natural species, the contrast between particular animals and their general class, along with the homologous relation between animal traits and human behavior thus provide the most obvious set of distinctions in terms of which primitive man imposes a systematic order on the natural world, as well as on his own social organization of kinship relations.

> Totemic classifications have a doubly objective basis. There really are natural species, and they do indeed form a discontinuous series; and social segments for their part also exist. Totemism [. . .] confines itself to conceiving a homology of structure between the two series, a perfectly legitimate hypothesis [. . .].[41]

But underlying the classificatory schema of totemism and myth is a logical relation, that of "binary opposition." Although nature itself exhibits obvious examples of op-posites, such as male/female, night/day, sky/earth, friend/foe, etc., Lévi-Strauss believes that experience alone of these empirical contrasts would be insufficient to account for the striking prevalence in primitive thought of "opposition" and "correlation" (and presumably for their occurrence also in the later speculations of such Presocratic philosophers as Anaximander, the Pythagoreans, and Heraclitus, as well as in the thinking of Hegel and Marx). Thus "opposition" and "correlation" are evidence of a particular logic, a "binary logic," which is innate and hence universal to all thought, but manifested in different ways in different cultures:

> [. . .] totemism is no more than a particular expression, by means of a special nomenclature formed of animal and plant names (in a certain code, as we should say today), which is its sole distinctive characteristic, of correlations and oppositions which may be formalized in other ways [. . .]. The most general model of this, and the most systematic application, is to be found perhaps in China, in the opposition of the two principles of Yang and Yin [. . .].[42]

Once the oppositions and correlations are incorporated into the classificatory schema, whether of totemism, myth, or philosophical speculation, the thinker can then form progressively more general relations on higher levels of abstraction. Thus totemism is interpreted by Lévi-Strauss as a manifestation of an innate logic which imposes order on the world in terms of successive abstractions.

> As medial classifier (and therefore the one with the greatest yield and the most frequently employed) the species level can widen its net upwards, that is, in the direction of elements, categories, and numbers, or contract downwards, in the direction of proper names [. . .]. Each system is therefore defined with reference to two axes, one horizontal and one vertical [. . .].[43]

In contrast to Cassirer who interprets myth and totemism as unique, necessary mental functions in the mind's striving for objectivity, or in contrast to Malinowski who finds in psychoanalysis a basis for interpreting myth, Lévi-Strauss finds such cultural expressions to be abstract, complex, taxonomic schemata which are natural expressions of an innate binary logic by means of which the primitive imposes an ordered network on experience.

There is, in this view, no essential difference between the thought of the primitive and that of the modern scientist. The difference lies in the fact that while the primitive utilizes abstractions derived from the most immediate sensory level of experience in his interpretation of phenomena, the modern scientist, as a result of a long tradition in the development of knowledge, incorporates into his theoretical explanations abstractions based on experimentation and mathematics, which abstractions are therefore at some remove from immediate perception. While the primitive will explain the birth of a child with a harelip by the fact that just before the birth the mother tore a rend in her garment, or that a child was born with a curled ear because in the last stages of the mother's pregnancy the father had unaccustomedly rolled up his sleeping mat, a scientist will reject these superficial resemblances as explanations and seek for more basic causes. For Lévi-Strauss, although one can distinguish "two distinct modes of scientific thought," depending upon the level of abstraction, this does not imply "a function of different stages of development of the human mind [. . .]."[44]

On this view, the primitive form of thought manifested in myth and totemism is not an example of a different *kind of thought from that of science*, but of a different *mode of scientific thought* itself: a type of scientific thought which seeks for explanations within the immediate sensory realm of experience rather than in terms of inferred entities lying behind this experience. As we shall find later, this differentiation of two modes of scientific thought essentially distinguishes Aristotelian science, based on immediate common sense experience, from the mechanistic science of the 17th century, based on abstract mechanical concepts. What Lévi-Strauss affirms regarding primitive thought will prove to be true of Aristotelian science: "This science of the concrete was necessarily restricted by its essence to results other than those destined to be achieved by the exact natural sciences but it was no less scientific and its results no less genuine."[45]

CONCLUSION

What, then, can be concluded from these various ethnological interpretations of primitive mentality? Are there different logics of explanation, depending upon whether the primitive is considering the secondary empirical causes of phenomena or the more occult, psychic causes, as Lévy-Bruhl maintained? Do the mental processes of the primitive man of action differ from those of the more reflective individual, so that only the thought of the latter approximates that of the modern thinker, as Radin argued? Should one distinguish between two domains of experience, the sacred and the profane, realizing that it is only with respect to the latter that the primitive displays genuinely scientific forms of thought, as Malinowski claimed? Are there specific innate modes of mythical intuition and forms of thought that the human mind necessarily manifests in its cognitive development from subjectivity to objectivity, as Cassirer believed? Or does the primitive display logical processes which are the same as, and just as rigorous as, our own, as exemplified in the intricate classificatory systems embedded in totemic organizations, as Lévi-Strauss thought he had discovered?

While it seems reasonable to countenance different levels or modes of thought, as well as different logics of explanation, depending upon the domain of experience and/or intellectual intent of the primitive, the evidence for there being a single logic, equally rigorous underlying all thought, seems less convincing. Although the cells in our cerebral cortex often discharge according to a simple on/off, either/or system, thereby providing a neural analogue for Lévi-Strauss' innate binary logic, it is not that apparent that all thought is governed by such a logic. The rationale of myths, for example, as dreams, is not controlled by the usual laws of logic, since contradictions are commonplace. If all thought reflected the same innate logic, then, as Locke maintained in arguing against Descartes' doctrine of innate ideas, this logic would have to be manifested in all thought. But the acceptance of contradictory attributes and fantastic occurrences is typical in the literature of primitive peoples, as well as in the fairy tales of children. Even in Western culture it was not so long ago that many people believed that the world was created from nothing, and that the deity consisted of three persons in one. Moreover, in his investigation of the development of logical and mathematical thinking in children, Piaget found that the manifestation of logical principles is not the same at all ages, but varies with different stages of development.[46] Why, then, should this not be true of mankind in general? After all, Parmenides was the first to employ the law of contradiction as a theoretical principle, while the formulation of the laws of logic had to await Aristotle in the 4th century B.C.

Nor need we account for the prevalence of opposites and the desire for classification in primitive thought solely on the basis of an inherent binary logic. As already suggested, nature itself exhibits numerous examples of opposites, while the need to impose order on experience in terms of a classificatory schema does not imply anything as specific as an innate binary logic — primitive man could hardly survive in a world that did not admit of any systematic organization. Lévi-Strauss would probably reply that as in structural linguistics, where the grammatical differences of radically dissimilar natural languages have not precluded the discovery of a universal, deep linguistic structure, so the widely different patterns of thought merely represent surface variations of a deeper logical structure. But apart from the question of the validity of the analogy with linguistics, the issue is not yet settled even in linguistics itself.

The main question, then, is whether all the diverse manifestations of thought reflect a common, underlying, universal logic, so that all human beings necessarily think alike, or whether a comparison of various peoples, cultures, and mental activities disclose different modes of thought, some developing in diverse ways. Expressed differently, do the apparent variations in human thought merely reflect an application of a common logic to alternative contents, relative to evolving cultural stages, or has the inherent form of thought, as well as the content, also undergone change?

It is significant that in opposition to Lévi-Strauss, historians of science often stress the fact that revolutionary developments in science do not depend simply on the discovery of novel data, but on encountering theoretical anomalies that require drastic new ways of thinking about traditional problems. Moreover, even the recognition of anomalies may itself mark a change in the intellectual outlook, or at least the beginning of such a change, which was the case with Copernicus. In his classic study of the origins of modern science, Butterfield states that "change is brought about, not by new observations of additional evidence in the first instance, but by transpositions that were taking place inside the minds of the scientists themselves [. . .].[47]

The fact that at times of revolutionary scientific change much of the fabric of past thought had to be radically modified, in terms of presuppositions, basic concepts, and conceptual implications or demonstrations, was also emphasized by Alexandre Koyré:

> [. . .] what the founders of modern science, among them Galileo, had to do, was not to criticize and to combat certain faulty theories, and to correct or to replace them by better ones. They had to destroy one world and to replace it by another. They had to reshape the framework of our intellect itself, to restate and to reform its concepts, to evolve a new approach to Being, a new concept of knowledge, a new concept of science — and even to replace a pretty natural approach, that of common sense, by another which is not natural at all.[48]

While it is reasonable to assume that thought processes, as physiological functions, must be based on certain innate structures, is it necessary to suppose that these structures are unchanging? Just as histology provides evidence for changing biological functions related to evolving physiological structures, so the history of science seems to provide evidence for changing forms of thought related to developing cognitive structures. This, at least, is one of the questions this study shall attempt to answer.

CHAPTER II

[1] Paul Radin, *Primitive Man as Philosopher* (New York: Dover Pub., Inc., [1927] 1956), p. 5.

[2] *Ibid.*, pp. 6-7.

[3] Cf. Claude Lévi-Strauss, *The Savage Mind*, trans. by George Weidenfeld and Nicolson (Chicago: The Univ. of Chicago Press, 1966), chs. 3, 5.

[4] Cf. Claude Lévi-Strauss, *Totemism*, trans. by Rodney Needham (Boston: Beacon Press, 1963), ch. 5.

[5] Heinz Werner, *Comparative Psychology of Mental Development* (New York: Science Editions, Inc., [1948] 1961), p. 39.

[6] Cf. Sir Edward B. Tylor, *Primitive Culture*, 2nd ed. (London: John Murray, [1871] 1872).

[7] Cf. Sir George Frazer, *The Golden Bough: A Study in Magic and Religion,* 3rd ed. (New York: The Macmillan Company, 1935), 12 vols.

[8] Sir George Frazer, *The New Golden Bough* (Pt. VIII, sec. 583), ed. by T. H. Gaster (New York: Criterion Books, 1959), p. 648.

[9] Cf. Emile Durkheim, *The Elementary Forms of Religious Life*, trans. by J. W. Swain (New York: Collier Books, [1915] 1961), p. 22.

[10] Lucien Lévy-Bruhl, *How Natives Think*, trans. by L. A. Clare (New York: Washington Square Press, Inc., [1910] 1966), p. 19.

[11] Cf. Bronislaw Malinowski, *Magic, Science and Religion* (Garden City: Doubleday & Co., 1955), p. 25.

[12] Lévy-Bruhl, *op. cit.*, pp. 63-64.

[13] *Ibid.,* p. 63.

[14] *Ibid.,* p. 23.

[15] *Ibid.,* p. 25.

[16] Lucien Lévy-Bruhl, *Primitive Mentality*, trans. by Lilian A. Clare (Boston: Beacon Press, [1922] 1966), p. 60.

[17] Radin, *op. cit.*, p. 373.

[18] *Ibid.,* p. 365.

[19] Malinowski, *op. cit.*, p. 34.

[20] *Ibid.,* p. 35.

[21] *Ibid.,* p. 33.

[22] *Ibid.,* p. 31.

[23] *Ibid.,* p. 87.

[24] H. and H. A. Frankfort, "Introduction," *Before Philosophy* (Baltimore: Penguin Books, Inc., 1949), p. 11.

[25] Malinowski, *op. cit.*, pp. 83-84.

[26] Jean Piaget, *The Language and Thought of the Child*, 3rd ed., trans. by Marjorie Gabian (London: Routledge & Kegan Paul, Ltd., [1926] [1932] 1959), p. 43.

[27] *Ibid.,* p. 159.

[28] H. and H. A. Frankfort, *op. cit.*, p. 11.

[29] Werner, *op. cit.*, p. 38.

[30] Lévy-Bruhl, *How Natives Think*, *op. cit.*, p. 89.

[31] Claude Lévy-Strauss, *Structural Anthropology*, trans. by Jacobson and Schoeph (Garden City: Doubleday & Co., Inc., 1967), p. 227.

[32] *Op. cit., The Savage Mind*, p. 268.

[33] Cf. *op. cit., Structural Anthropology*, p. 32ff.

[34] *Ibid.*, pp. 21-22.

35 Ernst Cassirer, *The Philosophy of Symbolic Forms*, Vol. II, trans. by Ralph Manheim (New Haven: Yale Univ. Press, 1955), p. 11.

36 Lévi-Strauss, *Totemism*, *op. cit.*, p.90.

37 Cassirer, *op. cit.*, pp. 13-14.

38 Lévi-Strauss, *Structural Anthropology*, *op. cit.*, p. 227.

39 Werner, *op. cit.*, p. 17.

40 Lévi-Strauss, *The Savage Mind*, *op. cit.*, p. 9-10.

41 *Ibid.*, p. 227.

42 *Op. cit.*, *Totemism*, pp. 88-89.

43 *Op cit.*, *The Savage Mind*, p. 149.

44 *Ibid.*, p. 15.

45 *Ibid.*, p. 16.

46 Cf. Jean Piaget, *Genetic Epistemology*, trans. by Eleanor Duckworth (New York: Columbia University Press, 1970).

47 Herbert Butterfield, *The Origins of Modern Science* (New York: Collier Books, [1957] 1962), p. 13.

48 Alexandre Koyré, "Galileo and Plato," in *Roots of Scientific Thought*, ed. by Philip Wiener and Aaron Noland (New York: Basic Books, 1957), p. 152.

CHAPTER III

PRIMITIVE FORMS OF COGNITION

The primitive mentality is unusually rich in 'condensations.'[1]
Werner

Regardless of the differences of opinion among ethnologists as to the degree of rationality manifested by primitive peoples, the one point on which all investigators of primitive forms of cognition seem to agree is that the difference between lower and higher levels of thought consists essentially of the greater concreteness and subjectivity of primitive thought. Whether one considers the mental capacity of primitive peoples, of children, or of mentally retarded individuals (i.e., those born defective or those who have suffered brain damage), one finds that the various levels are measured in terms of the development from more concrete to more abstract, from more subjective to more objective modes of perception and cognition. The purpose of this chapter, therefore, is to delineate the characteristics of a more elementary type of primitive mentality — whether that of undeveloped individuals, such as the primitive and the child, or mature individuals with a very low mental capacity — in contrast to a more advanced or conceptual form of thought. While this chapter does not claim to describe the mental processes of primitive peoples as such, it does propose to distinguish, utilizing both psychological and anthropological data, a form of primitive mentality. These distinguishing characteristics will be used later in identifying and describing the thought processes of Hellenic and Hellenistic thinkers in their intellectual effort to attain a more objective, comprehensive understanding of nature.

SYNCRETIC CHARACTER OF PRIMITIVE EXPERIENCE

Though it is natural to assume that everyone's perception and theoretical interpretation of the world, in terms of a natural order of independently existing physical objects and events, occurring within an objective framework of space and time, and exhibiting inherent causal relations, must hold for all individuals, this is easily shown not to be the case. At a certain level of mental development the world is not structured by means of such perceptual forms and theoretical categories. The achievement of such an objective interpretation required considerable mental development in terms of greater discrimination among subjective and objective contents, differentiation of cognitive functions, and conceptual refinements. Lower stages of mental functioning, in which the contents and the functions of the mind differentiated on higher levels are still fused, is called "syncretic perception" by Werner: "We have repeatedly emphasized the syncretic character of primitive perception. This means, in brief, that motor and affective elements are intimately merged in the perception of things."[2]

Neither the child nor the primitive draw as clear a distinction between subjective experience and the objective natural world as does the contemporary adult. On the level of syncretic perception, objective sensory impressions are not sharply differentiated from subjective feelings and kinaesthetic sensations, nor are dreams, hallucinations, and fantasy recognized as being merely subjective, nor is the association of images and concepts formed as much by the external sequence of events as by the heat of subjective emotions. The reality of an event is not determined primarily by its occurrence within a natural causal context, but by the effect it has upon the emotional life of the individual — the stronger the effect, the greater the reality. The literature is replete with examples of the primitive tendency to exaggerate martial and marital exploits, while the strangeness and vividness of images are taken (or mis-taken) as criteria of their greater objectivity.

> Everywhere in primitive society we find that visionary appearances not only are accorded an objective value, but are also invested with a superior significance, in that they may even supersede the common reality of life from day to day. One is proud of seeing things others do not.[3]

This explains the tendency among primitives to look to hallucinations as authenticating beliefs and rites, seeking individuals with unusual temperaments who claim extraordinary psychic powers for the role of medicine-man or shaman. As Radin states, "Throughout the world of primitive man some form of emotional instability and well-marked sensitivity has always been predicated as the essential trait of the medicine-man and shaman."[4]

The common primitive fear of the dead and of their possible effects on the living also reflects the import of affective experiences on his conceptions. The usual dread with which we all apprehend the deceased was evidence for primitive man of the lingering presence of the dead man's spirit and the potentiality it could have for harming the living. Moreover, when in dreams, fever, or religious ecstasy he saw phantom human forms, these experiences were taken as providing authentic evidence for the existence of such creatures.

> Just as there was no sharp distinction among dreams, hallucinations, and ordinary vision, there was no sharp separation between the living and the dead. The survival of the dead and their continued relationship with man were assumed as a matter of course, for the dead were involved in the indubitable reality of man's own anguish, expectation, or resentment. 'To be effective' to the mythopoetic mind means the same as 'to be.'[5]

In addition to the fusion of sensory, affective, and kinaesthetic elements at the level of syncretic perception, described by Werner, this level of perception has a further characteristic, as pointed out by Piaget, which accounts for two other commonly cited manifestations of primitive mentality. It is a well-known and occasionally annoying fact that at a certain age a child will constantly ask for the "reasons why" of things, having no conception that events occur by accident. Piaget found that for "lack of a definite idea of chance, he [the child] will always look for the why and wherefore of all the fortuitous juxtapositions which he meets in experience."[6] Similarly, the primitive also believes that nothing occurs accidentally but always because of some intention. If he suddenly is seized with a tremor of foreboding or fear, this is not interpreted as merely a subjective apprehension probably without objective significance, but is invariably attributed to some unseen (but felt) daemon or spirit. A terrifying dream is not merely a subjective phenomenon of no objective import, but is a portent of some dreadful event to occur. The fact that a native is surprised and wounded by a hostile war party, that a child is born deformed, or that a woman is dragged into the water and eaten by a crocodile is never merely an accidental matter, but due to some more basic power or magic. In the view of Cassirer,

> inability to conceive of an event that is in any sense "accidental" has, in any case, been called characteristic of mythical thinking. Often where *we* from the standpoint of science speak of "accident," mythical consciousness insists on a cause and in every single case postulates such a cause. For primitive peoples a catastrophe that descends on the land, an injury which a man suffers, sickness, and death are never "accidental;" they always go back to magical interventions as their true causes.[7]

This notion that one must be able to explain the "why" as well as the "how" of things was still prevalent in the thinking of the ancient Greek philosophers. Socrates told of his disappointment with the lectures of Anaxagoras because the latter failed to make enough use of mind or *"nous"* in his explanations. Plato's whole cosmology as stated in the *Timaeus* has as its primary intent to show that the universe, which exhibits order and purpose, could not have arisen by chance or necessity, but must have been formed by a Divine Craftsman. Aristotle would not consider any causal explanation complete which did not include an account of the end or purpose attained by the occurrence, his famous four causes serving to explain the "why" of things. Even today this mode of thinking is apparent when people ask why (rather than how) an unexpected tragic event, such as a death or an accident, occurred. Piaget attributes this mode of thought to the wholistic, interrelated nature of syncretic perception and understanding: "The illogical character of childish 'whys,' or the absence of the notion of chance revealed in primitive 'whys' are therefore due to syncretism of understanding and perception [. . .]."[8]

Furthermore, the fact that in syncretic perception the aspects or parts of things are not sharply distinguished and categorized leads to the equivalence of the parts with the object, thus explaining another commonly noted example of primitive mentality, termed *"pars pro toto."* This refers to a lack of differentiation between the part and the whole, such that the part retains the properties of the whole, regardless of its fragmentary nature or spatial or temporal separation from the whole. In magic, which exemplifies this mode of conception, whatever rite is performed on a part of the individual, his hair, excrement, piece of clothing, etc., is believed to have the same effect on the individual himself.

SYNCRETIC CONCEPTION OF SPACE AND TIME

The relatively undifferentiated, concrete character of syncretic perception is exemplified also in numerous accounts of the primitive's conception of space and time. As described by many investigators, the primitive notion of space and time is not that of an abstract, metrical, isomorphic coordinate system, but a configuration fused with the affective, social, and practical activities of the individual and the tribe.

> Space is postulated by us to be infinite, continuous, and homogeneous — attributes which mere sensual perception does not reveal [. . .]. The spatial concepts of the primitive are concrete orientations; they refer to localities which have an emotional colour; they may be familiar or alien, hostile or friendly. Beyond the scope of mere individual experience the community is aware of certain cosmic events which invest regions of space with a particular significance. Day and night give to east and west a correlation with life and death.[9]

Leucippus and Democritus were the first to formulate a conception of space as empty intervals separating the atoms, a necessary presupposition to account for the motion and compactness of the atom. Newton later gave space its modern (but not contemporary) form as a three dimensional, Euclidean, isotropic, infinite container enclosing all cosmic events. But even Newton had not quite freed space from earlier mythological associations in that he conceived of space (and time) ultimately as *"sensoria dei,"* an absolute framework of God's mind. In contrast to the isomorphic quality of Newtonian space, primitive space is structured in accordance with some concrete referent, a person's body, the earth, the totemic organization of the tribe, or the rising and setting of the sun. As Cassirer notes,

> the terms of spatial orientation, the words for 'before' and 'behind,' 'above' and 'below' are usually taken from man's intuition of his own body: man's body and its parts are the system of reference to which all other spatial distinctions are indirectly transferred.[10]

Thus primitive space is not conceived as logically independent of the objects and events located in it, as in Newtonian science, but is apprehended as qualified by whatever occupies it (remotely analogous to the Einsteinian notion of "space-time" as a metric varying with the density of matter and the strength of the gravitational field).

As late as Aristotle the various elements, the celestial element aether and the terrestrial elements, air, earth, fire, and water, were assumed to have their natural place such that they (or their motions) were affected by their positions in space, though space itself had no existence apart from the totality of places; moreover, up and down were absolute directions either away from or towards the center of the earth which was believed to be in the center of the universe. In medieval thought the "great chain of being" was associated with a hierarchical structuring of space. But even as late as the 17th century, space was still divided qualitatively into celestial and terrestrial realms.

However, when it comes to more practical considerations, relating spatial and temporal designations to actual activities, the primitive conception can approximate that of relativity theory. Whorf provides an interesting account of the operational meaning of time for the Hopi Indians — interesting because it approximates the Einsteinian notion of relational space-time as compared to the Newtonian conception of an absolute, independent space and time. In the Hopi language, as in relativity theory, the Newtonian (and common sense) conception of the absolute simultaneity of distant events has no meaning. Events which are spatially separate cannot be simultaneous because the time it takes to convey information about one of the distant events places its occurrence in the past in relation to the other event: the abstract conception of 'there-now' has no meaning, only the operational concept 'there-then.' As in relativity theory, only events occurring in the same place at the same moment can be truly simultaneous.

> The Hopi metaphysics does not raise the question whether the things in a distant village exist at the same present moment as those in one's own village, for it is frankly pragmatic on this score and says that any 'events' in the distant village can be compared to any events in one's own village only by an interval of magnitude that has both time and space forms in it [as in relativity theory]. [. . .] What happens at a distant village, if actual (objective) and not a conjecture (subjective) can be known 'here' only later. If it does not happen 'at this place,' it does not happen 'at this time'; it happens at 'that' place and at 'that' time.[11] (Brackets added)

When freed from an operational definition, time for the primitive usually is related to the periodic phases of biological, social, or natural events. The beginning of time is often dated from the legendary appearance of the founder of the tribe, just as before Darwin some people believed that the universe was created in 4004 B.C., and as our designations "B.C." and "A.D." mark the impact of a sacred historical figure on our own culture. As we shall find continually in the course of intellectual history, concepts initially derive their meaning from the most familiar and concrete areas of experience, gradually becoming more abstract and refined as they are subjected to further analysis and critical application.

PHYSIOGNOMIC PERCEPTION

In addition to acknowledging the concreteness of primitive thought, most accounts of primitives also agree on another point; namely, that the world of primitive peoples is a dynamic, animated, living reality in which natural phenomena are considered to be manifestations or embodiments of a kind of spiritual power. Codrington's definition of "mana" as "a force altogether distinct from physical power, which acts in all kinds of ways for good and evil,"[12] called attention to this dynamic aspect of the primitive's world. Tylor also traced the origin of religion to a pervasive "animism" that colors the mentality of primitive peoples contributing to an inherent disposition to "personify" natural phenomena as the embodiment of souls, spirits, ghosts of the dead, etc.: "[...] spiritual beings are modelled by man on his primary conception of his own soul [...] [and] their purpose is to explain nature on the primitive child-like theory that it is truly and throughout 'Animated Nature'."[13] While denying Tylor's thesis of primitive animism, H. and H. A. Frankfort also claim that primitive man encounters an animate world, a world that can be characterized as a "Thou" due to its living qualities, as compared to the impersonal "It" world of contemporary man. However, on their view primitive man does not "personify" nature — he directly apprehends it as animate.

> Primitive man simply does not know an inanimate world. For this very reason he does not 'personify' inanimate phenomena nor does he fill an empty world with the ghosts of the dead, as [Tylor's] 'animism' would have us believe [...]. Any phenomenon may at any time face him, not as 'It,' but as 'Thou.' In this confrontation, 'Thou' reveals its individuality, its qualities, its will.[14] (Brackets added)

Werner also denies that animism is the result of a personification of nature, but he goes further in providing a psychological interpretation according to which animism is rooted in a very early and primitive form of perception. On this level of perception entities and occurrences are not perceived as inanimate, objective, independent phenomena, but as "things-of-action" expressive of temperament, will, intention, etc. Like physiognomic characteristics which are defined as external bodily manifestations (e.g., facial expressions and physical gestures) of inner states of mind, such as moods, intentions, character traits, etc., Werner has called this level of perception "physiognomic."

> Things perceived in this way may appear "animate" and, even though actually lifeless, seem to express some inner form of life [...]. I have proposed the term *physiognomic perception* for this mode of cognition in general. There is a good deal of evidence that physiognomic perception plays a greater role in the primitive world than in our own [...].[15]

Phenomena perceived physiognomically are apprehended within an active interpersonal context and thus exhibit qualities which are a function of one's own actions, expectations, reactions, etc. Objects are perceived in terms of their subjectively felt effects upon the individual, such as threatening, friendly, cruel, protective, awesome, etc. A baby naturally responds to the subjective moods of the mother on the basis of physical cues, and young children have a great capacity for seeing things physiognomically. Werner has recounted a number of illustrations of this, but the following is one of the most charming:

> A five-year-old girl is asked by her mother during a thunderstorm: "What does the thunder look like?" The child replies: "He has a head, but no eyes and no nose or mouth." "Then how does he look?" "Oh, he looks like this . . ." and the child makes an angry face and draws her brows together. [16]

On this level objects are not perceived in terms of what we understand to be their permanent, independent, physical properties, as abstracted from their subjective effects upon the individual, but as dynamic configurations in which sensory qualities, kinaesthetic sensations, and feelings are fused. These affective dynamic properties of things are what stand out as the most essential properties, as compared to their more static objective properties.

It should be emphasized that in this mode of perception there is no *transference* of qualities from the personal context to the impersonal context; rather, the individual *first* perceives a thing as possessing *all those qualities* which he experiences when it affects him. Only later does he learn to differentiate and abstract his subjective, affective feelings and sensations from the more objectively permanent, sensory qualities and properties of the object. This mode of experience is not a deviation from a standard type of perception characteristic of the educated 20th century individual, but is itself a normal level of perception in both the child and the primitive. Werner claims that this capacity is so basic that it antedates the primitive tendency toward personification and anthropomorphism:

> So far as *primitive man* is concerned it is hardly necessary to labor the commonly known facts of animism and anthropomorphism in primitive civilizations. It is my belief that the anthropomorphic concept of nature is only a secondary phenomenon based on a deep-seated dynamic and physiognomic perception. Nature when known physiognomically is alive throughout, not because the soul, the vitality, is invested in the inanimate object, but rather because every thing is understood to behave dynamically [. . .]. [17]

In accordance with Werner's contention that primitive modes of cognition occur along with more advanced forms, both in better educated individuals as well as in more advanced societies, one can recall instances of physiognomic perception in himself and in others. Who has not seen a face in the clouds, a shadowy bush as a lurking form, or felt a storm as threatening and the sea as ominous? Also, as Werner has pointed out, it seems that physiognomic perception has played an important role in the creativity of certain artists. It has often been said that some of the charm and insight of poetry derive from the freshness and sensitivity with which the poet is capable of experiencing the world. The following description by Kandinsky indicates how, as an artist, he was aware of his own capacity for perceiving things physiognomically:

> On my palette sit high, round rain-drops, puckishly flirting with each other, swaying and trembling. Unexpectedly they unite and suddenly become thin, sly threads which disappear in amongst the colors, and roguishly skip about and creep up the sleeves of my coat [. . .]. It is not only the stars which show me faces. The stub of a cigarette lying in an ash-tray, a patient, staring white button lying amidst the litter of the street, a willing, pliable bit of bark — *all these have physiognomies* for me [. . .]. As a [. . .] boy I bought a box of oil colors with pennies slowly and painfully saved. To this very day I can still see these colors coming out of the tubes [. . .] jubilantly, festively, or

grave and dreamy, or turned thoughtfully within themselves, the colors came forth. Or wild with sportiveness, with a deep sigh of liberation, with the deep tone of sorrow, with splendid strength and fortitude [. . .] these curious, lovely things that are called colors [emerged].[18] (Italics and brackets added)

Although Werner does not mention Van Gogh, there can be little question but that his swirling stars, dynamic landscapes, and vividly animated colors were based on his own mode of perception affected by a very disturbed state of mind. In addition, the unusual character of sensory experiences reported by those who have taken hallucinogenic drugs would seem to be evidence of an augmenting of the physiognomic mode of perception. In fact, in some primitive societies hallucinatory drugs were used to heighten religious frenzy and induce visions. Radin, for example, describes the experience of an American Indian, and the effects on the religious practices of the tribe, due to eating peyote seeds which contain the hallucinogenic drug mescal. [19]

THE LABILE AND DIFFUSE CHARACTER OF PRIMITIVE CONCEPTS

Related to both the syncretic nature of primitive mentality and the physiognomic character of primitive perception is the "labile" and "diffuse" nature of primitive concepts. As we have seen, it is not the more objective, stable properties of objects that are of primary interest to the primitive mind, but the dynamic, emotive qualities accompanying the object's effects. The categorical analysis of experience into a relatively permanent substratum underlying changing attributes, which Whitehead regarded as the fundamental legacy of Aristotelian philosophy, and that since has become enshrined in the ordinary language, academic philosophy, and (until recently) natural science of the West, is not typical of a certain level of primitive mentality. For example, before the child has acquired the concept of an object and learned to structure his experience into a permanent, self-subsisting world, he lives in a fluid environment of diffuse images and forms. As described by Piaget: "The child's universe is still only a totality of pictures emerging from nothingness at the moment of the action to return to nothingness at the moment when the action is finished." [20]

At this developmental level the child's experience is impressionistic: it lacks the conceptual interpretation of images as independent, continuously existing things. Objects are real insofar as the child interacts with them but cease to exist when unperceived. Even after he has learned that the familiar objects around him continue to exist when unobserved, the child still may be unable to conceive of *distant* objects as having the same permanent existence — perhaps explaining why man's conception of remote objects or events has always been more fanciful than his thinking about his immediate environment.

The child who speaks, or even the adult, may alike bestow the quality of object on the things which surround them and yet find themselves incapable of so doing with regard to the stars or other distant bodies [. .] .[21]

One merely has to recall historically the tendency to personify the planets, and the reluctance to admit that the "celestial bodies" were made of the same material as the "terrestrial world," to realize how natural it is to endow remote entities with an enhanced status. In the 5th century B.C. Anaxagoras was prosecuted and then exiled from Athens because he "held that the moon was made of earth, and had plains and ravines on it." A generation later one of the (false) charges brought against Socrates was that he denied that the moon and the sun were gods. And opposition to Galileo's telescopic discoveries was largely due to the fact that if they were accepted as true, then one would have had to admit that the sun, moon, and other planets were bodies similar to the earth — still a heresy at that time — thus refuting the cherished distinction between the celestial and the terrestrial worlds. More recently, when the astronauts landed on the moon a number of simple people were apprehensive as to the ultimate consequences for man, as if there had been a trespass on forbidden or sacred domain.

Correlated with this lack of conceptual object constancy are certain labile and diffuse tendencies of primitive mentality, particularly evident in myths. Anyone familiar with myths is aware of the multiplicity of meanings, the fusion of images, the contradictions and inconsistencies, the bizarre associations, and the transposition of contents from diverse domains of experience inherent in the narrative. The images of different mythical personas melt into one another through the processes of historical, rhetorical, and unconscious condensation, attributes and functions derived from entirely different and incompatible domains of experience transferred indiscriminately from one to the other.

For example, in the Babylonian myth named "Enuma Elish,"[22] the genesis of the universe is traced from three primary gods which are personifications of elemental domains of the world. The primordial stage is one of watery chaos consisting of three intermingling elements personified as gods: Apsu, representing sweet or fresh water; Ti'amat, representing the sea; and Mummu, who seems to represent cloud banks and mist. From this original watery chaos a sequence of Gods are "begotten," representing the horizon, earth, sky, winds, etc. Thus the myth attempts to "penetrate the mystery of creation" by utilizing aspects abstracted from the alluvial setting of Mesopotamia. Just as silt is deposited by the two great rivers, the Tigris and the Euphrates, so out of the original watery chaos the earth was formed.

> In speculating about the origin of the world, the Mesopotamians thus took as their point of departure things they knew and could observe in the geology of their own country. Their earth, Mesopotamia, is formed by silt deposited where fresh water meets salt water; the sky, seemingly formed of solid matter like the earth, must have been deposited in the same manner and must have been raised later to its present lofty position.

Superimposed on this empirical aspect is the schema of sexual generation. The second pair of gods, Lahmu and Lahamu were begotten by Apsu, the sweet waters and born of Ti'amat, the sea, and would appear to represent silt which had formed in the waters. They in turn engender Anshar and Kishar, two aspects of the horizon who then give birth to Anu, the god of the sky. As the myth unfolds it would seem that the model of social strife and political organization becomes the dominant motif governing further speculation as to the differentiation and organization of the universe. Thus the younger gods "disturb" Ti'amat and Apsu, and when the latter opposes the younger gods, one of them, Ea-Enki, rises to the occasion and because Ti'amat neglects to come to the aid of her husband is able to slay him. The means by which Ea was able to kill Apsu was a spell, the power of command, which followed the Mesopotamian view that "authority" is "a power inherent in commands, a power which caused a command to be obeyed, caused it to realize itself, to come true."

But the real hero of the myth is Marduk, Ea's son who later vanquishes Ti'amat when she and her second husband Kingu decide to attack the younger gods. Marduk agrees to be the champion of the younger gods, but only if he is given "authority on a par with that of the powerful senior members of the community [. . .] [and thus] his demand foreshadows the coming state with its combination of force and authority in the person of the king." This authority granted, Marduk then does battle with Ti'amat. His weapons are those of a god of storm and thunder with arrows of lightning and a net held by four winds. "In addition, he fashions seven terrible storms, lifts up his mace, which is the flood, mounts his war chariot, 'the irresistible tempest,' and rides to battle against Ti'amat [. . .]." Thus armed he defeats Ti'amat by sending in the winds to hold her jaws apart as she opens them to swallow him. "The winds swell her body, and through her open mouth Marduk shoots an arrow which pierces her heart and kills her." He then crushes her skull with his mace, cuts her body in two, lifting up half to form the sky, the other half forming the earth.

Although a marvellously imaginative story, this hardly can be called a rational account of the origin of the universe. The figures and motifs are abstracted from such diverse areas of experience as the geographical, biological, social, and political. Similar to dreams, the content is not regulated by logical principles. "The diffuse structure allows the object to be determined by successive properties standing in logical contradiction."[23] A god of the sea can take on human form so that her skull can be crushed and her body then divided to form the sky and the earth. Even if, as in the case of the "Enuma Elish" myth, the condensation and transference of the central figures and themes were the result of historical transformations, the final account still illustrates essential characteristics of myths. It is not surprising, then, that H. and H. A. Frankfort concluded that peoples who could accept such an account must have been incapable of "rational speculation."

However, as was suggested earlier, myths may satify more than rational needs. To account for this, Joseph Campbell has posited innate, energy releasing, physiological mechanisms that can be triggered by archetypical images or symbols. Specifically, Campbell postulated the existence of "central excitatory mechanisms (CEMS)" and "innate releasing mechanisms (IRMS)" responsive to certain universal, mythical, sign stimuli that trigger the release of inherent sources of psychic energy.[24] Analogous to Jung, this conjecture would explain the recurrence in myths of certain universal figures and motifs (death and resurrection, the fall, the redemptive act of a saving hero, etc.) as due to a physiological and psychological makeup shared by all mankind throughout history. These innate mechanisms would account for the creation, prevalence, and cultural importance of mythic archetypes; consequently, myths would not be characteristic simply of a certain level of mental development, but would represent a cultural response to a universal human need.

> Man, apparently, cannot maintain himself in the universe without belief in some arrangement of the general inheritance of myth. In fact, the fullness of his life would seem to stand in direct ratio to the depth and range not of his rational thought but of his local mythology. Whence the force of these unsubstantial themes, by which they are empowered to galvanize populations, creating of them civilizations, each with a beauty and self-compelling destiny of its own? And why should it be that whenever men have looked for something solid on which to found their lives, they have chosen not the facts in which the world abounds, but the myths of an immemorial imagination — preferring even to make life a hell for themselves and their neighbors, in the name of some violent god, to accepting gracefully the bounty the world affords? [25]

But whether the pervasive prominence of myths in early cultures can best be explained by postulating certain neurological excitatory mechanisms, like the question whether certain forms of primitive thought depend upon an innate logic, must be left to future investigations to decide. In the meantime, this study shall proceed on the simpler assumption that man has the capacity to form various kinds of abstractions and that it is these abstractions that primarily determine the content of his beliefs and thought.

From this point of view, mythopoetic thought is not the result of a specific neurological need or capacity, but a natural way of thinking about the universe at a certain level of mental development. The fantastic contents of myth are not seen as indicative of a special origin but characteristic, as we have noted, of thought that relates to distant times and places. To the extent that one's thinking about the world refers to some remote episode, the more imaginative, fanciful nature of man's thought tends to predominate. Rigorous thinking apparently requires the continual opposition and reinforcement of stubborn empirical influences.

CONCRETE SYNCRETIC THOUGHT

As one would expect, primitive thought resembles primitive perception in being relatively concrete and syncretic. The younger the child, the more severely retarded the individual, or the more primitive the peoples the more one finds that the thinking of such individuals depends upon conceptual images, affective associations, and concrete generalizations. This thinking, according to Werner, "has as yet not achieved release from [. . .] imaginative-perceptual activities," but is "limited to and enclosed within concrete, picture-like forms."[26] Even Lévi-Strauss described "savage thought as a system of concepts embedded in images."[27]

The languages of earlier cultures illustrate this in that they are richer in imagery, allegory, and metaphor than modern languages which are more attuned to scientific thinking. As thought becomes more abstract and rational, the more the language loses in richness of content, vividness of description, and color of expression. Thus ancient literatures, as reflected in the major religious writings in the world, are vastly superior in metaphorical illusions, poetic symbolism, and lyrical expressiveness to our own. When the primary function of language changes from richness of expression to lucidity of explanation, the image

> originally rooted in the sensuous sphere of feeling and of fantasy, is gradually changed in functional character. It becomes essentially subject to the exigencies of abstract thought. Once the image changes in function and becomes an instrument in reflective thought, its structure will also change. It is only through such structural change that the image can serve as an instrument of expression in abstract mental activity. This is why, of necessity, the sensuousness, fullness of detail, the color and vivacity of the image must fade.[28]

This explains why, if one wants to enlarge his *experience* or increase his *sensitivity* to the rich qualities of the world he turns to art, but if he wants to *understand* or *explain* phenomena he turns to science. (In later life Darwin regretfully admitted that the demands of exact scientific research had diminished his capacity for aesthetic appreciation.) While literature conveys a much more direct experience of the marvellous diversity and subtle complexity of individual personalities than psychology, the latter provides a needed explanation of some of the general traits and causes of human behavior. But who would venture to say who understood human nature better, Shakespeare and Dostoevsky or Freud and Jung?

In any case, primitive thought utilizes concrete descriptions and explanations as narrative stories more than it does abstract concepts and theoretical interpretations. This is why "In the beginning [. . .] " is such a characteristic introductory phrase to both children's stories and myths (e.g., Hesiod's *Theogony* and the *Old Testament*).

The traditional *explanations* of primitive peoples are most often descriptive and narrative in form. The properties of natural objects are not derived from principles of a universal necessity, they are not conceived in any lawful sense; rather they stem from an individual history, a myth telling "how it came to pass." [29]

And H. and H. A. Frankfort concluded from their investigation of Eastern myths that

the ancients told myths instead of presenting an analysis or conclusions. We would explain, for instance, that certain atmospheric changes broke a drought and brought about rain. The Babylonians observed the same facts but experienced them as the intervention of the gigantic bird Imdugud which came to their rescue. It covered the sky with the black storm clouds of its wings and devoured the Bull of Heaven, whose hot breath had scorched the crops. [30]

The concreteness of primitive thought is illustrated also in the meaning the terms have for the primitive. Words are not understood to be conventional symbols used to denote or replace their referents in a symbolic sense; instead, words are identified with the objects they stand for thus taking on their properties. Moreover, words are hypostatized or substantialized, thereby acquiring an independent status and power. This, as we shall see later, partially underlies the Platonic conception of an independent realm of eternal meanings or Forms, though Plato's theory is much more sophisticated and on a much higher abstract level than that of the child or the primitive. Both the child and the primitive believe that things have acquired their names by some exalted person, animal, or god, and creation is often thought of as issuing from the spoken word. As Werner states, "in the primitive sphere a name is in no sense regarded as something imposed willfully, or as something fortuitous, a mere sign. A thing cannot be grasped until its name is known [. . .] it is part of the object itself."[31] Piaget's description of the child's conception of names is similar:

[. . .] children believe that every object has received a primordial and absolute name, which somehow is part of its being. When very young children ask about an unknown object "what is it? ", it is the name of the object they are enquiring for, and the name plays the part not only of a symbol, but of a definition and even of an explanation [. . .].[32]

In the Babylonian myth "Enuma Elish" described earlier, to determine whether the younger god Marduk had sufficient "magic quality" to overcome Ti'amat, he was confronted with the following test:

They placed a garment in their midst
And said to Marduk their firstborn:
'O Lord, thy lot is truly highest among gods.
Command annihilation and existence, and may both come true.
May thy spoken word destroy the garment,
Then speak again and may it be intact.'
He spoke — and at his word the garment was destroyed.
He spoke again, the garment reappeared.
The gods, his fathers, seeing (the power of) his word,
Rejoiced, paid homage; 'Marduk is king.' [33]

The oldest Gospel of the *New Testament*, Saint John, begins with the words, "In the beginning was the Word, and the Word was with God, and the Word was God. All things were made by Him [. . .]." Gradually, however, words came to lose their magic power, became freed from their attachment to things, and as vehicles of thought became the conveyors of meanings. As analyzed by Werner, this progression occurs in four stages: (1) the name is first thought of as a material property of the thing (e.g., primitive magic), (2) the name then becomes a concrete physiognomic picture of the thing (e.g., hieroglyphics), (3) the name is believed to have an essential though conceptual meaning (e.g., in the philosophy of Plato and Aristotle), (4) the name finally is recognized as a conventional sign of a concept. [34]

In addition, at this level of thinking entities are discriminated and classified in terms of their spontaneous groupings or relationships within a natural situation, rather than in terms of a more abstract differentiation of qualities, aspects, or forms. For example, different colors will often be designated by the same word because they are found together on the same object.

> With the expression *tu ku éng* the Bakaïré Indians of Brazil designate the following colors: emerald green, cinnabar red, and ultramarine. This surprising conjunction of colors so decidedly different in quality is explained by the fact that *tu ku éng* is the name of a parrot which bears all of them. Here the concrete togetherness is determined not by any actual similarity, but rather by a realistic configuration. [35]

Wood and fire, food and eating, thirst and water may also be designated by the same word because they are related aspects of a total situation, illustrating further that primitive groupings are not as determined by abstract considerations as they are by the natural tendency of similar elements to spontaneously configure and stand out against a background field. The configurations may involve either perceptual qualities or dynamic properties, but in any case they are less mediated by rational considerations than is true of more advanced thought.

> Colors do not stand alone self-subsistently, but are fused with the object they qualify. Thus the natives of New Pomerania use the expression *kott-kott* (black crow) for black; *gab* (blood) for red, etc. [. . .]. We may say that the function of abstraction is expressed here in a selection, a focussing, of qualities linked with a situative, concrete relation. Given such a relation, certain qualities stand forth of themselves. [36]

Even when, as pointed out by Boas,[37] a primitive language such as the Chinook is exceedingly rich in abstract terms (e.g., instead of saying "The bad man killed the poor child," this is expressed in Chinook as "The man's badness killed the child's poverty"), these abstractions are used in a very concrete, hypostatized sense.

The generalizations of primitive thought also tend to be more concrete and situative, more determined by the sensory-motor and affective-dynamic characteristics of things. In contrast, on the more advanced level the resemblances exhibited by qualities, forms, and aspects, even when the latter belong to different objects, are recognized as indicative of similarities among more abstract qualities.

> One who can shift his point of view in a purposeful grouping activity is no longer subject to the forces of sensory stimulation. He is able consciously to perceive that objects have different qualities, any one of which may be taken as the point of departure for an ordering process. In other words, this development indicates an immensely important step away from an abstraction closely allied to sensory organization and toward an abstraction guided by deliberately selected categories such as color, shape, number, size, etc. [38]

This ability to alter one's assumptions, to be selective in the choice of abstractions, to be imaginative in their synthesis within a general intellectual framework, and to think in terms of different perspectives brings us to a final differentiation of levels of mentality.

EGOCENTRIC THOUGHT

In the development of Western physical science (after the emergence of proto-scientific thought from myth), there have been two fundamental revolutions. The first was the well-known Copernican Revolution which occurred roughly during the 17th century, in the course of which the geocentric, homocentric, and organismic cosmology of Aristotle was replaced by the heliocentric universe of Copernicus and the mechanistic cosmology of Newton. The second revolution began just at the turn of the 20th century with Planck's introduction of the quantum of action in 1900 and Einstein's publication of the special theory of relativity in 1905. The latter developments, though less well-known and understood by the layman, constitute at least as fundamental a revolution in scientific thought as the more familiar Copernican Revolution.

Central to both of these revolutions was a radical shift in perspective requiring a drastic reshaping of our thinking about the universe and man's place in it. That is, in place of a more absolute, self-centered point of view one began thinking about the universe in terms of a context of relations that are less immediate but more adapted to physical reality itself. This change in orientation is analogous to Piaget's description of a crucial stage in the mental development of the child when he revises his basic system of reference from an absolutistic, self-centered point of view to one that is more flexible and objective in outlook. Piaget characterized the earlier, more rigid form of thought "egocentrism."

> If we may be permitted to make a somewhat daring comparison, the completion of the objective practical universe [by the child] resembles Newton's achievements as compared to the egocentrism of Aristotelian physics, but the absolute Newtonian time and space themselves remain egocentric from the point of view of Einstein's relativity because they envisage only one perspective on the universe among many other perspectives which are equally possible and real.[39] (Brackets added)

As Piaget uses the term "egocentrism," it means "the inability to differentiate between one's own point of view and other people's or between one's own activity and changes in the object."[40] By accepting himself as the center of his reference system the child refers all motions and changes of appearance to himself, rather than placing them in a more objective framework. In failing to distinguish those aspects of experience due to his own situation, the child automatically attributes them to the objective world, as prior to the Copernican Revolution one attributed motion to the planets (e.g., retrograde motion) that did not belong to them but which was merely an *apparent* change in position due to the observer's own unobserved motion. Although he has often been misunderstood, this is precisely what Kant meant when he termed his own philosophy a "Copernican Revolution," for he attributed to the innate structure of the mind basic perceptual forms and conceptual categories that previously had been attributed to the world itself.[41] And prior to Einstein's relativity theory one extended to the entire universe concepts which were appropriate only to the "limiting conditions" of the earth with its relatively slow velocities and limited dimensions.

The transition to a less egocentric point of view involves a number of different aspects: (1) the ability to think in terms of more than one perspective or assume more than one point of view; (2) the capacity to overrule common sense experience and assumptions in favor of more abstract theoretical considerations (e.g., conceptual coherence, logical consistency, theoretical simplicity); (3) the courage to face a stranger, more impersonal universe; and (4) the strength to withstand the professional criticism and personal abuse which often accompanies radical innovations in thought (e.g., Anaxagoras, Galileo, Darwin, and Freud).

CONCLUSION

In this and the previous chapter an attempt has been made to distinguish and delineate a form of primitive mentality discernible in all cultures, as contrasted with a type of mentality common to all peoples at a certain chronological stage of development. While rejecting the 19th century view that all primitive thought is determined by the evolved cultural level of the society within which it occurs, the opposite thesis of Lévi-Strauss that the cultural forms of all societies, primitive or modern, manifest the same type or level of logical thinking also seems too extreme. Rather than all human thought being determined by the same innate logic, and therefore being equally rigorous, it has been argued that one can distinguish between various modes or levels of thought in terms of their degrees of objectivity, abstractness, clarity, and flexibility.

It would seem, then, that in any society of sufficient size there is a correspondingly wide range of mental abilities. While there is considerable evidence that among primitive peoples there are individuals capable of a very high level of thought, as manifested within the cultural or intellectual limitations of that society, there is even greater evidence to show that in any advanced society there are individuals who exhibit a very low level of mental development, even when exposed to the education of that society. For example, many varieties of primitive religious sects exist along with the predominant scientific and technological strains in our society. The widespread institutionalization of science in the West, while influencing to some extent the mental outlook of all people exposed to it, does not necessarily mean a more highly *evolved* mental capacity among *all* the individuals of that society; for some, the exposure to science apparently results merely in a superficial modification of deeper, more elementary forms of thought closer to animism and myth than to science. Thus the progressive advance of science does not imply a more evolved mental capacity but the continued refinement and development of a form of thinking that has always been characteristic of certain individuals — just as there are people with greater inherent artistic, mathematical, athletic, or leadership abilities. The difference in the intellectual contributions of Anaximander, Archimedes, Newton, and Einstein is not due to a genetically evolved, higher level of scientific thought, but to the progressive development of scientific concepts and theories, improved methods of investigation and experimental apparatus, as well as accompanying advances in mathematics. What justifies the distinction between lower and more advanced societies is not the *evolution* of peoples with a genetically higher mental capacity, but the cultural development of, as well as the greater distribution and influence of, certain kinds of knowledge and modes of thought. How this cognitive development has come about is the intent of this study to describe.

CHAPTER III

[1] Heinz Werner, *Comparative Psychology of Mental Development* (*op. cit.*, ch. II), p. 153.

[2] *Ibid.*, p. 337.

[3] *Ibid.*, p. 146.

[4] Paul Radin, *Primitive Religion* (New York: Dover Pub., Inc., [1937] 1957), p. 106.

[5] H. and H. A. Frankfort, *Before Philosophy* (*op. cit.*, ch. II), p. 21.

[6] Jean Piaget, *The Language and Thought of the Child* (*op. cit.*, ch. II), p. 174. Brackets added.

[7] Ernst Cassirer, *Mythical Thought* (*op. cit.*, ch. II), pp. 47-48.

[8] Piaget, *op. cit.*, p. 155.

[9] Cf. Frankfort, *op. cit.*, p. 30.

[10] Cassirer, *op. cit.*, p. 90.

[11] Benjamin Whorf, *Language, Thought and Reality* (*op. cit.*, ch. I), p. 63.

[12] R. H. Codrington, *The Melanesians*, pp. 118-119. Quoted from Paul Radin, *Primitive Man as Philosopher* (*op. cit.*, ch. II), p. 254.

[13] Sir Edward B. Tylor, *Religion in Primitive Culture:* Part II of *Primitive Culture* (New York: Harper & Brothers Pub., [1871] 1958), p. 270. Brackets added.

[14] Frankfort, *op. cit.*, p. 14.

[15] Werner, *op. cit.*, p. 69.

[16] *Ibid.*, p. 77.

[17] *Ibid.*, p. 80.

[18] Wassily Kandinsky, "1901-1913," in *Der Sturm*, 1913. Quoted from Werner, *op. cit.*, pp. 71-72.

[19] Cf. Radin, *Primitive Man as Philosopher, op. cit.*, Appendix B.

[20] Jean Piaget, *The Construction of Reality in the Child*, trans. by Margaret Cook (New York: Basic Books, Inc., 1954), p. 43.

[21] *Ibid.*, p. 78.

[22] Cf. Thorkild Jacobsen, "The Cosmos as a State," in Frankfort, (*op. cit.*, ch. V). All the following quotations pertaining to this myth are from this chapter.

[23] Werner, *op. cit.*, p. 286.

[24] Cf. Joseph Campbell, *The Masks of God: Primitive Mythology* (New York: The Viking Press, 1959), p. 37ff.

[25] *Ibid.*, p. 4.

[26] Werner, *op. cit.*, pp. 352-353.

[27] Claude Lévi-Strauss, *The Savage Mind* (*op. cit.*, ch. II), p. 264.

[28] Werner, *op. cit.*, p. 152.

[29] *Ibid.*, p. 304.

[30] Frankfort, *op. cit.*, p. 15.

[31] Werner, *op. cit.*, p. 254.

[32] Piaget, *The Language and Thought of the Child, op. cit.*, p. 217.

[33] Frankfort, *op. cit.*, p. 193.

[34] Cf. Werner, *op. cit.*, p. 266.

[35] *Ibid.*, p. 225.

[36] *Ibid.*, p. 241. For a different analysis, see Lévi-Strauss, *The Savage Mind, op. cit.*, p. 63ff.

[37] Cf. Franz Boas, "Handbook of American Indian Languages," Part I, *Bulletin* 40, Bureau of American Ethnology, Washington, D.C., 1911, pp. 657-658.

[38] Werner, *op. cit.*, p. 240.

[39] Piaget, *The Construction of Reality in the Child, op. cit.*, p. 367.

[40] *Ibid., The Language and Thought of the Child, op. cit.*, p. 267, f.n. 1.
[41] Cf. Immanuel Kant, *Critique of Pure Reason*, sec. B xvi - B xvii.

CHAPTER IV

THE DAWN OF SCIENTIFIC RATIONALISM

With the first dawn of scientific insight the mythical world of dream and enchantment seems to sink into nothingness. [1]
Cassirer

At one time or another everyone must have had the experience either of feeling a cool breeze disperse a sultry mist or of finding the sun's rays dissipate a heavy fog. The first philosophizing of 6th century Greece has often been likened to a burst of sunlight or to a sweep of fresh wind dispelling the mythical haze that enveloped early cultures. In fact, there even have been attempts to attribute the brilliant clarity of Greek thought to the brilliant luminosity itself of the atmosphere of Greece. For anyone with a rational bent or a predilection for scientific reasoning, it is easy to see Greek thought as an entirely new beginning, to view Greek philosophy as a kind of "motherless Athena," an unaccountably new way of thinking about the universe. For, as Cornford states, "The philosopher's business is to dispel the veil of myth and penetrate to the 'nature of things', a reality satisfying the requirements of abstract thought." [2]

Because of the fragmentary nature of what is left of their writings, as well as the cultural, intellectual, and historical distance that separates us, a certain amount of conjecture must enter into any attempt to reconstruct the contribution to knowledge and to the growth of abstract thought of the Presocratic philosophers. If Aristotle and his student Theophrastus, two of our main sources of knowledge of the Presocratics, had difficulty barely two centuries later in understanding and reconstructing the thought of "the ancients," how much more difficult — and less likely to succeed — must the task be today. The difficulty comprises two main, though obviously not unrelated, problems: (1) the necessity of determining as accurately as possible what each of the Presocratic philosophers actually said; and (2) reconstructing what they might have meant by what they said.

While there is a certain amount of conjecture with regard to both problems, there is much more speculation about the second, for even though philologists may agree as to what the philosopher *said* (though they also disagree), they more often disagree as to what he could have *meant* by what he said, or as to what were the meanings and thought processes underlying the fragmentary statements attributed to him. This is due in part to the fact that in most cases what one has to work with (called the "doxography"[3]) are not direct quotations, but attributions, paraphrases, and interpretations contributed by ancient scholars throughout the succeeding centuries, or at best separate and even fragmentary statements which have to be reconstructed into a total picture of the philosopher's views — much as one would put together the pieces of a puzzle or draw a composite picture of someone from different descriptions. In addition, given the differences in culture and language that separate us from such early philosophers, a large amount of interpretation would in any case be required to provide an intelligible rendering of what was meant by what was said. Often in the past the two types of problems (though not unrelated) have not been kept distinct; today, however, we are fortunate in having a compilation of the statements of the Presocratics which clearly distinguishes these two problems. [4]

Given this careful evaluation of the doxography by recent scholars, the present reconstruction and assessment of Presocratic philosophy will be guided by the general aim of describing its emergence and development as an early form of scientific rationalism in comparison with the previous mythopoetic, theogonic traditions. However, as one might expect, there are considerable differences of opinion among classicists and historians as to the extent of the independence and rationality of the thought of the Presocratics in relation to the religious heritage and the mythopoetic writings of Homer and Hesiod. This problem is particularly acute as regards the earliest Presocratic speculations, those of the Milesian philosophers. [5] The key figure in the controversy is the brilliant scholar, Francis Cornford, who, in two separate works written in different periods of his life, attempted to trace out the origins of Presocratic philosophy and distinguish its original elements from those features inherited from the ancient traditions. As he expresses the problem in the later book, *Principium Sapientiae*, on which he was working when he died:

> Is the Milesian cosmogony the work of rational inference based on observation and rechecked by at least rudimentary methods of experiment? Or are its features to be referred to an attitude of mind [i.e., of the earlier mythopoetic tradition] uncongenial to natural science as we understand it? [6] (Brackets added)

In his earlier book written when he was a young man, Cornford was greatly influenced by the writings of some of the early 20th century anthropologists whom we have already discussed, namely the French School of Durkheim and Lévy-Bruhl. [7] Under the influence of this school Cornford then interpreted the physical speculations of the Presocratics not as spontaneous creations of the philosophic mind, but as speculations inadvertently influenced by the general problems and "collective representations" of the earlier mythical and religious traditions. Moreover, following Durkheim, he interpreted this earlier religious tradition as itself being unconsciously formed by the conventions and structures of society. As he states: "Here we touch at last the bedrock. Behind philosophy lay religion; behind religion, as we now see, lies social custom — the structure and institutions of the human group." [8]

According to this interpretation, while it often seems as if the Presocratics had completely detached themselves from the vestiges of the earlier mythical-religious outlook, they were merely pouring new wine into old bottles. That is, while they consciously eliminated the specific contents of such cosmogonic and theogonic accounts as those of Homer and Hesiod, and intentionally banished the Olympian gods from the heights of their philosophic speculations, nonetheless, they were speculating within the cognitive framework of these earlier mythical-religious traditions. Thus the cosmogonic and cosmological problems they initially faced were not problems of their own choosing, but problems unconsciously taken over from the earlier theogonic literature. As Charles Kahn states, agreeing to some extent with Cornford, "the early cosmologists present a theory which they consider to be much more satisfactory, but which is still of the same outward form: cosmogony is the heir of theogony." [9]

Moreover, the derivative ways of attacking these problems were not free creations of their own, but formulations and patterns of reasoning unintentionally derived from the earlier traditions. While this will be discussed more fully later in terms of particular doctrines, it is well to quote Cornford more generally at this point:

[. . .] philosophy rediscovers in the world that very scheme of representation which had, by a necessary process, been projected into the world from the structure and institutions of society in its earlier stages of development. The concepts and categories which the intellect brings with it to its task, are precisely those by which the chaos of phenomena had long ago been coordinated and organized into the significant outlines of a cosmos. [10]

In his later book Cornford seemed to emphasize much more the rationality of the Milesian outlook in comparison with the mythical aspects of the earlier theogonies and cosmogonies. Thus he says of Ionian philosophy: "Every reader is struck by the rationalism which distinguishes it from mythical cosmogonies [. . .]. The Milesians brought into the world of common experience much that had previously lain beyond that world." [11] However, in spite of this change of emphasis, Cornford is still very much convinced of the continuity of thought and of the extensive influence of the earlier mythopoetic traditions on the thought of the Milesian philosophers in particular, and of the Presocratics in general. Therefore, in addition to finding rational aspects in such earlier works as Hesiod's *Theogony* ("It has advanced so far along the road of rationalization that only a very thin partition divides it from the earlier Ionian systems." [12]), he finds mythical figures recurring in the later writings of the Presocratics.

But there is something to be added on the other side. If we give up the idea that philosophy or science is a motherless Athena, an entirely new discipline breaking in from nowhere upon a culture hitherto dominated by poetical and mystical theologians, we shall see that the process of rationalization had been at work for some considerable time before Thales was born. We shall also take note of the re-emergence in the later systems of figures which our own science would dismiss as mythical — the Love and Strife of Empedocles and the ghost of a creator in the Nous of Anaxagoras. And when we look more closely at the Milesian scheme, it presents a number of features which cannot be attributed to a rational inference based on an open-minded observation of the facts. [13]

As regards the latter "features," Cornford points out that the two primary assumptions of Presocratic thought, that (1) the present world order arose out of some preexistent state and thus had a beginning in time, and (2) that this world order occurred by a process of differentiation from a previous state, are not suggested by an unbiased observation of nature itself. That is, there is little in the observation of nature as such to suggest that the entire cosmos began in time or that the present state of things arose by a process of differentiation from a simpler previous state. He then goes on to elaborate more specifically the basic assumptions underlying Presocratic speculations which he believed were inadvertently taken over from the earlier mythopoetic tradition: (1) that the present universe began in time and arose as a kind of separation from a preexisting, relatively undifferentiated unity; (2) that this separation was generally explained as due to an antagonism between opposite qualities occupying different domains; (3) that this antagonism subsequently produced the cosmos along with the heavenly bodies; and (4) that this conflict of opposites was balanced or controlled by a contrary tendency to reunite and blend, thereby accounting for the transformation of natural processes and the existence of living organisms. [14]

While this conception of the influence of the earlier theogonic tradition on Milesian philosophy in particular is now generally acknowledged (e.g., by Kahn and Kirk and Raven), a few scholars disagree. For example, de Santillana, in *The Origins of Scientific Thought*, does not find Presocratic speculation to be predetermined by the earlier mythopoetic tradition. Perhaps because he brings to his study a scientific background in contrast to the classicist training of Cornford, he finds in Presocratic philosophy a highly developed form of scientific rationalism, one not so much dependent upon a religious heritage as one bred in the cultural climate of an early Greek renaissance marked by new explorations, interchange of ideas, engineering advances, and new physical discoveries and mathematical developments. Rather than Milesian speculation presupposing mythopoetic forms of thought, he sees it as continuous with an older rational tradition, but incorporating new principles of interpretation (such as "the principle of sufficient reason"), new abstractions derived from natural processes, and a new use of empirical models and analogies. In direct opposition to Cornford, he says of Thales, the first of the Milesian philosophers:

> It would be very wrong to imagine only medicine men behind him, or sad mythographers like Hesiod [. . .]. What we discern in their background are not priests and prophets, but legislators, engineers, and explorers. [15]

Even de Santillana's interpretation of myth is different from that of Cornford. Somewhat analogous to Lévi-Strauss' emphasis on the underlying rationality of myth, if properly decoded, de Santillana finds within the complex web of mythical, legendary accounts of the origin of the universe and of the dramatic movements of the celestial bodies a kind of code which, when deciphered, reveals a somewhat exact and technical understanding of astronomical phenomena. As he says:

> We can see then how so many myths, fantastic and arbitrary in semblance, of which the Greek tale of the Argonauts is a late offspring, may provide a terminology of image motifs, a kind of code which is beginning to be broken. It was meant to allow those who know: (a) to determine unequivocally the position of the given planets in respect to the earth, to the firmament, and to one another; (b) to present what knowledge there was of the fabric of the world in the form of tales about "how the world began." [16]

Even myths, therefore, are not just imaginative stories of the origin of the world, created out of the visions and legends of shamen, seer-poets, or prophets. Instead, they represent a feat of intellectual abstraction in which certain prominent empirical occurrences are rendered stable and intelligible by being fitted into the only available theoretical framework of the time — a story as to "how things came to pass." It is as if the conceptual thought of such early speculations was submerged in and thus hidden by the dramatic narrative because of the lack of an empirical symbolism and rational framework by which it could be expressed — but a quality of thought nonetheless truly abstract and highly developed.

> The variety of names given to each planet and the multiplicity of mythical motifs to accompany them were made necessary by the almost incalculable complexity of relations to be expressed [. . .]. The colossal intellectual effort, the abstraction, that this entailed are worthy of the greatest modern theorists. We must assume every age has minds of the order of Archimedes, Kepler, or Newton. In the last millenaries, as well as yesterday, those minds had to create within the context of their time. [17]

Thus de Santillana supports the thesis developed in the previous chapters: namely, that in all societies, however remote in time, there generally are the same range of mental capacities as in modern society, but that these capacities necessarily are manifested within the "context of their time." This in itself, however, does not preclude the unconscious influence of a more primitive form of thought (in addition to the more advanced form) on even the most developed or far-sighted thinkers of an epoch. Along with asserting that every society has its Archimedes, Keplers, and Newtons, we have found that each individual operates on different mental levels. This certainly was illustrated in the thought of Kepler whose mathematical formulation of the first astronomical laws was motivated by a neo-Pythagorean, neo-Platonic mysticism (and perhaps by a belief in astrology), and in the thought of Newton whose views of the order of the planets, the origin of motion, and of the nature of absolute space and time were definitely influenced by theological considerations. And it has been pointed out many times that in spite of the tremendous intellectual freedom exhibited by Einstein in his special and general theories of relativity, he remained very conservative as regards his attitude toward quantum mechanics.

The tendency to attribute to the Presocratics not only a complete break with the older theogonic tradition, but an anticipation of much later scientific concepts is particularly characteristic of scientists or scientifically oriented historians. To take only one example, in his account of the Presocratics, Sambursky finds in Empedocles' "poetic" conceptions of Love and Strife controlling cosmological transitions an "intuition" of the modern scientific conception of the attraction and repulsion of forces: "The modern physicist is amazed at the intuition which led Empedocles to propound the simultaneous existence of forces of attraction [Love] and repulsion [Strife]."[18] Underlying this interpretation is a conception of Presocratic philosophy diametrically opposed to that of Cornford: "With the study of nature set free from the control of mythological fancy, the way was opened for the development of science as an intellectual system." [19]

But was Presocratic philosophy "freed from the control of mythological fancy"? Was Empedocles' use of the terms "Love" and "Strife" an intuitive groping toward the scientific conception of attractive and repulsive forces or, as Cornford stated earlier, does the reliance on such concepts indicate "the reemergence in the later systems of figures which our own science would dismiss as mythical — the Love and Strife of Empedocles and the ghost of a creator in the Nous of Anaxagoras"? Is it merely a coincidence that in Hesiod's *Theogony* "Eros" and "Strife"[20] are principal causes of generation, or is Sambursky anachronistically reading into Empedocles' terms his own understanding of how certain processes occur? Is it reasonable to suppose that Empedocles' thinking was closer to that of the 18th century physicist, Charles Dufay, who differentiated the electrical properties of attraction and repulsion, than it was to Hesiod? As Kahn states (referring, however, to Theophrastus), such interpretations represent

> not so much a question of misunderstanding as of the involuntary projection of more specialized, abstract notions into a period where models of thought and expression were simpler and closer to the concrete language of poetry and myth. [21]

These questions illustrate the difficulty in attempting to assess the nature and uniqueness of Presocratic thought. While the answers must remain somewhat conjectural, any attempt to assess the originality, nature, and significance of Presocratic philosophy, as well as the origin and development of scientific thought, depends upon giving some reply to these questions, however tentative. As Cassirer states: "Our insight into the development of science [. . .] is complete only if it shows how science arose in and worked itself out of the sphere of mythical immediacy [. . .]." [22]

HESIOD'S *THEOGONY*

In order to try to answer the questions posed above, it will be helpful to have in mind what could very well be called the "Bible" of ancient Greece, Hesiod's *Theogony*. As

Kahn asserts, "The precedent for all Greek speculation is of course the *Theogony* of Hesiod [. . .]."[23] Heraclitus deplored the fact that "Hesiod is most men's teacher." Like the *Bible,* the cosmological and theogonic accounts in the *Theogony* were not "invented" by Hesiod, but represent a compilation of views based on a rich religious heritage, even borrowing from the older Babylonian cosmological myth, "Enuma Elish," discussed earlier.[24] As such, the *Theogony* presents a narrative account expressed in mythopoetic language of the origin and emergence of the present structure of the physical universe, and of the origin, genealogy, conflicts among, and present control over the universe of the gods. It contains a personified or anthropomorphic interpretation of the physical features of the universe, of social, artistic, and political functions, and of various human traits. Finally, it describes the acts of the gods that contributed to the present status of man and to their influence over him.

The cosmological account begins with the coming into being of Chaos (how or from what is not explained), usually described as a "yawning gap," and then of "broad-bosomed Earth" and Eros, the latter depicted as "the most beautiful of the immortal gods." From these primal entities Hesiod derives the rest of the universe, the history of the gods, and that of man. Generation is usually explained by the influence of Eros (the "passion of sexual love"), though in a few instances creation does not depend upon sexual generation. For example, from Chaos alone came Darkness (Erebos) and black Night, but from Night "conceived after union in love with Darkness," came her children Light and Day. Also, Earth alone produced starry sky, the tall mountains, and gave birth to the barren waters, "sea with its raging sources — all this without the passion of love." Thereafter, however, she "lay" with Sky and gave birth to Ocean and a diverse series of gods representing various kinds of personifications (e.g., law, memory, thunder, lightning, etc.).

As far as the narrative is concerned, the most important of these offspring is "cunning Cronus" who takes revenge on his father Sky for keeping all the Titan children hidden in the "bowels of Mother Earth." Following the stratagem of his mother, he is able to castrate his father thus freeing the other gods and becoming the head of the Titans. Later, however, the "reign" of Cronus also is overthrown when his son, Zeus, born of the union with Rhea, unites the Olympian gods against the forces of the Titans headed by Cronus and, after a volcanic battle, deposes his father. Zeus then consolidates his power and as the "lord of wisdom" reigns over both gods and men.

While this is no more than a capsule summary of the main outline of the *Theogony,* it does illustrate certain general features of Hesiod's way of construing the world: (1) that things originated by "separating apart from" or "differentiating from" an original gap or chasm (the etymological derivation of the word 'chaos' is 'to gape' or 'to yawn'); (2) that sexual intercourse (Eros) is the main model of generation; (3) that a "strife" among opposing powers is a common cause of historical change; (4) that the sequential stages of the myth are marked by the successive rebellion of a younger god against his father with the aid of his mother; and (5) that there is a kind of moral or legal order, a "retributive justice," and a sort of "just portioning" of the various roles and domains of the gods that underlies the political conflicts. With this schema in mind, we shall try to determine whether Presocratic philosophy represented a clean break with the older mythopoetic, religious traditions, or whether in fact the form of this philosophy, if not the content, was abstracted unconsciously from the earlier tradition. As Cornford posed the problem, was Presocratic philosophy a kind of "motherless Athena," conceived in the heads of certain philosophers, or was it really mothered by an older tradition — just as Athena herself was actually conceived by Metis.

THALES

As tradition has it, it was in the city of Miletus in Ionia where the dawn of rational speculation about the world began. This was in the 6th century B.C., perhaps a century or so after Hesiod and two centuries before Aristotle. According to the latter, Thales was the first of the philosophers, the first to pose and seek an answer to the question 'what is the first principle of things, the reality behind the phenomena.'[25] According to Simplicius, while Thales is 'traditionally the first to have revealed the investigation of nature to the Greeks,' he had many predecessors, 'but so far surpassed them as to blot out all who came before him.' Indeed, he was one of the seven wise men of antiquity and somewhat of a Renaissance figure: a statesman, astronomer, meteorologist, physicist, engineer, mathematician, businessman, and philosopher. He is even the prototype of the absent-minded professor having his attention so fixed on the heavens above that he stumbled into a well on the ground below — according to a tale told by a Thracian servant girl. He did study the heavenly bodies making note of star groups, measuring the occurrence and variations of the equinoxes and solstices, and predicting a solar eclipse in 585 (this prediction was made, apparently, not on the basis of any theoretical knowledge of eclipses, but with the aid of Babylonian tables — a "lucky prediction" according to Kirk and Raven, and "a *tour de force* [. . .] achieved on the basis of century-long Babylonian observations," according to Kahn). Thus far we have the image of an extremely accomplished person involved in various scientific, civic, and theoretical concerns. As de Santillana asserted, here is no "prophet, priest, or sad mythographer."

From Aristotle we learn of Thales' metaphysical and cosmological doctrines: (1) that the underlying substance or first principle of things is water; (2) that the earth floats on water like a log or a ship; (3) that inanimate objects that move and are moved (magnets and iron, and probably amber and wool) possess souls; and (4) that soul is 'intermingled' in the universe and that 'all things are full of Gods.' Not knowing any more about what he actually said, even Aristotle was forced to conjecture what Thales might have meant by these doctrines.

As regards the first, Aristotle conjectured that Thales may have arrived at this conclusion from observing that all living things depend upon water for nourishment and are themselves moist (e.g., sperm and seeds), and that heat itself is generated by moisture.[26] Burnet suggested that Thales' meteorological considerations may have influenced this conclusion in that he must have observed that water (liquid) can be transformed into the two other physical states, solid (ice) and gas (vapor). Moreover, anyone who has looked out onto the Aegean or Mediterranean Seas from the shores of Ionia or from the Temple of Sounion at the southernmost extension of mainland Greece, as Thales must have, could well have concluded that water must comprise the most extensive and basic element of the universe. In addition, he might have been influenced in his thinking, as Aristotle suggests and as his second thesis indicates, by near-eastern mythical cosmologies which attributed the origin of the earth and its resting place or support to water.

Aside from conjectures as to the possible influences on Thales' thought, an even more important philosophical question arises. What could Aristotle's formulation of Thales' primary thesis that the original substance (*physis*) or first principle of things (*archê*) was water have meant for *Thales*? Did Thales mean that everything is 'composed of' or 'made up of' water (even fire and earth), or did he mean that various phenomena 'originated from' water — that though earth and fire are not now water, yet in some way (no account being given or remaining as to how), they and the multiplicity of phenomena originated from water? In other words, at this time had Thales clearly differentiated the two distinct cosmological and physical problems: (1) the question as to an original indestructible substance out of which everything originates and returns; and (2) the question of a substratum underlying currently existing physical properties and transformations? As Aristotle clearly states the problem, it relates to

> the original source of all existing things, that from which a thing first comes-into-being and into which it is finally destroyed, the substance persisting but changing in its qualities [. . .] there [being] no absolute coming-to-be or passing away [. . .] for there must be some natural substance [. . .] from which the other things come-into-being, while it is preserved. [27] (Brackets added)

However, while Aristotle had behind him two centuries of philosophical speculation during which this problem was analyzed, clarified, and finally stated in precise conceptual terms and logical alternatives (particularly by and after Parmenides), is it likely that at the time of its first formulation the problem received the precision of its final form? Probably not. In any case, one can agree with Kirk and Raven that the most that can be concluded regarding Thales' thesis is the following:

> (i) 'all things are water' is not necessarily a reliable summary of Thales' cosmological views; and (ii) even if we do accept Aristotle's account (with some allowance, in any event, for his inevitably altered viewpoint), we have little idea of *how* things were felt to be essentially related to water. [28]

When it comes to the two remaining fragments, those pertaining to souls and gods, there is even less to guide one's conjectures. For those predisposed to see in the rational speculations of the Presocratics a more or less complete break with the earlier mythical-religious tradition, there is a tendency to minimize the significance of these fragments (for example, they are not even mentioned by Sambursky). Yet, as Cornford asserts, for all we know these views were just as important to Thales as the more scientific ones:

> [. . .] if our aim is to abandon our own standpoint and to regain that of the ancients, we cannot afford to discard all the elements which seem foreign to our own way of thinking [. . .]. Rather we should fix attention on elements which strike us as strange and unaccountable. We may find in them a clue to the attitude of mind we are trying to recover. [29]

In fact, such pre-scientific notions are just what we should expect to find in the thinking of the first outstanding philosopher-scientist following the earlier mythopoetic tradition and inaugurating the scientific outlook. Recent scholarship has shown that there are no complete breaks or abrupt transitions in intellectual traditions. Copernicus was preceded by Philolaus, Heraclides, and Aristarchus, and Einstein also had predecessors in Leibniz's relational view of space and time, in Poincaré's critique of the concept of simultaneity, in the contraction theories of time and space of Lorentz and FitzGerald, and in Clifford's prescient theory of a varying space-time continuum.

Accordingly, it is natural to find in the thinking of Thales vestiges of the older animistic, mythic, religious tradition. The very meaning of *'physis'* connotes life, vitality, and divinity. Cornford even suggests that *physis* is an abstraction from the dynamic aspects of *mana, wakanda,* etc.[30] That Thales should have explained the power of a magnet to "attract" iron or of amber to "move" wool, as due to the possession of souls, is not at all surprising. As late as Plato and Aristotle souls were thought to be the only adequate source or cause of motion (e.g., in the *Phaedrus* Plato describes the soul as being the self-moved source of motion, and for Aristotle, the motions of the planets were guided by "divine movers" or souls). At this time, as Cornford states, "the primitive assumption is that whatever is capable of moving itself or anything else, is alive — that the only moving force in the world is Life, or rather soul-substance."[31]

Recalling the discussion in Chapter II of the pervasiveness of animism in the outlook of early peoples, we should expect to find traces of it in the thinking of Thales. As Kirk and Raven have pointed out:

> It is a common primitive tendency to regard rivers, trees and so on as somehow animated or inhabited by spirits: this is partly, though not wholly, because they seem to possess the faculty of self-movement and change, they differ from mere stocks and stones. Thales' attitude was not primitive, of course, but there is a connection with that entirely unphilosophical animism [...]. Thus Thales appears to have made explicit, in an extreme form, a way of thinking that permeated Greek mythology but whose ultimate origins were almost prearticulate.[32]

Thales' outlook "was not primitive," or we should not begin our Western scientific heritage with a study of his views, but neither was he entirely removed from the heritage of the older mythopoetic, theogonic traditions. If one finds this difficult to reconcile with Thales' reputation and achievements as an engineer, mathematician, meteorologist, astronomer, etc., then this is because one still does not recognize sufficiently the difference between concrete practical investigations as controlling factors on one's thought and speculations on more remote and difficult topics. Recall again that even Newton when speculating on the ultimate nature of space and time attributed them to God's mind.

The striking originality of Thales consists in discarding the personification of natural phenomena, in rejecting the anthropomorphic explanations of events as found in the *Theogony,* and in abstracting from the content of experience a natural observable element as the basic constituent or principle of things. This in itself was a tremendous achievement worthy of one of the seven wise men of antiquity. But to expect him to have avoided recourse to souls and gods in all explanations would be unrealistic. How, for example, would Thales have explained life and consciousness in terms of water alone? While it is true that sperm and seeds are usually moist, in what does their living vitality consist? Since at this time consciousness and life were universally attributed to the soul, if one were to explain their origin, why not assume that souls are mingled with the primary substance water, and that it is full of gods? Again this is conjecture, but one which has the advantage of construing intellectual history as a continuity, rather than as a series of abrupt beginnings and endings. While Cornford may be correct in saying that only "a thin veil" separated Ionian philosophy and Hesiod, it nevertheless took a wise man to penetrate that veil.

What, then, were Thales' specific contributions to the emergence of a higher level of empirical-rational thought? First, keeping in mind the contrast between Hesiod's *Theogony* and the fragmentary description we have of Thales' system, and Simplicius' assertion that he had "many predecessors," one can conclude that Thales partially broke the mold of the earlier mythical-religious framework, rejecting both personifications of natural phenomena and gross anthropomorphic explanations. Secondly, he gave to the earlier cosmological problem taken over from Hesiod, as to how things first began, a clearer formulation in terms of an original substance (*physis*), or first principle (*archē*) of things, thus posing one of the fundamental problems for later philosophers.[33] Thirdly, in place of the vague concept of Chaos, he isolated a natural substance as the primary stuff of reality. Fourthly, his thinking shows a greater recognition of empirical processes (e.g., meteorological, magnetic, electrical, and probably physiological) and the first recorded use of physical models in explanations (e.g., earthquakes caused by disturbances in the supporting water). Thus there seems to be good evidence to show that Thales' system manifests the characteristics of a higher level of rational thought described in the last chapter: greater differentiation in the contents and forms of thought, less egocentrism and more objectivity in the understanding of nature, and use of empirical concepts woven into a more rational framework.

ANAXIMANDER

The next in the tradition of Milesian philosophy, variously called a 'pupil,' 'associate,' or 'successor' of Thales, advanced the development of empirical-rational thought to a remarkable degree. Impressed by the striking originality of Anaximander, de Santillana goes so far as to assert that "it is not a paradox to say that Anaximander's system is as much an innovation on the way of thinking that came before as the whole of science has been since, from Anaximander to Einstein."[34] This statement will undoubtedly strike the reader as somewhat exaggerated, and insofar as it seems to imply that Anaximander's philosophy (as much as we know of it) is a radical "innovation" of a new way of thinking, it probably is. Not only was Anaximander preceded by Thales who himself "had many predecessors," he also was the heir to the cosmogonic speculations of Hesiod.

And yet there is an element of truth in de Santillana's assertion in that nothing is more difficult than the articulation of a new way of thinking about the world. Whether this new way ought to be attributed particularly to Anaximander, or whether he should be used as a symbol for the flowering of a new intellectual point of view, the tradition of scientific rationalism initiated by the Milesians not only set the stage for later developments in Greek philosophy but also laid the foundation for the scientific outlook toward the universe. For it is to the Milesians that history ascribes the first systematic attempt to produce *natural* explanations of the whole range of physical processes, from the origin of the universe and formation of the heavens, to the causes of meteorological phenomena and generation of living creatures on the earth.

Like his predecessor Thales, Anaximander apparently was extremely accomplished, being referred to as 'the most outstanding of his generation.' He led an expedition which founded a colony at Appolonia, he drew the first map of the known ancient world, and though he probably did not invent the gnomon, a vertical rod whose shadow could be interpreted as indicating the varying positions and heights of the sun, he may have been the first to introduce it into Greece.

Judging from the accounts we have of his writings, derived mainly from Aristotle, Theophrastus, and Simplicius, Anaximander began his speculation with the problem bequeathed by Thales, that of defining the original substance (*physis*) or first principle (*archē*) from which the universe began. But though Anaximander began with the question posed by Thales, he rejected the latter's answer that water is the primary substance, perhaps as Aristotle conjectured on the grounds that if it were, it would be difficult to understand how the other substances (especially such a contrary substance as fire) could be generated from it, or why they would not be 'destroyed' by it. We are fortunate in having a statement by Simplicius, based on the text of Theophrastus, which scholars agree contains an authentic fragment from Anaximander:

Anaximander [. . .] said that the principle and element of existing things was the *apeiron* [indefinite or infinite], being the first to introduce this name of the material principle. He says that it is neither water nor any other of the so-called elements, but some other *apeiron* nature, from which come into being all the heavens and the worlds in them. And the source of coming-to-be for existing things is that into which destruction, too, happens 'according to necessity: for they pay penalty and retribution to each other for their injustice according to the assessment of Time', as he describes it in these rather poetical terms. [35]

The above statement contains the primary source for the interpretation of Anaximander's philosophy. Needless to say, there has been much speculation as to just what the various assertions contained in this statement mean. As Kahn states: "As the oldest document in the history of Western philosophy, the brief fragment of Anaximander has been the object of endless discussion." It is clear, however, that Anaximander rejected Thales' notion of water as the original *archē*, as well as any of the other visible substances, probably for the reason added by Simplicius: '[. . .] having observed the change of the four elements into one another, he did not think fit to make any one of these the material substratum, but something else besides these.' Thus in order to account for the endless transformation of phenomena in the world, as well as for the origin of all that exists, Anaximander turned to something inexhaustible and indestructible, apparently introducing a new name for the generating principle, the '*apeiron*.' The etymology of this term connotes that which is "endless," something which "cannot be passed over or traversed from end to end," and thus came to mean an inexhaustible and endless entity, usually translated as the 'Indefinite,' the 'Infinite,' or the 'Boundless.'

This interpretation is supported also by Aristotle who undoubtedly was referring to Anaximander in the following passage:

> For there are some who make the *apeiron* not air or water, but a thing of this sort, so that the other elements should not be destroyed by the one of them that is infinite. For they are characterized by *opposition* to one another; air, for instance, is cold, water is wet, fire is hot; if one of these were infinite, the others would now have perished. Hence, they say, the *apeiron* is something else, from which these things arise. [36] (Italics added)

As both the quotations from Simplicius and Aristotle indicate, none of the observable elements can be the beginning and source of all change, hence a new *archē* is needed, one which is boundless in extent and prior in time. Moreover, as Aristotle suggests, the *apeiron* must be inexhaustible and indestructible, for 'only so would generation and destruction not fail, if there were an infinite source from which that which is coming-to-be is derived.' Thus the Boundless not only is imperishable in contrast to the empirical world, it is also that inexhaustible stuff out of which and into which the transformations of the world take place. Just as Thales' *archē* was 'full of Gods,' so Aristotle claims that the 'deathless and imperishable' nature of the *apeiron* indicated its divine nature: '[. . .] as it is a beginning, it is both uncreatable and indestructible [. . .]. Further, they identify it with the Divine, for it is "deathless and imperishable" as Anaximander says [. . .].'[37]

But what precisely does Aristotle's 'infinite source' mean? Does it refer to temporal infinity or spatial infinity or both? As 'uncreated and indestructible' the Boundless would certainly be everlasting and hence always in existence. But it is doubtful that Anaximander could have formulated an explicit conception of spatial infinity at this early date. The Boundless was indefinite in the sense of having no boundary, but this itself is not an explicit conception of spatial infinity. The question of temporal and spatial infinity pertains also to Simplicius' assertion that the *apeiron* is that 'from which come into being all the heavens and the worlds [*kosmoi*] in them.' This implies a plurality of worlds existing in the heavens, but are these worlds coexistent in space or successive in time — one world succeeding another from time immemorial? The existence of innumerable worlds, both coexistent and successive, was a view later formulated by the Atomists, but could it have been held by Anaximander? The question has been extensively discussed by Cornford, Kahn, and Kirk and Raven, the consensus being that it was unlikely that Anaximander held the view of separate world systems, either as coexistent or as successive. As Kirk and Raven state: "It would be entirely contrary to the whole mythical background of Greek thought, and to the dictates of common sense, to believe in a cycle of separate worlds [...]."[38] Moreover, it is extremely unlikely that the kinds of problems and reasons which led the Atomists, a century later, to posit infinite worlds would have occurred at this early date.

In addition to the attributes of the Boundless already discussed, its 'eternal motion' was mentioned by Theophrastus as 'the cause of the generation of the heavens,' and by Simplicius as the source of the 'separating out of the opposites' (to be discussed shortly). That an explanation of the ceaseless transformations of the world would require an 'external motion' seems evident, but no explanation is given either as to the nature or the origin of this motion. However, since the *apeiron* was said to be divine, Cornford conjectured that no explanation was needed: "Motion was inherent in the divine stuff because it was alive, and eternal because that life was immortal."[39] This equating of motion with soul, life, and the divine, noted earlier, was characteristic of all primitive peoples and still a prevalent belief among the ancient Greeks. Thus the earlier traditions are still apparent in the thought of Anaximander, as Kirk and Raven indicate: "Anaximander seems to have applied to the Indefinite the chief attributes of the Homeric gods, immortality and boundless power (connected in this case with boundless extent) [...]."[40]

Anaximander's explanation of how the cosmos arose out of the Boundless includes one of the basic presuppositions of Greek thought, namely, that nature represents a dynamic interaction among antagonistic powers. More specifically, this presupposition can be broken down into the following assumptions: (1) that the visible world is composed of various substances such as earth, water, mist, air, wind, fire, etc.; (2) that the disparate qualities of these substances (the dense, cold, wet, dry, rare, hot, etc.) result in their being in opposition to one another; and (3) that these opposing powers account for the phenomena of nature. So embedded is this schema in Greek thought that we shall find it appearing in different forms in each of the Greek philosophers from Anaximander to Plato and Aristotle.

In part, at least, this schema must have had an empirical origin. When one observes the panorama of nature, especially atmospheric and meteorological phenomena, it might well seem that it represents a conflict among opposite qualities, such as hot and cold, wet and dry, light and dark. We associate these contrasting qualities with the various seasons, hot with summer, cold with winter, damp with spring, dry with autumn, and dark with night and light with day. Moreover, atmospheric changes are often accompanied by wind and storms, thunder and lightning, as if exhibiting the wrath of the gods. Something like this conflict of opposites must have impressed itself on the mind of the ancient Greeks, for they were described already by Hesiod in the *Works and Days*.

However, in addition to this empirical aspect, it seems that Anaximander again was influenced by Hesiod's *Theogony* (even de Santillana suggests that the Boundless was "certainly meant to develop the Hesiodic idea of Chaos, the 'yawning gap' "[41]). In the earlier summary of the *Theogony* we noted the following thematic developments: (1) that the various domains or aspects of the world arose by "coming out of" the original Chaos; (2) that sexual generation is the main model of creation; (3) that an "opposition" or "Strife" is a cause of change; (4) that pervading this change there is a moral or legal order, a kind of "retributive justice;" and (5) that there is a "just portioning" of the various "domains" of the gods underlying the political conflicts. In Anaximander's explanation of the origin and formation of the visible world, one cannot but see a somewhat more abstract version of this schema being imposed on the empirical qualities just described to account for the creation and transformation of natural phenomena. Kahn asserts that the poetic "division of the world into Earth, Sea, Underworld, and Heaven," mentioned by Hesiod in the introduction to the *Theogony*, and "familiar to every Greek schoolboy from the recitations of Homer and Hesiod," helps to explain "how the later doctrine of the four elements [. . .] could meet with such rapid and widespread approval." [42]

If what has just been said regarding the influence of the *Theogony* on Amaximander's thought is true, then it seems that he began his cosmological account by abstracting from Hesiod's vague notion of Darkness and black Night "coming out of" Chaos, the more explicit conception of opposite qualities "separating off" the Boundless. According to Simplicius there were several pairs of such opposites, 'hot, cold, moist, dry, and the rest.' These opposites were not mere qualities, but substantial 'powers,' since it is due to their antagonism or 'aggression' that the universe is formed. Moreover, just as Hesiod explained generation as due to Eros, Anaximander may have described the first act of generation as due to "the secretion of a pregnant seed or germ out of the Boundless, which thus became the parent of the universe." [43] That there was such an intermediate germinal principle is suggested by a fragment of Pseudo-Plutarch: 'Something capable of generating Hot and Cold was separated off from the eternal [Boundless] in the formation of this world [. . .].'

Again, this is conjecture, and it is possible that Anaximander had a more empirical model in mind, either in place of or in addition to the schema just described. Aristotle suggests an image of a vortex, but there is no additional doxographical evidence to support this. De Santillana points to the model of a "winnower's sieve," a model used by later Presocratics and by Plato:

> In that world-eddy, he thought, what is mixed in the uniform boundless must come to separate out; and the familiar image that came to his mind must have been the Winnower's sieve, rotating and shaken, where the heavy grain remains in the middle and the chaff wanders toward the rim. For he said that contraries "separated out," earth going to the center and fire to the outside, water and air remaining in between. [44]

It is unlikely that we shall ever know which of these interpretations is correct, or whether even Anaximander could have told us, the influences on one's thought being so diverse and often less than explicit.

In any case, the eternal motion of the Boundless generated something from which separated off the primary opposites, the Hot and the Cold. These opposites might have been contained in the Boundless in an indistinct state before they were separated off by the germinal principle, perhaps existing in the Boundless the way dissolved substances exist in a solvent without being apparent. This view is supported by Cornford but denied by Kahn. [45] The primary opposites then form a mixture in which air, mist, water, and earth separate from the Cold. It is at this point that Simplicius' statement (previously quoted) containing the original fragment of Anaximander enters in: 'And the source of coming-to-be for existing things is that into which destruction, too, happens "according to necessity: for they pay penalty and retribution to each other for their injustice according to the assessment of Time [. . .]." ' Apparently, the various qualities or powers, once having become distinct, are by virtue of their opposite natures engaged in perpetual strife and, infringing on each other's domain, must pay eternal retribution.

Here again we see the influence of the *Theogony* in which a "Strife" among opposing powers is a common cause of historical change, notwithstanding the fact that the various domains and roles of the Gods are regulated by a kind of moral or legal order, a "retributive justice." It is also likely that this conception reflects the more archaic totemic structure of society. Kahn implies as much in his discussion of the origin of the concept of *kosmos*.[46] Like the primitive, Anaximander apparently discerned a continuity between the continuous strife that plagues human society and the eternal transformations of nature, with a kind of compensatory moral order or harmony regulating both.

A generating principle having separated off the Boundless, the Hot and the Cold separate from it, the Hot forming 'a sphere of fire' surrounding the air in the center of which will condense the earth from the moist core of the air. The sphere of fire, Pseudo-Plutarch tells us, 'grew around the air about the earth like bark around a tree.' The heat from this fire apparently caused the enclosed air to expand until rings of fire separated off the primordial fire forming the circular bands of the sun, moon, and stars. Aëtius describes these rings or bands of fire, surrounded by a concealing mist, as 'similar to the wheel of a chariot, which has a hollow rim filled with fire, letting this fire appear through an aperture at one point, as though through the mouthpiece of a bellows [. . .].' The sun and moon are each such round apertures in their separate rings of fire, the moon having its own firelight, though fainter than that of the sun which consists of 'purest fire.' The similes of 'wheel-apertures,' 'breathing holes,' and 'nozzles of bellows' allow Anaximander to explain the phases of the moon and lunar and solar eclipses as partial or total blocking of the apertures.

If these similes really are due to Anaximander, and not to some "Hellenistic popularizer" as Kahn finds "more likely," they represent a remarkable use of empirical models and deductive reasoning to account for astronomical phenomena. Not the least noteworthy is the fact that celestial phenomena are being explained with concepts derived from terrestrial experience — such bold speculation, though carried further by Hellenistic astronomers, will not appear again in the West until Galileo's astronomical arguments, based on his telescopic observations, refute the qualitative distinction between the celestial and the terrestrial worlds.

But Anaximander does not stop there. Again according to Aëtius, he asserted that the circles of the sun and moon 'lie ascant,' indicating that he must have been familiar with the inclination of the ecliptic. There is some evidence also that he was interested "in solstices, equinoxes, and the measurement of the diurnal 'hours.' "[47] He assigned dimensions to the size of the sun (the sun's aperture being the same size as the circumference of the earth) and a mathematical ratio between the diameter of the wheels of fire of the sun and moon and the diameter of the earth: the sun-wheel being twenty-seven (or twenty-eight) earth diameters, the moon-wheel being eighteen (or nineteen) earth diameters. As described by Aëtius, 'The sun is set highest of all, after it the moon, and beneath them the fixed stars and planets.' No dimension is given for the size of the rings of the stars and planets, nor reason why Anaximander thought the orbit of the fixed stars was closer than that of the sun and the moon, even though the latter can be seen to pass in front of the stars.

Given the paucity of detail, there has been much speculation as to the rationale behind this mathematical model. As Kahn says, "Many conjectures have been made in order to fill some of the yawning gaps in this account." Kahn himself provides a plausible conjecture as to why Anaximander believed that the orbit of the fixed stars was closer to the earth than that of the sun and the moon: if one assumes that the brightness of the celestial bodies is due either to the purity or to the intensity of the fire composing them, and the purest and most intense fire is at the circumference, which makes the sun the brightest, then the faintest celestial bodies must be closest to the earth.

A more difficult problem of interpretation pertains to the dimensions assigned the lunar and solar orbits. Not only is there a discrepancy in the reported size of the orbits (eighteen or nineteen and twenty-seven or twenty-eight), there is no evidence explaining how Anaximander arrived at the proportions. Cornford states that the

> variants 28 for the sun and 19 for the moon have been explained by the supposition that each ring is one earth-diameter in thickness, so that the figure would be 18 or 19 according as measurements were made to the inside or the outside of the ring. [48]

The advantage of this explanation is that it would enable one to reduce the inner dimensions of the sun and moon rings to 27 and 18 times the diameter of the earth, i.e., some multiple of three or nine. And since the diameter of the earth is three times its height, if one divides the other dimensions by three, this gives a factor of nine which would lead to the neat progression 9, 18, 27, suggesting nine earth diameters as the missing dimension for the band of the fixed stars.

While this supposition is probably too neat to be true, it still leaves unexplained the choice of the factor nine. Cornford brusquely states: "However the figures may be explained, it is obvious that they are *a priori* and cannot be based on any kind of observation." Kahn, on the other hand, takes a different view. After dismissing Heath's suggestion of "three Visnu-steps" as being a possible source of the dimensions, he concludes:

> I suggest that quite a different inspiration is likely to have led him to these results. In the first place, Anaximander clearly believed that the universe was governed by mathematical ratios, and by ratios of a simple kind. Furthermore, his Babylonian predecessors must have provided him with considerable data concerning the periods in which the various bodies return to the same relative position [...]. Our fragmentary knowledge of Anaximander's system does not permit us to reconstruct the method by which his numerical results were reached; we are not even sure of the figures he obtained. But that the inspiration was essentially mathematical seems to me beyond doubt. [49]

But even here one can discern perhaps the influence of the *Theogony* in which it is stated that "a bronze anvil falling from the sky would fall nine days and nights and reach earth on the tenth; a bronze anvil falling from the earth would fall nine days and nights and reach Tartarus on the tenth."[50] This reference would not give the specific ratios, of course, but it might account for the significance of the multiples of nine and for the discrepancy in the additional number as given by Aëtius.

Aside from these conjectures, what is of primary significance is Anaximander's attempt to fit astronomical phenomena into a geometrical model based on some mathematical ratios — it hardly being necessary to emphasize or document the importance of this attempt on later Hellenic, Hellenistic, and Renaissance astronomical developments. An added consequence of this model is that the east-to-west movement of the heavens could be explained as due to the rotation of the celestial wheels, while the particular 'wanderings' of the sun, moon, and other planets could be attributed to the wind. While there is only indirect evidence that Anaximander actually utilized these explanations, he can be credited with distinguishing between the planets and the fixed stars. Here we see for the first time the modern conception of our solar system beginning to take shape. In the center of this geometric model is the earth, maintained in this position by a kind of 'equilibrium,' there being no reason for it to move in any particular direction. As Aristotle says,

> there are some, Anaximander, for instance, among the ancients, who say that the earth keeps its place because of its indifference. Motion upward and downward and sideways were all, they

thought, equally inappropriate to that which is set at the centre and indifferently related to every extreme point; and to move in contrary directions at the same time was impossible: so it must needs remain still. [51]

This conception of the earth positioned in equipose in the center of the spherical heavens is significant for a number of reasons. First, at least Aristotle's statement of the view certainly implies a definition of the circle as having a center from which 'every extreme point' is 'indifferently related,' adding further evidence of Anaximander's conception of a geometrical model of the heavens. Secondly, the reasoning behind the assertion of an unsupported earth being held in equilibrium in the center of the heavens is the first known application of "the principle of sufficient reason," a principle that has played such an important role in the development of Western thought. Actually, as de Santillana defines the principle, it is a negative application of the principle that every phenomenon must have a sufficient reason for being as it is: the negative application being, if there is no reason to the contrary, one may assume in the interest of simplicity a certain feature of the world to be thus and so, rather than otherwise. If there is no reason for the earth to move from the center in any particular direction, one may assume it remains there in a state of equilibrium; if there is no reason for there to be a boundary, one may assume that the world is boundless. As stated by de Santillana,

if a theory cannot deduce for a given feature a sufficient reason why it should be thus and so and not otherwise, that feature has no place in the theory. This is the *negative* aspect of the principle, as it was applied by Anaximander [. . .]. [52]

In a sense, the principle of sufficient reason is a precise expression of the belief in an intelligible universe: everything has a sufficient explanation and the simpler the explanation the better. It is because of Anaximander's use of this principle that de Santillana claimed, as noted earlier, that his achievement was "as much an innovation on the way of thinking that went before as the whole of science has been since [. . .]." And Kahn also affirms:

For the history of ideas, Anaximander's theory of the earth's position is of an entirely different order of importance. Even if we knew nothing else concerning its author, this alone would guarantee him a place among the creators of a rational science of the natural world. [53]

Anaximander goes on to describe the earth, according to Hippolytus and Pseudo-Plutarch, as 'moist, rounded like a stone column,' with 'a depth one-third of its width.' Thus the earth resembles in shape the truncated marble columns that one sees among the ruined splendors of Greece and which must have been familiar to Anaximander. No evidence remains as to how he arrived at the ratio of the earth's diameter to its length. There is a reference to the antipodes or the opposite surface of the earth, although nothing is known as to what the reference might have signified.

The formation of the heavens and the earth having been described, the doxography next records Anaximander's attempt to explain various meteorological phenomena. Somewhat surprisingly, Kahn is rather disparaging of this attempt: "Although we must recognize its rational quality, the meteorology of the ancients fails to awaken our enthusiasm." [54] On the contrary, however, the rationality of Anaximander's explanations deserves our admiration, as we shall see.

The earth having been condensed at the center of the heavens from the air-mist, at first both the earth and the surrounding region are moist. But the heat from the sun dries the moist earth producing dry land, some of the moisture remaining as the sea while that which is evaporated produces clouds and winds. 'Rainfall arises from the vapor emitted by the earth under the action of the sun.' Thunder and lightning are caused by wind caught in 'thick black clouds' bursting out with violent force, the bursting of the cloud causing the noise of thunder, the sudden contrast between 'the lightness and the fineness of the parts' of the escaping wind and the blackness of the containing cloud causing the appearance of a flash. According to one source, the evaporating moisture, 'the vapors and exhalations,' which form the wind and the clouds also cause the solstices and the 'turnings' of the moon (and perhaps of the sun). Why the turning of the moon and the sun should depend upon such 'vapors and exhalations' is not made clear. Recalling that the Boundless which continues to 'surround all the heavens' is divine, there may be an aspect of anthropomorphic nourishment in all of this. Be that as it may, what an extraordinary use of terrestrial analogies to explain in relatively clear rational concepts celestial or meteorological phenomena. Here, finally, Zeus and his thunderbolts have been banished to the netherworld of mythology.

In his derivation of life in general, and of human life in particular, Anaximander turned his back on the genealogical descriptions typical of such mythical accounts as the *Theogony* and the creation myth of the *Old Testament.* Instead, we find Anaximander speculating that as the earth was dried by the sun living things arose from the moisture. These first creatures were covered by prickly bark similar to the protective covering of sea urchins. As Aëtius states: 'The first animals were generated in moisture and enclosed in thorny barks; as they grew older, they came out onto the drier [land] and, once their bark was split and shed, they lived in a different way for a short time.' Moreover, taking note of the fact that infants are defenseless and cannot feed themselves, Anaximander reasoned that the first men could not have originated from babies (the first appearance of the chicken and egg controversy). Instead, 'man was generated from living things of another kind, since the others are quickly able to look for their own food, while only man requires prolonged nursing.' Thus, according to Hippolytus, 'In the beginning man was similar to a different animal, namely, a fish.' In what way man was 'similar' to a fish, or how he developed from a fish, is not recorded. Naturally Anaximander did not have, as we do, the embryological evidence that the human fetus displays gill slits during its development. Nonetheless, this account presages the later theory of evolution.

In summary, while the evidence clearly supports Cornford's thesis regarding the continued influence of the older mythopoetic and religious traditions, especially Hesiod's *Theogony*, on the general form of Anaximander's way of construing the universe, the content itself discloses a remarkable degree of empirical awareness and rational thought. As de Santillana says:

> We may discern poetic ambiguity in Anaximander's words, charged with all the themes of the past. But the light of reason has dispelled Chthonian darkness. If what emerges is not physical thought, one may ask what is. [55]

Assuming the authenticity of the doxography, Anaximander's clearer formulation and solution of the cosmological problem, his bold use of inductive and deductive reasoning and of the principle of sufficient reason, his attempt to provide a geometrical representation of astronomical phenomena, his imaginative use of empirical analogies and models to explain meteorological occurrences, and his endeavor to account for organic life in purely naturalistic terms mark him as a thinker of the first rank and a founder of scientific rationalism. In Kahn's estimation:

What the system of Anaximander represents for us is nothing less than the advent, in the West at any rate, of a rational outlook on the natural world. This new point of view asserted itself with the total force of a volcanic eruption, and the ensuing flood of speculation soon spread from Miletus across the length and breadth of the lands in which Greek was spoken. [56]

Even modern descriptions of the origin of the universe sound remarkably similar to Anaximander's view. For example, in a newspaper report of a new creation theory according to which matter is continuously created within galaxies by cells of creation called 'irtrons,' the following summary occurs.

In each [of 12 galaxies so far investigated] [...] both matter and anti-matter are being created, then annihilating each other — and the resulting debris is continuously sprayed out to form all the stuff that fills the universe, making stars, making planets, making new worlds. [57] (Brackets added)

As in Anaximander, we are presented with an account of creation in terms of opposite substances arising from some unknown background material which, in their self-destructive conflict, give rise to the material which fills the universe and forms the planets, stars, and new worlds. Thus so contemporary is ancient thought, and so ancient are modern theories.

ANAXIMENES

The last in the trilogy of Milesian philosophers is Anaximenes. As Anaximander lived one generation after Thales, so Anaximenes was a younger compatriot of Anaximander and probably his 'pupil' and 'associate.' Though not as innovative as Anaximander's, his system illustrates the progressive clarification, refinement, and extension of explanatory concepts that usually accompany critical reflection on previous theories. His views are important also because of their influence on later philosophers such as Anaxagoras, Leucippus, and Plato.

As in the cosmologies of Thales and Anaximander, the central problem was that of defining the *archē* or the nature of *physis*. Apparently finding Anaximander's Boundless too indefinite, he selected a more specifiable substance, 'Air,' as the original stuff or first principle of things, although this 'Air' was not the common atmospheric air whose presence is indicated by qualities such as dampness or coldness. Instead, his conception was more abstract and theoretical, a kind of inexhaustible Air-substrate from which the present atmospheric qualities and different substances originate. Simplicius, following the account of Theophrastus, describes his system in this way:

Anaximenes [...] of Miletus, a companion of Anaximander, also says that the underlying nature is one and infinite like him, but not undefined as Anaximander said but definite, for he identifies it as air; and it differs in its substantial nature by rarity and density. Being made finer it becomes fire, being made thicker it becomes wind, then cloud, then (when thickening still more) water, then earth, then stones; and the rest come into being from these. He, too, makes motion eternal, and says that change, also, comes about through it. [58]

Although retaining the eternal motion and infiniteness of Anaximander's *apeiron*, Anaximenes' explanation of how the ordinary physical world came to be is considerably clearer. Discarding the vague concept of "separating off," he adopted the simpler, more intelligible theory that all the various kinds of phenomena are merely different 'appearances' of the underlying 'Air.' The contrasting elements are simply 'Air' in different states of density or rarity, the latter produced by the eternal motion of the 'Air' itself. As described by Hippolytus, this 'Air'

> through becoming denser or finer [. . .] has different appearances; for when it is [. . .] finer it becomes fire, while winds, again, are air that is becoming condensed, and cloud is produced by air from felting. [59]

In an 'equable' state, one of complete uniformity, air is invisible though in constant motion. 'Felting' apparently was a new term introduced to designate the process of condensation.

This explanation was so clear and seemed to fit the cyclical transformations of nature so well that it was adopted by Plato in the *Timaeus* and expressed in the same way:

> [. . .] we see that what we just now called water, by condensation, I suppose, becomes stone and earth; and the same element, when melted and dispersed, passes into vapour and air. Air, again, when inflamed, becomes fire; and again fire, when condensed and extinguished, passes once more into the form of air [. . .] [which] when collected and condensed, produces cloud and mist; and from these, when still more compressed comes flowing water, and from water comes earth and stones once more [. . .]. [60] (Brackets added)

A major theoretical asset of this type of explanation is that it allows for functional correlations among the variables, motion, density, and temperature, with the possibility (not actualized by Anaximenes, but attempted by Plato) of quantifying these correlations. For example, an increase in motion causes a decrease in density and an increase in temperature, while a decrease in motion causes an increase in density and a decrease in temperature. Theoretically, therefore, qualitative changes could be reduced to quantitative relations, again showing the superiority of this explanation over that of Anaximander.

Using breath as an example, Anaximenes tried to illustrate how variations of density can result in differences in temperature: 'for the breath is chilled by being compressed and condensed with the lips [when one blows outward], but when the mouth is loosened the breath escapes and becomes warm through its rarity.' Aristotle scornfully dismissed this illustration as due to 'the man's ignorance,' pointing out that the warm air we exhale is not due to its rarity but to its having been heated in the body, while the coolness of the blown air is not a result of its density, but of the cooler air outside being propelled by our breath. While the choice of example may not have been apt, it does indicate a tendency to use empirical examples to support his theory.

As in the cosmologies of Thales and Anaximander, there still is a visible trace of the older theogonic tradition and of primitive animism. Anaximenes' 'Air,' like the Water of Thales and the *apeiron* of Anaximander, is the divine begetter of all that exists. Hippolytus asserts that 'Anaximenes [. . .] said that the *arché* was infinite air, from which arise what is becoming, what has become, and what is to be, and gods and things divine; the rest arise from the progeny of the air.' Other sources maintain that the 'Air' itself is a god, so that it may be reasonably inferred that whatever gods Anaximenes believed to exist were derived from the primal 'Air' which itself was divine. The phrase 'progeny of the air' recalls Hesiod, as Kahn maintains: "[. . .] the stylistic echoes of Hesiodic theogony are [. . .] unmistakable in the case of Anaximenes." [61]

The element of animism is indicated in a statement by Aëtius identifying the infinite 'Air' with breath and so with life or soul: 'As our soul, he says, being air holds us together and controls us, so does wind [or breath] [. . .] enclose the whole world.' It is not clear what the above expression means, yet, since this concept of breath was used earlier to illustrate how different densities of air could give rise to the different qualities hot and cold, it apparently played some role in Anaximenes' cosmology. An ambiguous reference to 'exhalations' occurred in Anaximander's system, while the concept of a breath-soul will be used by the Pythagoreans in their account of creation.

The conception of the soul as air or breath [. . .] is founded in the age-old sense of the link between respiration and life that is contained in the very etymology of the word *psyche*: the power by which we breathe, and which we "expire" with life itself. Such a view of the *psyche* as a breath-soul is still prevalent at the time of Plato. [62]

The condensation or 'felting' process eventually produces the earth, which in turn generates the sun, moon, and remaining heavenly bodies by its exhalations. Hippolytus provides an explicit description: 'The heavenly bodies have come into being from earth through the exhalation arising from it; when the exhalation is rarefied, fire comes into being, and from fire raised on high the stars are composed.' The earth and the heavenly bodies, though fiery, are all flat, supported by the air as leaves float in the breeze: 'the sun is flat like a leaf,' and 'the stars are fiery leaves like paintings.' Aëtius adds that 'the stars are implanted like nails in the crystalline,' though no explanation is given as to the nature or status of the latter. Anaximenes denied that the heavenly bodies move under the earth, claiming that they turn around it as 'a felt cap turns round the head,' the sun being hidden at night 'by the higher parts of the earth,' presumably a reference to the mountains in the north. He corrected Anaximander's notion that the fixed stars are closer to the earth than the planets, asserting that they produce no heat and less light than the sun and moon because of their greater distance from the earth.

In his explanations of hail, snow, rainbows, and earthquakes Anaximenes expanded on the meteorological contributions of Anaximander. Aëtius reports that 'Anaximenes said that clouds occur when the air is further thickened; when it is compressed further rain is squeezed out, and hail occurs when the descending water coalesces, snow when some windy portion is included together with the moisture.' He too attempted to explain lightning, claiming that a flash of lightning is similar to 'what happens in the case of the sea, which flashes when cleft by oars.' His explanations of rainbows and earthquakes were particularly sophisticated.

A rainbow arises from the radiation of the sun in contact with a cloud which is dense, thick, and hard, since the beams become entangled with the cloud and are not able to break through [. . .]. The first part [of the air or cloud] appears as brilliant red, burnt through by the sun's rays, the rest as dark overpowered by moisture. [63]

Earthquakes occur in periods of draught and excessive rains, 'for in draughts [. . .] it dries up and cracks, and being made over-moist by the waters it crumbles apart.'

CONCLUSION

As stated earlier, while not as brilliantly innovative as Anaximander, Anaximenes' conception of *physis*, and particularly his account of the generation of the empirical world in terms of the processes of condensation and rarefaction, were significant advances in the handling of these problems. In terms of theoretical implications, if not in actual conceptual explicitness, there is only a lack of refinement separating his views and those of the later Pluralists. But conceptual worlds, like actual worlds, are not created overnight. His conceptions of creation and motion still exhibit earlier animistic features, associations with biological generation, life, breath, soul, and the divine. And even though he discarded the vestigial totemic framework of a moral-political order that haunted Anaximander's cosmology, there is more than a hint of anthropomorphism in his phrase 'progeny of the air.' Although it is clear that we are further along the way to a more explicit scientific rationalism, the legacy of the past has not been completely outgrown.

Nevertheless, storm clouds are no longer explained in terms of the extended wings of the gigantic bird Imdugud, as in Babylonian myths, and lightning is no longer a thunderbolt thrown by Zeus. Moreover, a definite attempt is made to explain the origin of the world in terms of substances and processes abstracted from nature, rather than exclusively in terms of biological and social models. According to Cornford's assessment:

> It was an extraordinary feat to dissipate the haze of myth from the origins of the world and of life. The Milesian system pushed back to the very beginning of things the operation of processes as familiar and ordinary as a shower of rain. It made the formation of the world no longer a supernatural, but natural event. Thanks to the Ionians, and to no one else, this has become the universal premiss of all modern science. [64]

Thus the primary legacy of the Milesians might well be the innocent but bold assumption that, however complex or remote, natural phenomena can be understood in terms of empirical concepts and rational principles.

NOTES

CHAPTER IV

[1] Ernst Cassirer, *The Philosophy of Symbolic Forms, Mythical Thought,* vol. II, trans. by Manheim (New Haven: Yale Univ. Press, 1955), p. 14.

[2] F. M. Cornford, *Principium Sapientiae,* ed. by W. K. C. Guthrie (New York: Harper & Row, Pub., [1952] 1965), p. 145.

[3] For an excellent summary of the doxography pertaining to Milesian philosophy see the work of Charles H. Kahn cited below, pp. 5-71.

[4] Cf. the following works on the Presocratic philosophers:

Hermann Diels, *Die Fragmente der Vorsokratiker,* ed. by Kranz, 7th ed., 3 vols. (Berlin: Weidmann, 1954). This work is often referred to by the hyphenated name, Diels-Kranz.

John Burnet, *Early Greek Philosophy,* 4th ed. (New York: The Macmillan Co., 1930).

G. S. Kirk and J. E. Raven, *The Presocratic Philosophers* (Cambridge: At the University Press, 1957).

Kathleen Freeman, *The Pre-Socratic Philosophers,* 2nd ed. (Oxford: Basil Blackwell, 1959).

Charles H. Kahn, *Anaximander And The Origins of Greek Cosmology* (New York: Columbia Univ. Press, 1960).

Kathleen Freeman, *Ancilla To The Pre-Socratic Philosophers* (Oxford: Basil Blackwell, 1962).

Special mention should be made of the comprehensive study by W. K. C. Guthrie of the *History of Greek Philosophy,* vols. I-IV (Cambridge: At the University Press, 1967-1975).

The author is particularly indebted to the authoritative study of Milesian philosophy by Charles H. Kahn for the interpretation of Anaximander presented in the present chapter, and especially to the excellent work of Kirk and Raven for the interpretation of Presocratic philosophy in general. Finally, the influence of F. M. Cornford will be apparent throughout.

[5] The terms Milesian and Ionian philosophy or 'school' refer to the oldest tradition of Presocratic philosophy consisting of the speculations of Thales, Anaximander, and Anaximenes. The 'school' derives its name from the city of Miletus in Ionia where each member lived and taught.

[6] Cornford, *op. cit.,* p. 186.

[7] Cf. F. M. Cornford, *From Religion to Philosophy* (New York: Harper & Brothers, [1927] 1957), p. viii.

[8] *Ibid.,* p. 54.

[9] Kahn, *op. cit.,* p. 156.

[10] Cornford, *From Religion to Philosophy, op. cit.,* p. 126.

[11] Cornford, *Principium Sapientiae, op. cit.,* p. 187.

[12] *Ibid.,* p. 198. Kahn on the other hand takes an opposite view: "Serious speculation seems to me to play very little part in Hesiod's *Theogony.*" Kahn, *op. cit.,* p. 95. I tend to agree with Kahn.

[13] *Ibid.,* p. 188.

[14] *Ibid.,* pp. 188-189.

15 Giorgio de Santillana, *The Origins of Scientific Thought*, vol. I (Chicago: The Univ. of Chicago Press, 1961), pp. 23-24. The reference to "medicine men," "sad mythographers like Hesiod," and "priests and Prophets" is to Cornford's discussion in *Principium Sapientiae*, chs. V, VI, VII, XI.

16 *Ibid.*, p. 14.

17 *Ibid.*, p. 17.

18 S. Sambursky, *The Physical World of the Greeks,* trans. by Merton Dagut (New York: Collier Books, [1956] 1962), p. 34.

19 *Ibid.*, p. 39.

20 Hesiod uses the term "Strife" several times in the *Theogony*, at one point saying "Hateful Strife gave birth to [. . .]." Also, he talks of "the Strife which is good for men" in *Works and Days*, lines 174-178.

21 Kahn, *op. cit.*, p. 75.

22 Cassirer, *op. cit.*, p. xvi.

23 Kahn, *op. cit.*, p. 200. However, it was Cornford more than anyone else who called attention to the influence of the *Theogony* on Presocratic philosophy. Cf. *Principium Sapientiae*, ch. X. See also Kirk and Raven, *op. cit.*, pp. 24-32.

24 Cf. Chapter III. For a comparison of the *Theogony* and the "Enuma Elish," see Norman O. Brown's "Introduction" to *Hesiod's Theogony* (New York: The Liberal Arts Press, 1953), pp. 37-49. The following summary of the *Theogony* is based on Professor Brown's translation and "Introduction."

25 As this information is common knowledge which can be had from any of the standard works on the Presocratics listed under footnote 3 above, I shall not indicate sources except where the view expressed is of special significance or of a particular interpretation. I shall identify original source material by single quotation marks.

26 Cf. Aristotle, *Met.*, 983b19-27.

27 Aristotle, *Met.*, 983b8-13.

28 Kirk and Raven, *op. cit.*, p. 93.

29 Cornford, *Principium Sapientiae, op. cit.*, p. 5,

30 Cf. Cornford, *From Religion to Philosophy, op. cit.*, p. 87.

31 *Ibid.*, p. 131.

32 Kirk and Raven, *op. cit.*, p. 95.

33 It is not known who was the first specifically to use the terms *'physis'* and *'arche;'* they were used by Aristotle and Theophrastus to describe Thales' and Anaximander's first principles. For the earliest use of *'physis,'* see Kahn, *op. cit.*, p. 4, and for *'arche,'* pp. 30-31; 235-236.

34 de Santillana, *op. cit.*, p. 36.

35 Simplicius, *Phys.* 24, 13; DK 12A9, based on Theophrastus, *Phys. Opin.*, fr. 4. Quoted from Kirk and Raven, *op. cit.*, pp. 105-107. Brackets in the original.

36 Aristotle, *Phys.* 204b22.

37 *Ibid.*, 203b7-14. (Hardie and Gaye trans.)

38 Kirk and Raven, *op. cit.*, p. 123.

39 Cornford, *Principium Sapientiae, op. cit.*, p. 181.

40 Kirk and Raven, *op. cit.*, p. 116.

41 de Santillana, *op. cit.*, p. 38.

42 Kahn, *op. clt.*, pp. 136-137.

43 *Ibid.*, pp. 86-87.

44 de Santillana, *op. cit.*, p. 28.

45 Cf. Cornford, *Principium Sapientiae, op. cit.*, pp. 160-162; Kahn, *op. cit.*, p. 41.

46 Cf. Kahn, *op. cit.*, pp. 222-223.

47 Cf. Kahn, *op. cit.*, p. 88.

48 Cornford, *Principium Sapientiae, op. cit.*, p. 164, f.n. 3.

49 Kahn, *op. cit.*, pp. 96-97.

50 Brown, *Hesiod's Theogony, op. cit.*, p. 73.

51 Aristotle, *De Caelo,* 295b10-17. (Stocks trans.)

52 de Santillana, *op. cit.*, p. 36.

[53] Kahn, *op. cit.*, p. 77.

[54] *Ibid.*, p. 98.

[55] de Santillana, *op. cit.*, p. 40.

[56] Kahn, *op. cit.*, p. 7.

[57] Victor Cohn, "New Theory of Universe [by Dr. Frank J. Low] Says It Is Created Continuously," in *International Herald Tribune*, Paris, January 3-4, 1970, p. 3. Brackets added.

[58] Simplicius, *Phys.* 24, 26. Quoted from Kirk and Raven, *op. cit.*, p. 144, f.n. 143.

[59] Hippolytus, *Ref.* 1, 7, 1. *Ibid.*, pp. 144-145, f.n. 144.

[60] Plato, *Timaeus*, 49. (Jowett trans.)

[61] Kahn, *op. cit.*, pp. 156-157.

[62] *Ibid.*, p. 114. *Psyche* has been substituted twice in the text for the Greek lettering.

[63] Quoted from Kahn, *op. cit.*, p. 161. The Greek synonyms for some of the translated words have been omitted.

[64] Cornford, *Principium Sapientiae*, *op. cit.*, pp. 187-188.

THE FIRE AND THE LOGOS

> [. . .] *the eternal but ever-changing processes of motion* [. . .] *is just the process of* becoming, *first described by Heraclitus* [. . .].[1]
>
> Bohm

The philosopher to be discussed in this chapter was neither from Miletus nor a pupil or companion of Anaximenes. About forty years younger than the latter, he spent his life in Ephesus, a city close to Miletus on the coast of Ionia, and prided himself on not being a disciple of anyone. Indeed, he decidedly did not have the scientific outlook and rational cosmological interests of the Milesian philosophers discussed in the previous chapter. The reason for describing his views is the selection of fire as the *physis*, thus rounding out the choice of primary substances for the *archē* and, more particularly, to contrast his mode of thinking with that of the Milesians. While in this study we shall be concerned primarily with the growth of a mode of thought called scientific rationalism, it is well to be reminded occasionally that there are other intellectual outlooks and forms of thought that have also played an important role in the development of Western intellectual history: e.g., the poetic, mystical, theological and metaphysical. As a thinker, Heraclitus belongs to this latter group.

He was not a "wise man" in the scientific tradition of Thales and Anaximander, but more of a prophet, sage, or metaphysical poet. Had he lived in an earlier era or in a different cultural milieu he undoubtedly would have been a shaman or a priest — again attesting to the fact that the same diversity of minds occurs in almost all societies, though manifested within the intellectual context of the culture. Instead, like Xenophanes,[2] he retained the typical critical attitude of Greek philosophers toward the conventional religion ('they pray to these statues, as if one were to carry on a conversation with houses, not recognizing the true nature of gods or demi-gods'), though he believed in the all-pervading presence of the divine. He adopted the oracular style of the Delphic oracle and the Sibyl (at a time when they were losing favor among the philosophers), which gave to his philosophical pronouncements an enigmatic, cryptic, and profound air. In fact, he was known as an exceptionally 'haughty,' 'obscure,' and 'prophetic' individual.

As one would expect from this kind of personality, there is a strong ethical, paradoxical, and somewhat mystical flavor to his thought: the kind of mind that gives rise to moral imperatives, visionary insights, and metaphysical systems, rather than empirical discoveries, physical explanations, or mathematical laws. One is reminded less of the outlook of the Milesians and more of the inspirational writings of such great religions as Hinduism, Confucianism, and Taoism. He addresses himself more to man's condition in the world and moral outlook than to a scientific understanding of natural phenomena. What scientific instruction he does give is not primarily for its own sake, but as an aid in guiding man's life: 'The wise is one thing, to be acquainted with true judgment, how all things are steered through all.' Though somewhat obscure, his writings do exhibit a comprehensive unity characteristic of metaphysical systems in contrast to scientific inquiry which often gives the impression of teaching one more and more about less and less (although powerful scientific theories, such as the atomic theory, also unify one's understanding of diverse phenomena). As de Santillana says:

> He is and remains an unscientific thinker. But he comes back at all times to haunt the thoughts of the scientist. He has focused attention on how mysterious and ambiguous the objects of nature really are, however familiar and obvious they may appear to be. [3]

According to Diogenes Laertius, Heraclitus wrote a book entitled 'On Nature' which was 'divided into three discourses: On the Universe, Politics, and Theology.' The work was said to have had 'so great a reputation that from it arose disciples, those called Heracliteans.' None of this information is too reliable, however. The later Stoics were much influenced by his views and so, to some extent, was Plato, especially by his thesis that 'all things are in flux.' As indicated in the last chapter, the aspects of nature that particularly caught the attention of Presocratic philosophers were the rhythmic processes of change and transformation. This awareness became articulated in the fundamental theoretical questions of "The One and the Many" and of " Change and Identity;" i.e., from whence do the diverse kinds of things originate and how do they maintain their identity while undergoing change? We have seen how these questions were answered by the Milesians in terms of an original *arché* such as 'Water,' 'Air,' or the 'Boundless,' and by such processes as 'separating off,' and 'condensation and rarefaction.' Thus in the system of the Milesians a conception of an eternal, inexhaustible, underlying substance out of which the opposite qualities and elements of the world arise, and into which they return in a balanced process, was dominant.

In the system of Heraclitus (as in such 20th century process philosophers as Whitehead, Bergson, Čapek, etc.,), however, it is not the selection of a primordial element, but the *aspect of Change itself* that is abstracted from experience and made the fundamental *arché* — further illustrating the creative variation in the use of abstractions among the Presocratic philosophers. Process and change are thus the basic principles of things. As Plato says in the *Cratylus*, 'Heraclitus is supposed to have said that all things are in process and nothing at rest; he compares this to the stream of a river, into which he says you could not step twice into the same water.'[4] And though Aristotle does not mention Heraclitus by name, it is obvious to whom he is referring in the *Physics* when he says, 'the view is actually held by some that not merely some things but all things in the world are in motion and always in motion, though we cannot apprehend the fact by sense-perception.'[5]

Since ordinary observation, as Aristotle notes, does not indicate that at any one time all phenomena are undergoing constant change, Heraclitus probably did not derive his thesis from an analysis of the imperceptible changes continuously occurring in observable things. Instead, he seemed to have been struck, as we all are at times, by a kind of intuitive awareness of the transiency of all existence, however momentarily permanent things may appear to be. Especially as one ages, a lifetime seems to contract into a fleeting moment submerged in the onrush of time. What better fits this insight than the image of a river continuously flowing from its source to its estuary — as experience or time flows from the future to the past? Just as one can occupy only one moment of time, the vanishing present, one cannot 'step twice into the same water' of a river.

But even though the image of a flowing river conveys an appropriate representation of the various tenses of time, it is not quite adequate to represent change as such, since a river can become still and stagnant. How then can one embody the abstract concept of continuous change in a more concrete image? One can imagine Heraclitus putting the enigmatic question to himself: What continuously changes while ever remaining the same? In answer to this question only one image is appropriate, that of fire. What else maintains its identity while continuously undergoing change? Who has not experienced the hypnotic spell of fire leaping and cavorting, sometimes in a wild frenzy, other times in a steady rhythm, as it consumes matter? So it must have seemed to Heraclitus who chose 'Fire' as the primordial stuff embodying Change itself: 'This world-order did none of the gods or men make, but it always was and is and shall be: an ever-living fire, kindling in measures and going out in measures.' 'All things are an equal exchange for fire and fire for all things, as goods are for gold and gold for goods.'

In addition to the image of 'Fire' as the *archê*, Heraclitus also invokes the Hesiodic concept of Strife as the generating principle underlying changes in the world: 'War is the father of all and king of all [. . .].' 'It is necessary to know that war is common and right is strife and that all things happen by strife and necessity.' Just as Hegel later would stress that all the tragic conflicts and suffering of mankind were "necessary" for the self-revelation and development of the Absolute Spirit, so Heraclitus rebukes those who would deny opposition and strife: 'for there would be no musical scale unless high and low existed, nor living creatures without female and male, which are opposite.' Thus the Milesian notion of opposites occurs in Heraclitus, but in a sublimated form. Cosmic generation is not explained as a result of the conflict of primary opposites such as Hot and Cold, but opposition remains as the aspect of plurality and diversity inherent in things.

This is difficult to comprehend and express, but again it is similar to Hegel's notion of synthesis in which the thesis and the antithesis are *"aufgehoben."* For Heraclitus, too, the plurality of opposites is a unity of contrasts, just as 'the path up and down is one.' According to Aristotle he maintained that 'Things taken together are whole and not whole, something which is being brought together and brought apart, which is in tune and out of tune; out of all things there comes a unity, and out of a unity all things.' He apparently identified this unity of opposition with the Divine: 'God is day night, winter summer, war peace, satiety hunger [. . .].' No wonder he thought that 'The real constitution of things is accustomed to hide itself,' and that 'An unapparent connection is stronger than an apparent one.'

This overriding sense of a pervasive unity underlying diversity reappears in a somewhat different form in his cosmology as a harmony underlying change. Like the Milesians, he stressed the continuous but 'measured' transformation of world processes, wherein fire is 'extinguished' to form the sea and the sea the earth, which in their turn are 'exchanged' back into fire. It is the measured interchange of these "world-masses" (fire, sea, and earth) in Heraclitus' system that underlies and preserves the order and unity in the world, as it did in the cosmologies of Anaximander and Anaximenes. Perhaps it was to this continuous cyclical process that Heraclitus was referring when he spoke of the path up and down as being one.

Thus while accenting the aspect of change in experience, Heraclitus also stressed the 'measured' or proportioned nature of this change. As Kirk and Raven observe,

all Presocratic thinkers were struck by the dominance of change in the world of our experience. Heraclitus was obviously no exception, indeed he probably expressed the universality of change more clearly than his predecessors; but for him it was the obverse idea of the *measure* inhering in change, the stability that persists through it, that was of vital importance. [6]

Not satisfied with just recognizing this measured element inherent in change, or merely attributing it to fire which, as 'Thunderbolt,' 'steers all things,' he designated the controlling principle inherent in the fire the 'Logos.' It is this 'Logos' that determines the equable interchange of natural phenomena: 'Of the Logos men always prove to be uncomprehending [. . .]. For although all things happen according to this Logos men are like people of no experience [. . .] [and hence fail to understand it].[7] And just as earlier he referred to the unity comprising the opposition of things as a god, so the measured or proportionate exchange of things, namely the Logos, is also divine.[8]

This, obviously, is not the type of thinking characteristic of scientific rationalism, a mode of thought that aspires to explain phenomena in terms of empirical models and rational theories. Heraclitus' system is characteristic more of the earlier mythopoetic mode of thought, though expressed in the secular language of prose and on a more abstract intellectual level. Compared to the Milesians, his use of concepts is diffuse, syncretic, and concrete with little concern for logical distinctions and conceptual clarity. As in myth, Fire, Strife, and Logos all melt together into one dramatic image.

> Heraclitus, as Aristotle found, did not use the categories of formal logic, and tended to describe the same thing (or roughly the same thing) now as a god, now as a form of matter, now as a rule of behaviour or principle which was nevertheless a physical constituent of things.[9]

This condensing of images or concepts was found in Chapter III to be particularly characteristic of mythopoetic thought.

This raises the usual questions of interpretation and evaluation. Does Heraclitus' use of language indicate that he was talking nonsense, as some of the doxographers and 20th century positivists and ordinary language philosophers would have claimed? Or was he aspiring to express an insight which transcends the logical form of ordinary language (using Wittgenstein's terminology), as most mystics and metaphysicians would claim? According to Čapek:

> Since the time of Heraclitus, philosophers who insist on the dynamic nature of reality struggle with the extreme difficulties of expressing in adequate linguistic form this paradoxical "unity of opposites" which every temporal process realizes. No wonder that Heraclitus, like Bergson many centuries later, was called "obscure."[10]

When one turns to Heraclitus' meteorological explanations, it is almost as if he disdained the attempt to find serious explanations of astronomical phenomena (one can almost hear him say, "Of what use are they in life?"). While he must have had some idea (or should have had) of the meteorological speculations of his predecessors, he claimed that the sun is the actual size it appears to be ('the breadth of a man's foot') and 'is new each day.' Instead of utilizing more sophisticated empirical models, he apparently fell back on the popular myth of the sun being carried around the river Okeanos in a golden bowl, describing the heavenly bodies as bowls of fire nourished by exhalations from the sea and having to be 'rekindled' each day. According to the testimony of Diogenes Laertius, the heavens contain 'bowls turned with their hollow side towards us, in which the bright exhalations are collected and form flames [. . .]. Brightest and hottest is the flame of the sun [. . .].'

The bowl-image enabled Heraclitus to explain eclipses and the phases of the moon as due to different 'turnings' of the bowls: 'and sun and moon are eclipsed when the bowls turn upwards; and the monthly phases of the moon occur as its bowl is gradually turned.' Again we encounter that strange theory, found previously in the systems of Anaximander and Anaximenes, that the heavenly fires are nourished by moist exhalations or evaporations from the sea. Heraclitus maintained that the fixed stars were further from the earth than the sun and the moon, agreeing with Anaximenes. But coming after the Milesians, his views appear very primitive. Compared to the successive advances made by each of the Milesians on the thought of his predecessor, one can appreciate to what extent Heraclitus regressed in his meteorological explanations.

When one considers his reflections on man, however, a different evaluation is required. As already indicated, he equated wisdom with an understanding of the measured processes and unity of nature identified with the divine Logos. Thus Fire, Strife and Logos are different ways of referring to and regarding the divine reality within which all the contradictory aspects of the world seemed to be unified. Apparently he was not quite willing either to affirm or deny that this divine reality was identical with the Zeus of popular religion: 'One thing, the only truly wise, does not and does consent to be called by the name of Zeus.' He believed that the soul of man is akin to the divine fire, thus 'a dry soul is wisest and best.' Excessive drinking leads to a moist soul and thus to a reduction in man's cognitive capacities: 'A man when he is drunk is led by an unfledged boy, stumbling and not knowing where he goes, having his soul moist.'

One's degree of consciousness and intelligence seemed to depend upon contact with the divine Fire or Logos through breathing and perception, so that at night when these functions are reduced one falls into a deep slumber approaching the status of the dead.

> According to Heraclitus we become intelligent by drawing in this divine reason [logos] through breathing, and forgetful when asleep, but we regain our senses when we wake up again. For in sleep, when the channels of perception are shut, our mind is sundered from its kinship with the surrounding [...]. But in the waking state it again peeps out through the channels of perception as though through a kind of window, and meeting with the surrounding it puts on its power of reason [...]. [11]

Perhaps this is the source of the expression that we "perceive the world through the window of our senses."

Other fragments of Heraclitus sound remarkably similar to those Biblical or Shakespearean quotations which have been the major fonts of practical wisdom in the West: 'A lifetime is a child playing with draughts;' 'Man like a night light is kindled and put out;' 'Man's character is his destiny;' 'It is hard to fight desire; what it wants it buys with the soul;' 'To God all things are beautiful and good and just;' 'The wisest of men appear like an ape beside God, in wisdom, beauty, and everything else;' 'Nature loves to hide;' 'If one does not expect the unexpected one will not find it out;' and 'Men who love wisdom must inquire into many things.'

As the maxims indicate, Heraclitus, like Socrates after him, was more concerned with what man should be or how he should live than with what he could know. But there also was a strain of mysticism and paradox in his writings that is more typical of Eastern sages. Like mystics in general, he claimed that the strife and conflicts of life, along with the seeming permanence of the empirical world, are merely transitory manifestations of a deeper order of things, an underlying harmony and unity pervading reality. Rejecting the type of approach introduced by the Milesians that has since gained universal scientific acceptance, that phenomena can be explained by analyzing them into more fundamental elements from which they can be deduced, Heraclitus, like prophets, mystics, and metaphysical poets before and after, believed that scientific analysis and deduction can take mankind only so far — that behind the phenomenal world described by science there is a more fundamental unity indifferent to the methods of scientific rationalism. For some untold reason, this reality is best unveiled by the enigmatic, cryptic, poetic idiom of the mystic. Not too surprisingly, the current paradoxes and limitations inherent in contemporary knowledge, the present impasse in the mind-body problem, the mysteries of parapsychology, and the anomalies of quantum mechanics commend something of this point of view even to some modern physicists. [12]

NOTES

CHAPTER V

[1] David Bohn, *Causality and Chance in Modern Physics* (Philadelphia. Univ. of Pennsylvania Press, [1957] 1971), pp. 152-153.

[2] Xenophanes was an older contemporary of Heraclitus who, though born and brought up in Ionia, was forced to leave when a young man, perhaps spending most of his life in Sicily. He is particularly noted for his attacks on the anthropomorphic conception of God: 'The Ethiopians say their gods are snub-nosed and black, the Thracians that theirs have light blue eyes and red hair.' He also said that if animals could draw they would picture God in their own likenesses. He held that God was one, unlike mortals in body or thought and that 'without toil he shakes all things by the thought of his mind.' Perhaps Xenophanes' most important scientific contribution was the fact that he called attention to the existence of inland fossils, deducing that at one time the earth must have been covered by water and that it still remains a mixture of earth and sea.

[3] de Santillana, *The Origins of Scientific Thought* (*op. cit.*, ch. IV), p. 49.

[4] Plato, *Cratylus*, 402A.

[5] Aristotle, *Phys.*, 253b9-12. (Hardie and Gaye trans.)

[6] Kirk and Raven, *The Presocratic Philosophers* (*op. cit.*, ch. IV), pp. 186-187.

[7] Fr. 1, Sextus, *Adv. Math.*, VII, 132. Quoted from Kirk and Raven, *op. cit.*, p. 187, f.n. 197. Brackets added.

[8] Cf. Kirk and Raven, *op. cit.*, p. 192.

[9] *Ibid.*, p. 187.

[10] Milič Capek, *The Philosophical Impact of Contemporary Physics,(op. cit., Introduction)*, p. 372.

[11] Sextus, *Adv. Math.*, VII, 129. Quoted from Kirk and Raven, *op. cit.*, p. 207, f.n. 237. Brackets in the original.

[12] For example, the quantum physicist David Bohm advocates "a new description of reality: the *enfolded* order" in which "appearances are abstracted from an intangible, invisible flux that is not comprised of parts," but has "an inseparable interconnectedness." See the *Brain/Mind Bulletin*, 1977, p. 2(16). Also cf. Fitjof Capra, *The Tao of Physics* (Boulder: Shambhala Publications Inc., 1975).

CHAPTER VI

THE PYTHAGOREAN NUMBER COSMOLOGY

> *The Science of Pure Mathematics* [...] *may claim to be the most original creation of the human spirit.* [1]
> Whitehead

It is one of those ironies of history that while the founder of Pythagoreanism has achieved everlasting recognition because of the mathematical theorem named after him, much less is known about his thought than that of any of the other Presocratics, with the possible exception of Thales. Although he passed his early years on the island of Samos off the coast of Ionia, Pythagoras later left Samos to escape from the tyrant Polycrates, settling first in Croton and then Metapontium in southern Italy. In Croton during the last quarter of the 6th century he founded a religious order or fraternity the members of which were sworn to preserve the secrecy of the rituals, mathematical discoveries, and philosophical doctrines of the order. Because of this secrecy much of Pythagoreanism is "enveloped in a mist of legend."

As was customary with revered religious leaders at this stage of cultural development (e.g., Christ and Buddha), various miracles were attributed to Pythagoras, including his divine birth by Apollo from a mortal woman. But due to the imposed secrecy of the order little is known about the individual contributions to Pythagoreanism until Philolaus broke the vows in the 5th century (the exact date is a matter of dispute: de Santillana places his writings "about 460 B.C." while Kirk and Raven place them "at the end of the 5th century B.C.") by publishing accounts of the Pythagorean system. Tradition claims that these "notorious books" were bought by Dion, the ruler of Syracuse, at Plato's urging. In any event, most of our information about the 6th and 5th century Pythagoreans comes not from Philolaus but from Aristotle, particularly the latter's *Metaphysics* and *De Caelo.* Because of the difficulty in making exact attributions of discoveries we shall discuss what is generally believed to be the system common to the Pythagoreans of the 6th and 5th centuries, confining our attention to their mathematical, cosmological, and astronomical contributions.

MATHEMATICS AND SCIENCE

It is a well-known fact that the development of science has depended upon concurrent developments in mathematics. The keeping of accounts, as well as the exchange of goods, required some kind of numeration in terms of cardinal and ordinal numbers with the representation of divisible amounts in terms of fractions. While arithmetic includes the progression of integers (or rational numbers) which can be used to represent numerically the members of a group or set (one, two ... ten entities, etc.), and the kinds of computations that can be performed with these integers (e.g., addition and multiplication), geometry is concerned with formalizing the relationships embedded in geometrical figures (e.g., angles, triangles, circles, plane surfaces, etc.). Prior to the Presocratics, the necessity of measuring boundaries, surveying tracts of land, and performing such engineering feats as raising obelisks and constructing temples and pyramids led to the discovery by the Babylonians and the Egyptians of individual geometrical axioms and theorems. [2]

The Greeks in turn discovered additional mathematical theorems (such as the "Pythagorean theorem") and operations which they applied to astronomical calculations, e.g., Erathosthenes'determination of the earth's circumference and Aristarchus'calculation of the relative distances from the earth of the sun and the moon, and to the computation of the volumes and areas of certain geometrical figures. Plato states in the *Timaeus* that a major stimulation to the development of mathematical astronomy was the need for an accurate calendar to mark the religious holidays and to tell the time, claiming that the heavens were created for the express purpose of teaching men the conception of numbers needed for measuring the days, months, and years. As if to support Plato's contention, Copernicus wrote in the prefatory letter to the *De Revolutionibus* that the need for calendar reform was one of the major influences that led him to search for a better theory for describing the motions of the planets:

> For not long since, under Leo X, the question of correcting the ecclesiastical calendar was debated in the Council of Lateran. It was left undecided for the sole cause that the lengths of the years and months and the motions of the Sun and Moon were not held to have been yet determined with sufficient exactness. From that time on I have given thought to their more accurate observation, by the advice of that eminent man, Paul, Lord Bishop of Sempronia, sometime in charge of that business of the calendar. What results I have achieved therein, I leave to the judgment of learned mathematicians and of your Holiness in particular. [3]

Kepler, in turn, derived his three laws of planetary motion from the investigation of conic sections initiated in the 4th century B.C. by Menaechmus and in the 3rd century by Apollonius; i.e., the basis of Kepler's laws was the discovery that the planets describe elliptical rather than circular orbits, an ellipse being a plane closed figure formed by a diagonal slice through a cone. Newton in turn invented the infinitesimal calculus (discovered independently by Leibniz), utilizing the derivative, to improve upon the Greek "method of exhaustion"[4] for approximating the instantaneous rate of change of the motion of a planet due to inertial momentum and gravitational forces. In the 20th century Einstein not only incorporated a non-Euclidean geometry, Riemannian geometry, in the general theory of relativity, he also developed tensor calculus to express his physical theories. In quantum mechanics the differential equations of Newtonean dynamics were superseded by Schrödinger's wave and Heisenberg's matrix mechanics resulting in a probabilistic interpretation of natural laws.

Except for the description and the classification of phenomena, and the verbal formulation of inductive generalizations or laws, advances in the physical sciences have gone hand in hand with developments in mathematics. In brief, the basis of this partnership is the fact that given certain quantitative information, the scientist wants to utilize this information to extend his knowledge of the world. Insofar as mathematicians discover the formal principles underlying various kinds of valid mathematical operations, the scientist utilizes these formal techniques in his theoretical investigations. For example, in the theorem attributed to Pythagoras, if one knows the length of the sides containing the right angle of a right-angled triangle one can calculate the length of the hypotenuse. Similarly, by determining the angular distances of the sun and the moon (when exactly half full) to the earth, and utilizing certain geometrical axioms or theorems, Aristarchus was able to compute the relative distances of the sun and the moon from the earth. Given the geometrical properties of an ellipse, Kepler was able to formulate his Second Law that the orbital speed of each planet varies in such a manner that a line connecting the planet to the sun sweeps through equal areas of the ellipse in equal intervals of time. He then went on to discover his Third Law that the ratio of the orbital periods of the planets squared ($\frac{T^1}{T^2}$)^2 is equal to the ratio of the cubes of their average distances from the sun ($\frac{R^1}{R^2}$)^3. These ratios then provided Newton with the clue for his formulation of the universal law of gravitation, that the strength of the gravitational force varies inversely with the square of the distance.

Having obtained certain quantitative information the scientist wants to utilize this information to extend his knowledge of the universe. He determines the quantities involved by measurement, while the mathematician proves various rules or laws and formal operations which the scientist then uses in his mathematical deductions. Thus we now realize that mathematics alone can tell us nothing about the real world: in itself, mathematics is a formal system the truth-value of which depends upon its internal consistency or deductive validity. It is only in its *application* to the world that it can be used to further knowledge; i.e., only when its variables are interpreted in terms of measurable empirical magnitudes does an equation allow of being *empirically* true or false. As Einstein said, "as far as the propositions of mathematics refer to reality, they are not certain; and as far as they are certain, they do not refer to reality."[5] Or as Whitehead expressed the same thesis: "The certainty of mathematics depends upon its complete abstract generality."[6]

Another way of describing the applications of mathematics to science is to say that the physical scientist is usually investigating functional relationships among measurable magnitudes. In his investigations the scientist looks for a pattern in nature that can be expressed as a functional relationship or mathematical correlation among certain variable magnitudes. For example, given all the different rates of motion, speed or velocity can be defined more precisely as a function of the units of distance traveled per units of time. Having arrived at this definition, Galileo then discovered that the constant acceleration of free falling objects (in the earth's gravitational field) is a function of the units of time: $\frac{1}{2}gt^2$, where 'g' is the gravitational constant of the earth and 't' equals time. As these examples illustrate, the laws of the physical sciences are expressions of functional correlations, of certain numerical values being correlated with, varying with, or a function of other numerical values. In Newton's equation $F=MA$, force (F) is a function of the product of the numerical values of mass (M) and acceleration (A); or conversely, acceleration varies directly with the force and inversely with the mass. In Einstein's equation $E=MC^2$, energy (E) is a function of the product of mass (M) and the velocity of light squared (C^2); conversely, one can deduce that all mass is equivalent to the quotient of energy and the velocity of light squared.

Accordingly, knowledge of mathematics enables the scientist to do four things: (1) determine different magnitudes of phenomena (lengths, times, weights, velocities, etc.) by measurement; (2) derive a formal picture of the structural interrelations of empirical phenomena by expressing the functional relations in the form of an equation; (3) give an empirical significance to the equation by "plugging" empirical magnitudes into the variables; and (4) solve the equations to determine the values of the uninterpreted variables thereby extending one's knowledge of nature from what is known to what was unknown. Thus the application of mathematics to observable phenomena depends upon there being patterned, quantifiable correlations among the phenomena, while the expression of these correlations mathematically provides an insight into the structural interrelations and interdependencies of physical processes, along with the means of expanding or enlarging one's knowledge of nature.

Given our present understanding of the nature of mathematics and its application to the sciences, it is much easier to judge what was both extremely farsighted and exceedingly primitive in the conception of mathematics held by the 5th century Pythagoreans. What was farsighted was the realization, however dimly or obscurely conceived, that somehow the structure of physical reality embodies numerical relationships. For unlike the Milesians who defined the *archē* in terms of some primary stuff (*physis*), the Pythagoreans asserted that 'number is the substance of all things;' that 'numbers are the causes of the reality of other things.' This was a tremendous feat of intellectual abstraction.

However, what kept this insight exceedingly primitive was the fact, pointed out by Kirk and Raven, that among the Greek philosophers of this period no distinction had yet been made between the mode of existence of concrete, spatial, material objects, and that of abstract theoretical entities. Instead of numbers being conceived as formal units with a status of their own, they were thought to be substantial entities extended in space, for *substantial spatial existence was the only conceivable type of existence at that time.* This is an excellent example of syncretic thought in which distinctions that are clearly explicit and taken for granted on a higher level of thought are submerged in the concrete diffuseness of a lower level of primitive conception. As Kirk and Raven warn:

> It must constantly be remembered that no firm distinction between different modes of existence had yet been envisaged, and that what to us is obviously non-concrete and immaterial, like an arrangement, might be regarded before Plato as possessing the assumed ultimate characteristic of 'being,' that is, concrete bulk. To put it in another way, the arrangement would not be fully distinguished from the thing arranged, but would be felt to possess the same concreteness and reality as the thing itself. [7]

While the least educated person in our society would not be apt to confuse the numerical arrangement or classification of a group of objects with the objects themselves, "the earlier Pythagoreans, like all the rest of the Presocratics, failed to distinguish between the corporeal and the incorporeal."[8] It is this lack of *conceptual differentiation* between the spatial and the non-spatial, the concrete and the abstract that makes the views of the Pythagoreans so fascinating and so obscure. One is continually tempted to ignore in their thought the absence of a distinction which is commonplace today, and therefore attribute to their philosophy a degree of sophistication wholly alien to their thinking.

The basis of the Pythagorean number cosmology was the discovery, probably by Pythagoras himself (although even this is not known for certain), that the intervals of musical scales, the consonances and successive octaves, can be expressed in numerical ratios comprising the first four integers. Using a monochord, an ancient instrument consisting of a single string stretched over a sounding board with a movable bridge set on a graduated scale, Pythagoras discovered a correlation between the ratios of the lengths of the string and the musical tones: namely, that the ratio of the octave is 1:2 (i.e., by dividing the string into halves he successively produced the next higher octave of notes), that of the major fifth is 2:3, and that of the major fourth 3:4. Moreover, the division of the monochord into successive partials indicates that division by *even* fractions results only in higher octaves, while division by *odd* fractions results in "new tone individualities."[9]

Thus were discovered the intervals common to all Greek scales and, further, in what must be the boldest generalization in the entire history of thought, the insight that the harmonic ratios underlying the musical scales must somehow underlie the structure of the universe itself. According to Aristotle's account,

> the so-called Pythagoreans, who were the first to take up mathematics, not only advanced this study, but also having been brought up in it they thought its principles were the principles of all things. Since of these principles numbers are by nature the first, and in numbers they seemed to see many resemblances to the things that exist and come into being — more than in fire and earth and water [. . .] [and] since, again, they saw that the modifications and the ratios of the musical scales were expressible in numbers [. . .] [and] since, then, all other things seemed in their whole nature to be modelled on numbers, and numbers seemed to be the first things in the whole of nature, they supposed the elements of numbers to be the elements of all things, and the whole heaven to be a musical scale and a number. [10] (Brackets added)

From their discovery that mathematical proportions underlie the musical scale, they then *abstracted* the basic set of concepts in terms of which they interpreted the entire universe. As the musical scale was produced by imposing graduated limitations on the unlimited continuum of sound on the monochord, the tonal variations correlated with division by even or odd quantities, so the universe was thought to arise from the imposition of the "Limiting" on the "Unlimited," and to separate into various opposites under the categories of the odd and the even.

Thus we find appearing in the system of the Pythagoreans traces of Milesian philosophy, the "Unlimited" of Anaximander and the characteristic Milesian doctrine of opposites. Now, however, these doctrines have a mathematical significance. The *abstracted* aspect of oddness and evenness clearly characterizes the number system, while the *abstract aspect* of Limit and Unlimit applies to geometry, since the closed figures generated by points and lines represent bounded space. [11] In one fell swoop the Pythagoreans brought together arithmetic and geometry and attributed to them a cosmological significance. Again, as Aristotle tells us:

> Evidently, then, these thinkers also consider that number is the principle both as matter for things and as forming both their modifications and their permanent states, and hold that the elements of number are the even and the odd, and that of these the latter is limited, and the former unlimited [. . .]. [12]

In addition to these primary opposites, Aristotle lists eight other pairs of contraries correlated with the first two: under the principles of the Limited and the odd there is contained the one, right, male, resting, straight, light, good, and square; under the Unlimited and even there is plurality, left, female, moving, curved, darkness, bad, and oblong. This is certainly a strange organization of 'principles' for which, except for the arithmetical groupings, as we shall see, there appears to be little rationale. It is difficult not to attribute a strong element of syncretic or diffuse thought to such groupings as even, left, female, moving, darkness, bad, etc. Aristotle, too, was perplexed by the conjoining of such diverse kinds of abstractions, asking how 'white,' 'fire,' 'marriage,' 'justice,' etc. could be 'connected with numbers.'

Underlying this group of opposites is a unity reminiscent of the 'steering' principle of Anaximander, or the 'Logos' of Heraclitus, now defined in terms of Pythagoras' initial discovery, that of musical 'Harmony.' In the words of Philolaus,

> it would be impossible for any existing thing to be even recognized by us if there did not exist the basic Being of the things from which the universe was composed, (*namely*) both the Limiting and the Non-Limited. But since these Elements exist as unlike and unrelated, it would clearly be impossible for a universe to be created with them unless a harmony was added [. . .] through which they are destined to endure in the universe [. . .]. [13]

There is little clue, however, as to the nature of this Harmony, apart from the analogy with music. In de Santillana's opinion: "No use asking what it could mean precisely, in the moment of creative insight. There were in it centuries of thoughts and feelings unknown to us; the precise delicate twanging of the lyre bursting like stars upon the soul [...]."[14]

As regards the grouping of the mathematical configurations, here there is more to guide us. To literally 'see' why Limit, odd, one, and oblong are conjoined, one must consider the custom prevalent at the time of representing numbers spatially. That is, the typical manner of representing numbers diagrammatically by physical marks (similar to those on dice or dominos) or with pebbles in the sand, led to an identification of the *spatial* properties of numbers with what we would call their *formal* properties. Thus depending upon their spatial shapes, numbers were classified as triangular (Fig. 1 below), square (Fig. 2), or oblong (Fig. 3). The odd number 3, because of its representation by three dots (or *alphas*), was defined as having 'a beginning, a middle, and an end,' but even numbers have no 'middle' (or central dot), being equally divided into pairs of numbers. The number 10 is odd-even because division by two (which defines even numbers) results in two odd numbers. The following three diagrams illustrate the way in which the spatial characteristics of numbers contributed to their classification under the Table of Opposites.[15]

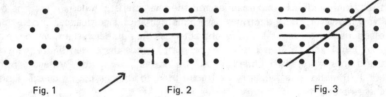

Fig. 1   Fig. 2   Fig. 3

The diagram in Fig. 1 is the most important for the early Pythagoreans because it represents 10, the perfect number, generated by the successive addition of the first four integers, those numbers included in the momentous discovery of the ratios of the musical scale (1:2, 2:3, 3:4). This triangular representation was known to the Pythagoreans as the '*Tetractys* of the Decad' and had such significance for them that they swore their oath by it. As Aëtius, the 2nd century A.D. compiler, relates:

Ten is the very nature of number. All Greeks and all barbarians alike count up to ten, and having reached ten revert again to the unit. And again, Pythagoras maintains, the power of the number ten lies in the number four, the tetrad. This is the reason: if one starts at the unit and adds the successive numbers up to four, one will make up the number ten; and if one exceeds the tetrad, one will exceed ten too. If, that is, one takes the unit, adds two, then three and then four, one will make up the number ten. So that number by the unit resides in the number ten, but potentially in the number four. And so the Pythagoreans used to invoke the tetrad as their most binding oath: 'Nay, by him that gave to our generation the tetractys, which contains the fount of eternal nature.'[16]

The significance of the tetrad of the *Tetractys* is illustrated in the equilateral triangle (Fig. 1): not only is it generated by the addition of 1+2+3+4=10, but each of the three sides is itself a tetrad. The Decad is the perfect number, thus the Table of Opposites consists of ten contraries. In addition, according to Aristotle, as the visible celestial bodies are only nine in number including the earth (the five planets, the moon, the sun, the sphere of the fixed stars, the earth), the Pythagoreans 'invented' a tenth, the invisible 'counter-earth,' to make up the perfect number. The importance of these numerical proportions or harmonies is revealed further in the crisis that was created by the discovery (perhaps by Pythagoras himself) of the incommensurability of the diagonal and the side of the square; i.e., the discovery that certain geometrical relationships or ratios of magnitudes (e.g., √3) cannot be expressed in terms of whole numbers. Tradition has it that the person who gave away the secret of the shocking discovery of what are now called irrational numbers was put to death.

The second and third figures, constructed by placing additional 'gnomons' or carpenter's squares around successive odd or even numbers, provide the clue as to why Limit, odd, one, and square are grouped together in contrast to Unlimit, even, plurality, and oblong. In Fig. 2, where each gnomon encloses, beginning with *one*, the series of *odd* numbers, we have the one, odd, and square all represented (for after one, imagining each previous gnomon removed, each gnomon encloses a *square* pattern of dots). In Fig. 3, where each gnomon encloses a series of *even* numbers, we have the even and the oblong represented together (for each gnomon encloses an *oblong* configuration of dots).

The equation of the odd with the Limit and the even with the Unlimit is more difficult. As Kirk and Raven explain this, whereas in Fig. 2 each addition of the gnomon results in the same figure, a square, in Fig. 3 each addition of the gnomon results in an *infinitely variable* figure, since with each addition the ratio of the length to the height in the oblong figure changes.[17] Thus the series of odd numbers in Fig. 2 produces a uniform or "limited" figure, while the series of even numbers produces a series of "unlimited" changing figures. This interpretation is supported by Aristotle:

> Further, the Pythagoreans identify the infinite with the even. For this, they say, when it is cut off and shut in by the odd, provides things with the element of infinity. An indication of this is what happens with numbers. If the gnomons are placed round the one, and without the one, in the one construction the figure that results is always different, in the other it is always the same.[18]

A further argument suggested by Kirk and Raven is illustrated by the contrasting arrows in Figs. 2 and 3. In Fig. 3 the arrow portrays the division of even numbers into equal parts *ad infinitum* (since they have no 'middle' dot), while Fig. 2 shows that odd numbers (having a 'middle') cannot be divided into whole integers (one would have to divide the central dot into two halves), thereby adding additional support for the identification of the even with Unlimit and the odd with Limit. As Aristotle states: 'Even is that which admits of division into halves without the interposition of the unit, odd is that which does not admit of division into halves because the unit is interposed as described.'

Though curious to us, this spatial representation of numbers allowed the Pythagoreans to equate arithmetic with geometry, "a 'geometry of numbers' or arithmogeometry of a rather fanciful kind,"[19] and to explain the generation of the universe from numbers. That is, not only were numbers *represented* spatially, they were thought of as actually having *spatial magnitude*. Again as Aristotle points out:

> And the Pythagoreans, also, believe in one kind of number — the mathematical; only they say it is not separate but sensible substances are formed out of it. For they construct the whole universe out of numbers — only not numbers consisting of abstract units: they suppose the units to have spatial magnitude. But how the first 1 [the original unit one] was constructed so as to have magnitude, they seem unable to say.[20] (Brackets added)

Ignoring the difficulty pointed out by Aristotle and starting with a unit-point, one can construct a line between two unit-points, then with several lines construct plane figures or 'surfaces' (triangles, rectangles, etc.), and finally by combining these plane figures or surfaces one can construct a three-dimensional representation of an object (see Figs. 4 and 5 below).

Fig. 4   Fig. 5

Fig. 4: The generation of the three-dimensional pyramid from points and lines.
Fig. 5: The generation of a rectangular solid from rectangular planes or surfaces.

In a similar manner, the Pythagoreans conceived the construction of the physical universe from "arithmogeometric units," or point-units, much as one conceives the lattice structure in crystallography or the arrangement of atoms in stereochemistry. All matter was thought to be composed of these three-dimensional point-units (later in the *Timaeus* Plato will attempt a similar reconstruction of the four elements in terms of triangles and squares, as a result of the influence of the Pythagoreans). Speusippus provides a succinct description of this construction:

> For 1 is the point, 2 the line, 3 the triangle and 4 the pyramid [Fig. 4 above] [. . .] and the same holds in generation too; for the first principle in magnitude is the point, the second the line, the third the surface and the fourth the solid.[21] (Brackets added)

There is in this conception considerable syncretic fusion and diffuseness of content characteristic of primitive thought, yet this attempt to conceive of the origin and structure of the universe in numerical spatial units, identifying the spatial properties of geometry with the elements of arithmetic, was no slight achievement. If nothing else, it exemplifies the primary aim of science to reduce the complexity of phenomena to structural magnitudes. Moreover, without anachronistically attributing more meaning or sophistication to these early thinkers than is warranted, there is a *slight* hint of similarity between this equivalence of matter with point-units and Einstein's "reduction" of matter and force fields to a "geometrization of space." As Whitehead states, "when Einstein and his followers proclaim that physical facts, such as gravitation, are to be construed as exhibitions of local peculiarities of spatio-temporal properties, they are following the pure Pythagorean tradition."[22]

There are, of course, many theoretical problems in this view, as pointed out by Aristotle. How, Aristotle, asks, can we conceive of 'weightiness,' and 'lightness' or of sensory qualities such as 'white,' 'sweet,' and 'hot,' as an 'assemblage of units?' However, while this is precisely what modern science (chemistry and physics) has attempted to do, are we any closer today to answering these questions? It has been argued that in spite of the tremendous progress made by the physical sciences, they have not been able to explain the 'emergence' of ordinary physical properties and qualities from a conception of the world essentially devoid of both. For example, introducing a spark in a gas composed of two portions of hydrogen and one of oxygen produces water, but how does one explain the 'emergence' of the qualities of water? How do light waves and sound waves by interacting with an organism become 'transformed' into colors and sounds? More generally, how do the inferred microstructures posited by the physical sciences give rise to ordinary macro-properties and qualities? Aristotle's questions were essentially the same, and he attempted in his own system to overcome these problems, as we shall see later.

Having described the generation of physical matter from spatial point-units, there is still the question as to how the point-units themselves arose. Probably because of its greater difficulty, the answer to this question is more obscure and primitive. As indicated earlier, Aristotle could not explain how the Pythagoreans accounted for the first unit's coming into being ('how the first 1 was constructed so as to have magnitude they seem unable to say'), but once it existed the Unit or Limit began to 'inhale' the surrounding Unlimit, and thereby 'breathed in' or 'inhaled' the void, the principle of distinction and separation. By 'inhaling' the void from the Unlimit, the Unit or Limit apparently brought into the universe the aspect of numerical distinction and spatial separation so essential to this arithmogeometric cosmology. For without the element of extension, the Unit could not generate the geometrical figures (points, lines, plane surfaces, and three-dimensional figures) that presuppose space, which in turn constitute physical matter. As such, the account shows both regressive and progressive aspects: it appears regressively to borrow the anthropomorphic analogues of 'breath' and 'inhalation' from Anaximenes to account for the consequent separation of the unit into the dyad, triad, tetrad . . . decad, etc., and progressively to introduce the concept of the void or pure extension that will play such an important role in later theories, especially atomism.

SPECULATIVE ASTRONOMY

In their astronomical system, the Pythagoreans again disclose their unique combination of abstract and syncretic reasoning. An essential component of scientific progress has been the willingness to reject, revise or replace empirical observations as a result of deductive reasoning. Given certain anomalous data and/or theoretical considerations, a scientist usually finds it necessary to supplement his observations with inferred entities or theoretical constructs, the existence of which is not directly observable. A dramatic example of this was the independent deduction by Adams and Leverrier, from Newton's laws of dynamics, of the existence of the (at the time) unobservable planet Neptune to explain the orbital irregularities of Uranus. The procedure has become so commonplace in the sciences, as attested to by such concepts as 'neutrons,' 'pulsars,' 'enzymes,' 'synaptic discharges,' etc., that it is difficult to understand how scientific progress could be made otherwise.

As regards the rejection or emendation of our common sense observations as a result of theoretical considerations, the most striking example of this was the Copernican Revolution. We do not feel the earth rotating on its axis nor does it appear to revolve in its orbit in reference to the clouds or even to the fixed stars, and yet for considerations based on mathematical symmetry (e.g., the fact that Ptolemy had to incorporate "equants" and "deferents" in his system in an *ad hoc* manner), Copernicus was willing to deny the direct evidence of our senses and attribute both an axial rotation and an orbital revolution to the earth. In this bold venture he was preceded by the Pythagoreans who followed their *a priori* deductions at the expense of common observations and beliefs, but who made the *right* deductions for the *wrong* reasons. In doing so they brought on themselves the condemnation of Aristotle, but exhibited a pattern of reasoning long since validated by science.

It was pointed out earlier that to make the number of celestial bodies equal ten the Pythagoreans invented the 'counter-earth,' a body similar to the earth but hidden from it because of its opposite position. Then, believing fire to be of greater importance than earth, they removed the earth from the center of the universe and replaced it by a central fire, calling it the 'Hearth of the World' and the 'Throne of Zeus.' Aristotle describes the system as follows:

> At the centre, they say, is fire, and the earth is one of the stars, creating night and day by its circular motion about the centre. They further construct another earth in opposition to ours to which they give the name counter-earth [...]. Their view is that the most precious place befits the precious thing: but fire, they say, is more precious than earth [...]. Reasoning on this basis they take the view that it is not the earth that lies at the centre of the sphere, but rather fire. The Pythagoreans have a further reason. They hold that the most important part of the world, which is the centre, should be most strictly guarded, and name it, or rather the fire which occupies that place, the 'Guard-house of Zeus' [...].[23]

Thus for the wrong reasons the Pythagoreans (and apparently "many others") made the farsighted correct deduction that the earth moved. Moreover, they also came to the remarkable conclusion (which Galileo had such opposition in establishing over two millenia later) that 'the earth is one of the stars, creating night and day by its circular motion about the centre.' However, the specific, *a priori* reasons for which they arrived at this conclusion were certainly unscientific and forced, as criticized by Aristotle.

> In all this they are not seeking for theories and causes to account for observed facts, but rather forcing their observations and trying to accommodate them to certain theories and opinions of their own. But there are many others who would agree that it is wrong to give the earth the central position, looking for confirmation rather to theory than to the facts of observation.[24]

As we now know, however, although Aristotle's criticism was correct, so was the Pythagorean position. Even the view that the central location should belong not to the earth, but to the more perfect element, fire, was one of the major reasons both Copernicus and Kepler favored the heliocentric theory.

Moreover, the attribution of two motions to the earth by the early Pythagoreans played a crucial role in the history of heliocentric astronomy. It was because of such prior conjectures that Copernicus himself felt free to entertain the heliocentric view, as he says in the prefatory letter to the *De Revolutionibus*, after quoting Plutarch's description of the contributions of the Pythagoreans.

> "The rest hold the Earth to be stationary, but Philolaus the Pythagorean says that she moves around the fire in an oblique circle like the sun and the moon. Heraclides of Pontus and Ecphantus the Pythagorean also make the Earth to move, not indeed through space but by rotating round her own center as a wheel on an axle from West to East."
>
> Taking advantage of this I too began to think of the mobility of the Earth; and though the opinion seemed absurd, yet knowing now that others before me had been granted freedom to imagine such circles as they chose to explain the phenomena of the stars, I considered that I also might easily be allowed to try whether, by assuming some motion of the Earth, sounder explanations for the revolution of the celestial spheres might be discovered.[25]

A further example of the soaring speculation of the Pythagoreans was their doctrine of the 'Music of the Spheres.' Having made their momentous discovery of numerical ratios underlying the musical scale, in a brilliant poetic generalization of this discovery they attributed a mathematical harmony to the motion of the planets. Apparently following Anaximander, they asserted that the orbit of the stars was nine times the radius of the earth, that of the moon eighteen times, and that of the sun twenty-seven times (they mistakenly followed Anaximander in holding that the stars were closer to the earth than the sun and the moon). Presumably the other planets were believed to be the same distance as the stars. Again, however, this conception of a mathematical astronomical order was more concrete and primitive than it first appears and was made for the wrong reasons. Believing that because of their size and speed the planets should emit sounds, they asserted that 'their speeds, as measured by their distances, are in the same ratios as musical concordances [. . .].'[26] Thus was conceived the alluring idea of the "Music of the Spheres," a celestial music to which only the gods were attuned, for its constant refrain was inaudible to human ears.

As premature and *a priori* as this notion of the "Harmony of the Spheres" was at that moment in time, it would be partially vindicated two millenia later by Kepler's Third Law published in a book in 1619, the title of which could have been given by the Pythagoreans: *The Harmonies of the World*. As mentioned earlier, Kepler's Third Law climaxed his search for a mathematical harmony among the celestial bodies by revealing a proportional correlation between the orbital speeds of the planets and their average distances from the sun. Thus the visionary, daring, but fanciful deductions of the Pythagoreans were given precise mathematical formulation later by Kepler, who considered himself a neo-Pythagorean.

CONCLUSION

It remains to summarize both what was extraordinarily farsighted in the thinking of the early Pythagoreans, along with what was still exceedingly primitive. Regarding the latter, it was the inability to conceive of numbers and numerical ratios as distinct from their empirical representation in spatial configurations that was particularly indicative of the syncretic thought of the early Pythagoreans. Also, their categories of opposites were extremely diffuse, as was their conception of the first unit "inhaling" extension from the Unlimited.

Yet in spite of these limitations, one can detect in their system several faint but discernible aspects essential in later scientific reasoning. First, as previously mentioned, one of the primary aims of science is to disclose functional correlations or patterns in nature that can be expressed in mathematical equations. The discovery that numerical ratios underlie the musical scale and the extension of this insight to "the harmony of the spheres" mark the first explicit realization of this crucial program of science. Consequently, later Pythagoreans were in the forefront in the attempt to reduce astronomical observations to a mathematically coherent system, explaining why Renaissance astronomers such as Copernicus and Kepler, who were primarily mathematicians, were referred to as Pythagoreans.

Secondly, the history of science shows that ordinary observations or presumed 'facts' cannot be accepted on face value, because when placed within another set of relations or interpreted by a different theoretical framework they can take on quite a new significance. As Kuhn has pointed out, this is what happens when a major paradigm shift or revolutionary change in point of view occurs. In their correct deduction that the earth moves, thereby causing the apparent rotation of the heavens, and in their willingness to accept the revolution of the earth around a central fire, in spite of the conflicting testimony of the senses, the Pythagoreans were not only illustrating an essential aspect of scientific thought, they were preparing the way for one of the most decisive revolutions in science. It was for this reason that they aroused such deep admiration in Galileo. As he has Salviati reply to Sagredo, who expresses surprise that the Pythagoreans had so few adherents in the course of the centuries:

> No, Sagredo, my surprise is very different from yours. You wonder that there are so few followers of the Pythagorean opinion, whereas I am astonished that there have been any up to this day who have embraced and followed it. Nor can I ever sufficiently admire the outstanding acumen of those who have taken hold of this opinion and accepted it as true; they have through sheer force of intellect alone done such violence to their own senses as to prefer what reason told them over that which sensible experience plainly showed them to the contrary.[27]

Thirdly, and finally, even though their attempt to construct the universe from arithmogeometric units depended upon very primitive, anthropomorphic concepts such as "inhalation," they still presaged the modern conception of chemical and physical elements as structures whose essential properties and configurations are quantitative. Even their reduction of the physical universe to arithmogeometric units is not without its contemporary adherents. Thus to have isolated from the innumerable kinds of qualities, substances, and relations in the world numerical units, however spatial and quasi-corporeal, as the ultimate *arché* of things, and then envisioned the entire universe as a manifestation of numerical harmonies, marks one of the milestones in the growth of abstract thought.

NOTES

CHAPTER VI

[1] Alfred North Whitehead, *Science and the Modern World* (New York: Mentor Books, 1925), p. 25.

[2] For a summary of the contributions to the development of mathematics of the Babylonians and the Egyptians see the final chapter of this book.

[3] Thomas Kuhn, *The Copernican Revolution* (*op. cit.*, Introduction), p. 142.

[4] Cf. Carl B. Boyer, *The History of the Calculus and Its Conceptual Development* (New York: Dover Pub., Inc., [1949] 1959), ch. V.

[5] Albert Einstein, "Geometry and Experience," a lecture before the Prussian Academy in 1921, pub. in *Ideas and Opinions*, new translations and revisions by Sonja Bargmann (New York: Bonanza Books, 1954), p. 233.

[6] Whitehead, *op. cit.*, p. 22.

[7] G. S. Kirk and J. E. Raven, *The Presocratic Philosophers* (*op. cit.*, ch. IV), pp. 188-189.

[8] *Ibid.*, p. 247, f.n. 1.

[9] Giorgio de Santillana, *The Origins of Scientific Thought* (*op. cit.*, ch. IV), p. 60.

[10] Aristotle, *Met.*, 985b24-986a3. (Ross trans.)

[11] Cf. Kirk and Raven, *op. cit.*, p. 242.

[12] Aristotle, *Met.*, 986a15-19. (Ross trans.)

[13] Kathleen Freeman, *Ancilla To The Pre-Socratic Philosophers* (*op. cit.*, ch. IV), p. 74.

[14] de Santillana, *op. cit.*, p. 58.

[15] These Figures are reproduced in all the anthologies of the writings of the Presocratics. The arrows in Figs. 2 and 3 are due to Kirk and Raven, *op. cit.*, p. 245. The present discussion owes much to their excellent account of the Pythagoreans.

[16] *Ibid.*, pp. 230-231, f.n. 280.

[17] Cf. *Ibid.*, pp. 243-245.

[18] Aristotle, *Phys.*, 203a10-15. (Hardie and Gaye trans.)

[19] de Santillana, *op. cit.*, p. 65.

[20] Aristotle, *Met.*, 1080b16-22. (Ross trans.)

[21] Speusippus, *ap. Theologumena Arithemeticae*, p. 84, 10 de Falco. Quoted from Kirk and Raven, *op. cit.*, p. 253-254, f.n. 319.

[22] Whitehead, *op. cit.*, p. 33.

[23] Aristotle, *De Caelo*, 293a21-293b6. (Stocks trans.)

[24] *Ibid.*

[25] Kuhn, *op. cit.*, p. 141.

[26] Aristotle, *De Caelo*, 290b12. Quoted from Kirk and Raven, p. 259, f.n. 330.

[27] Galileo Galilei, *Dialogue Concerning the Two Chief World Systems*, trans. by Stillman Drake, second edition (Berkeley: Univ. of California Press, 1967), pp. 327-328.

CHAPTER VII

PARMENIDES AND PURE REASON

Reason is impelled by a tendency of its nature [. . .] to venture in a pure employment [. . .] to the utmost limits of all knowledge [. . .].[1]

Kant

Parmenides was born in Elea on the west coast of southern Italy in the last quarter of the sixth century. Elea was not far from Metapontium where Pythagoras died, and though there is no evidence that Parmenides ever met Pythagoras, he is said to have been a Pythagorean in his youth. Indeed, one can clearly discern a knowledge of Pythagoreanism in his thinking, as well as of the philosophies of Anaximander and Anaximenes. Because one tends to think of these philosophers so remote from us in time as somewhat less (or more) than real individuals, the following description by Plato, of Parmenides and his disciple Zeno, reminds us that they were not so different from ourselves.

> According to Antiphon's account, Pythadorus said that Parmenides and Zeno once came to Athens for the Great Panathenaea. Parmenides was well advanced in years — about sixty-five — and very grey, but a fine-looking man. Zeno was then nearly forty, and tall and handsome; he was said to have been Parmenides' favourite. They were staying at Pythadorus' house outside the city-wall in the Ceramicus. Thither went Socrates, and several others with him, in the hope of hearing Zeno's treatise; for this was the first time Parmenides and Zeno had brought it to Athens. Socrates was still very young at the time.[2]

The impression made on the young Socrates by Parmenides must have been striking because Socrates later refers to him as "august and terrible in his greatness."

In fact, Parmenides is a truly singular kind of thinker. Unlike the philosophers already discussed, he does not take the experienced world as the point of departure for his philosophical speculations, but the speculative framework itself. Rather than offer another answer to the kinds of problems posed within the conceptual frame of the former Presocratics, he adopts a different approach, that of *examining the logical and linguistic presuppositions of the framework itself.* While his predecessors constructed answers to problems pertaining to the *archē*, to change and identity, and to the one and the many, Parmenides examined the logic underlying the questions; i.e., the "logical form"[3] underlying the linguistic expressions of the questions and the proposed answers. In particular, he examined the logic governing the use of such concepts as 'being and not being,' 'coming to be and ceasing to be,' 'change,' 'motion,' 'time,' etc.

Thus the question of the *archē* was transformed by Parmenides from what is the first principle of things to the relation of knowing to being. Having isolated the form of the arguments from the content, he undertook a critical examination of the form itself; namely, of the logical categories in terms of which reality can or cannot be described. As such he was raising formal questions which today are treated by the recently developed techniques of existential, predicate, and relational logic within the discipline of symbolic or mathematical logic. At the time of Parmenides, however, the uses of the Greek verb 'to be' did not distinguish (1) the assertion of the existence or non-existence of something (that Pegasus does not exist but horses do) from (2) the attribution or denial of a quality to something (that Parmenides was austere but not unkind) from (3) the assertion or denial of a relation between things (that Parmenides was younger than Pythagoras). As we shall see, Parmenides' analysis suffers from a lack of distinction among these different linguistic uses and logical operations. Admitting only the logical function of the assertion of the existence or non-existence of something, Aristotle's law of excluded middle (Pv-P), he was forced to deny logical validity to predications and relations.

Parmenides wrote only one work, a poem written in an obscure oracular style. Beginning with a "Prologue," an allegorical description[4] of the manner by which he was conveyed into the realm of truth (one is reminded somewhat of *The Divine Comedy*), the poem then divides into two parts, "The Way of Truth" and "The Way of Opinion." As the title indicates, "The Way of Truth" describes the true way of knowledge, the truth that can be attained by the use of pure logic or what Kant called "pure reason." "The Way of Opinion" provides a complementary view of the world, a more empirical cosmology based on ordinary perceptual experience. The latter "Way" is not a true account of reality but one that most people, because of the persuasive influence of the senses, will find more convincing. Thus the two "Ways" exhaust the possible accounts of the world though they are not offered as intellectual options: only "The Way of Truth" provides a true account of Reality or Being, but those who are incapable of following the rigorous argument in the "Way of Truth" can accept the "Way of Opinion," a true account, one might say, of the world of appearance.

THE WAY OF TRUTH

The "Way of Truth" begins by asserting what can and cannot be known. Knowledge and discourse can pertain only to that which is, Being, while that which is not, Not-Being, is nonexistent, unknowable, and inexpressible: 'For it is the same thing to think and to be.' Thus Parmenides establishes a correlation between knowing and being, thereby determining the logical bounds of existence, knowing, and discourse. Just as the logical form of our language (Heidegger being an exception that proves the rule) excludes the possibility of meaningfully attributing existence to what is not, Not-Being, so, Parmenides seems to be asserting, the logical form of our language also excludes the possibility of thinking of it or of uttering anything about it. It is true that one can form grammatically correct sentences (as Heidegger does) by which one can say "Not-Being is," or that "Nothingness exists," or that "Being comes to be from Not-Being," but the underlying *logical form* of these grammatically correct sentences excludes their meaningfulness. Although not clearly explicit, this seems to me to be an important aspect of Parmenides' approach. As he states at the beginning of the "Way of Truth:"

2. Come, I will tell you — and you must accept my word when you have heard it — the ways of inquiry which alone are to be thought: the one that IT IS, and it is not possible for IT NOT TO BE, is the way of credibility, for it follows Truth; the other, that IT IS NOT, and that IT is bound NOT TO BE: this I tell you is a path that cannot be explored; for you could neither recognize that which IS NOT, nor express it.

3. For it is the same thing to think and to be.

[. . .]

6. One should both say and think that Being Is; for To Be is possible, and Nothingness is not possible [. . .] .

7,8. For this (*view*) can never predominate, that That Which Is Not exists. You must debar your thought from this way of search, nor let ordinary experience in its variety force you along this way [. . .] .[5]

Having established the logical equivalence of knowing with Being, along with the nonexistence of Not-Being as well as the logical impossibility of knowing it and expressing anything about it, Parmenides then deduces further logical implications from this. Particularly, he denies the validity of the logic underlying the kinds of explanations that characterized the systems of the previous Presocratics. As already noted, the Presocratics were primarily concerned with change and with the problem of explaining how the visible world arose from an underlying, indestructible substance; or, alternatively, how the underlying permanent *physis* gave rise to the various substances and states of the visible world. Once the origin of the visible world was accounted for further change was generally explained as due to strife among the basic opposites. When Parmenides considers these problems he interprets all change within the logical categories of existing (P) or not existing (-P) and the law of excluded middle, either P or not P (Pv-P),[6] or its disjunctive equivalent, not both P and -P (P/-P). As he maintains, 'The decision on these matters depends on the following: IT IS, OR IT IS NOT.'

The problem having been formulated in these logical categories, Parmenides is forced to interpret all change as a coming to be or ceasing to be, so that what IS becomes what IS NOT, and what IS NOT becomes what IS. But the *logical form* of these expressions is itself inadmissible in terms of his logical principle: what IS cannot NOT BE, and what IS NOT cannot BE. Therefore, if change involves passing from what IS into what IS NOT, or from what IS NOT into what IS, change and becoming cannot be real. Parmenides reiterates this thesis in different ways: 'How could Being perish? How could it come into Being? If it came into being, it IS NOT; and so too if it is about-to-be at some future time. Thus Coming-into-Being is quenched, and destruction also into the unseen.'[7] Forced into the unnaturally tight logical schema of Being or Not-Being (Pv-P), the logic governing the concepts of change and becoming does not fit, and so the concepts are rejected.

As the above statement implies, not only is change logically impossible, but so is time, for time implies a succession of moments, the coming to be and ceasing to be of events, and thus is subject to the same devastating logical critique. Moreover, Parmenides adds to his critique of time Anaximander's *negative* application of the principle of sufficient reason: if one cannot deduce why anything should come to be at one time rather than another, there is no justification for supposing that it did. Hence there is no justification for supposing that anything ever comes to be or ceases to be in time; hence there can be no temporal distinctions. As he says,

7, 8 [. . .] . Being has no coming-into-being [. . .] for what creation of it will you look for? How, whence (*could it have*) sprung? Nor shall I allow you to speak or think of it as springing from Not-Being; for it is neither expressible nor thinkable that What-Is-Not Is. Also, what necessity impelled it, if it did spring from Nothing, to be produced later or earlier?[8]

The same form of arguments and questions will be used by Leibniz and others during the 17th and 18th centuries regarding the justification and time of God's creation of the world.

Accordingly, this application of the logic of 'IT IS, OR IT IS NOT' enables Parmenides to dispense with all previous explanations in terms of coming to be, whether from Water,

Air, or the Unbounded. Moreover, this applies obviously to the form of explanation which tries to account for the visible changes in the world in terms of the 'strife' or 'opposition' among contrasting qualities or substances (Anaximander, Anaximenes, and Heraclitus), or of dualistic categories (the Pythagoreans). As the opposition involves a ceasing to be and coming to be ruled out by Parmenides' logic of excluded middle, this form of change also is logically impossible. The doctrine of opposition that 'in everything there is a way of opposing stress,' he claims, is 'two-headed,' for those who hold it imply that 'To Be and Not To Be are regarded as the same and not the same.'

Further application of this logic rules out motion, degrees of existence (e.g., Anaximenes' doctrine of different degrees of density), and spatial separation or differentiation (the conception essential to the Pythagorean notion of unit-points and geometrical configurations). The logical refutation of the possibility of change and time implies the impossibility of motion, for motion is a form of change in time. However, motion differs from change in that it involves an alteration of position rather than of quality, hence an added refutation is necessary. But since change in position would imply ceasing to be at one place and coming to be at another place, change of position or motion also is eliminated: 'Being has no coming-into-being and no destruction, for it is whole of limb, without motion [. . .].'

Furthermore, spatial separation and differentiation also are rejected, for spatial separation would imply the existence of gaps or of a void within Being, and spatial differentiation would imply degrees of Being, both of which are logically impossible. Gaps in Being would imply the existence of Nothing and spatial distinctions (according to the logic of Pv-P) would also imply Not-Being, which we know cannot Be.

> Nor is Being divisible, since it is all alike. Nor is there anything (here or) there which could prevent it from holding together, nor any lesser thing, but all is full of Being. Therefore it is altogether continuous; for Being is close to Being. [9]

In short, Parmenides' logic of 'IT IS, OR IT IS NOT' allows of only one ontological category, that of the attribution of *existence*, hence Being is and Not-Being is not (P·-[-P]). No other forms of attribution are possible: neither change, becoming, motion, time, degrees of being, spatial distinctions and separation, nor *qualities*. Any such attributes are *nominal* but not real; i.e., they involve the application of names to what does not truly exist:

> [. . .] all things that mortals have established, believing in their truth, are just a name: Becoming and Perishing, Being and Not-Being, and Change of position, and alteration of bright colour. [10]

Having denied the validity of any qualitative attributions or relations to Being one is left with the timeless, changeless, qualityless, nearly empty category of Being itself. As Hegel said of Schelling's philosophy, "It is the night in which all cows are black." But this is not the last time when this type of logic will be applied to a critique of our experience and knowledge of reality. In the latter part of the 19th and first quarter of the 20th centuries, the absolute idealist F. H. Bradley also will deny validity to the logic of predication and relations, thus arriving at a "bloodless" conception of reality as a homogeneous unity without distinctions. That is, having 'proved' that all qualities and relations involve internal contradictions, Bradley concluded that they must be relegated to appearance as compared with the self-consistent unity of Reality.[11] As de Santillana says, paraphrasing Bradley's logic without mentioning him:

If we ask for the natures of the Many, we get only a set of internal contradictions — or of external relations. If we try to understand their motion, we get only a set of relations, that is, again, nothing in itself. The logic of objects turns out to be a logic of relations. Everything shows up as inconsistent against the absolute background of the One, the locus of all relations, where all contradictions are solved. [12]

In fact, it was to avoid being swallowed up by the Absolute (among other reasons) that Russell and other modern logicians developed the techniques of predicate and relational logic; i.e., an expansion of logical operations to include the valid predication of qualities and relations to the world.

But having said what *Being is not*, and merely established *that It Is*, is Parmenides' conception of Being otherwise empty? Can he deduce any additional properties belonging to it? If one has followed Parmenides' statements closely one can infer that Being does have certain positive attributes, namely, those implied by the denial of other characteristics. Thus having denied change, motion, and time to Being one can infer that it is 'changeless,' 'immobile,' 'imperishable,' and hence 'eternal.' Moreover, having denied spatial separation, distinctions, and degrees to Being, Parmenides asserts that it is 'continuous,' 'without distinctions' or homogeneous, and without gaps or 'full.' But can these attributes exist of themselves or do they in turn imply the existence of something else?

The only answer to this seems to be that they do imply the existence of something else, *that of pure spatial extension itself*; i.e., *the property of extension abstracted from its embodiment in objects*. For Parmenides is not thinking of corporeal extension, but three-dimensional fullness in the abstract. Thus he carried the apprehension of the abstract a step beyond the Pythagoreans who, as Aristotle said, identified numbers 'with sensible substances.' Parmenides, having abstracted Being from the conception of corporeality, could still not abstract it from spatiality; existence could not yet be conceived as non-spatial even if it could be conceived as non-corporeal. The relentless application of his logic led Parmenides to the brink of discovering pure *geometrical space*: the characteristics of continuity, homogeneity, and indifference of position and direction (i.e., isomorphism, although as we shall see Parmenides' space is not quite isomorphic) being precisely those later implied by Euclidean geometry and exemplified in Euclidean space. As de Santillana states:

> If we accept that word "Being," not as a mysterious verbal power, but as a technical term for something the thinker had in mind but could not yet define, and replace it by x in the context of his argument, it will be easy to see that there is one, and only one, other concept which can be put in the place of x without engendering contradiction in any point, and that concept is pure geometrical space itself (for which the Greeks had no word, but only such related terms as "place," "air," "breath," "emptiness"). [13]

At first it appears as if there *must* be a contradiction in identifying pure space, the essence of which is *emptiness*, with pure Being, the essence of which is *existence and fullness*. And yet the more one studies the words of Parmenides the more it appears that he was reaching for this identification: 'all is full of being [. . .] for Being is close to Being.' Thus the criterion of the 'fullness of Being' is spatial continuity: 'Therefore it [being] is altogether continuous; for Being is close to Being.' Nor will this be the only time in history when the attributes of space and Being are identified. In the 17th century, Henry More, finding the attributes of space coinciding for the most part with those of God, identified the two. [14] For similar reasons, Newton also was led to the spatialization of the Divine or the Divinization of space. Parmenides does not refer to the divine, but the rationale for identifying space and Being is essentially the same that led to the identification of space and God.

If there were any further doubt as to Parmenides' identification of spatial extension with Being, it would seem to be completely eliminated by the final attribute he assigns to Being — its *spherical shape*. Having implied so often the notion of continuous space, it seems that Parmenides should have embraced the concept of an infinite extension — and in fact his follower Melissus will draw this conclusion. But the notion of infinite extension was too close to the notion of Not-Being for him to equate it with Being. Arguing that if Being were without a boundary it would be spatially infinite and hence lacking in Being, he concludes that it must be 'bounded' or 'limited,' giving it the shape of a sphere, geometrically the most self-contained configuration. The logic of the argument is not unlike Einstein's who also concluded that the universe has a spherical shape and thus is both finite and infinite: finite in the sense that the dimensions are not infinite, but infinite in the sense that the universe is *continuous* so that one could never come to its limits, there being nothing beyond it to limit it. [15] As Parmenides states the argument, Being

> is motionless in the limits of mighty bonds, without beginning, without cease [. . .]. And remaining the same in the same place, it rests by itself and thus remains there fixed; for powerful Necessity holds it in the bonds of a Limit, which constrains it round about, because it is decreed by divine law that Being shall not be without boundary. For it is not lacking; but if it were (*spatially infinite*), it would be lacking everything. [16]

Parmenides therefore adds to the spatial properties of Being already mentioned the additional attributes of remaining 'fixed,' 'at rest,' and 'in the same place,' and being held by 'powerful Necessity' in the 'bonds of a Limit.' There must be a Limit (as in the Pythagorean system) or else everything would be 'lacking' or non-existent. He then describes the particular form of this Limit:

> But since there is a (*spatial*) Limit, it is complete on every side, like the mass of a well-rounded sphere, equally balanced from its centre in every direction; for it is not bound to be at all either greater or less in this direction or that; nor is there Not-Being which could check it from reaching to the same point, nor is it possible for Being to be more in this direction, less in that, than Being, because it is an inviolate whole. For, in all directions equal to itself, it reaches its limits uniformly. [17]

The argument that Being is 'like the mass of a well-rounded sphere,' in that it is 'equally balanced from its centre in every direction,' is a further application of Anaximander's principle of sufficient reason. So even though Parmenides is critical in general of the logical form of the systems of his predecessors, he is not beyond borrowing some of their arguments. Since the notion of a sphere implies a center, as he states, his space cannot be isotropic (the same in every direction), and as it is limited it cannot be infinite, two characteristics of later Euclidean space. Because these characteristics appear inconsistent with his conception of a qualityless, homogeneous Being, they were particularly attacked by his critics.

THE WAY OF OPINION

Such is the outcome of Parmenides' "Way of Truth." Having logically sliced away from Reality or Being all qualities except pure spherical extension, one is left with a surrealistic image of the world as an immense, timeless, static sphere seeming to be reflected in the pale light of pure reason itself. Small wonder that he added the "Way of

Opinion" to console those who could not follow the cold logic of his vision of Being. His own reason for adding the "Way of Opinion" is that 'no thought of mortal man shall ever outstrip him.' Thus, as mentioned earlier, the two "Ways" are thought to be logically exhaustive. Those formalists inclined to follow the severe path of pure logic will be brought to the domain of truth; lesser mortals can follow the second "Way," a kind of true opinion based on appearances rather than on Being. He therefore reintroduces the sensible contraries, Light and Darkness (Pythagorean opposites), the 'airy and the dense' (Anaximenes), and describes the world as a manifestation of these opposites. He then proceeds to astronomical and physiological considerations incorporating many of the doctrines of the previous philosophers: the heavens are divided into rings or 'crowns' of fire similar to Anaximander and the Pythagoreans; earth is grounded in water similar to Thales; in the center of the 'crowns' of fire a goddess 'steers' all, holding the 'keys of Justice and Necessity' (Anaximander), but who as Eros also controls mating and generation (Hesiod). Along with these eclectic views he may have introduced two new astronomical tenets: that the moon shines by reflected light and, according to Theophrastus, that the earth is a sphere. If in fact these latter were his discoveries they are quite remarkable and surprising considering his unempirical orientation.

SUMMARY AND EVALUATION

Surely we have in Parmenides' philosophy one of the strangest, most original cosmologies ever developed. But then, how many others have tried to base a theory of reality on one logical principle? For it should be evident now that Parmenides' system rests upon the discovery of the logical law of excluded middle: 'IT IS, OR IT IS NOT.' Embedded in this discovery, however, was a fundamental confusion between logic and ontology. Just as we found that the Pythagoreans were unable to differentiate the status of abstract numbers from that of spatial existence, and thus combined the two, so Parmenides could not differentiate the categories of logic and ontology, thereby conflating the two. While logic can be applied to theoretical frameworks as a criterion of consistency and valid inferences, its use cannot prejudge what is to be found in the world. Before the development of non-Euclidean geometries, most mathematicians held that such geometries were impossible because they would imply theorems which would contradict those in Euclidean geometry and some properties of Euclidean space. As a result of their development, however, one now separates the formal implications of a particular geometry from the physical properties of space.

Similarly, today one distinguishes between the purely formal characteristics of an abstract logical calculus and its application to particular arguments. While logic establishes the correct formal rules of *inferences* within arguments, it tells us nothing about the *contents* of the arguments: it can establish whether logical inferences are valid or not, but not whether the content of the argument is factually true or false. Like pure mathematics, logic itself tells us nothing about the world; but Parmenides, like the Pythagoreans, was unable to make this differentiation. Having discovered the purely logical principle that something either is or is not, he confused this with a discovery about reality. This confusion was enhanced probably by the verbal expression of his discovery, 'IT IS, OR IT IS NOT,' which implies that *something* is the subject of 'IT IS,' namely, that 'IT' and 'Being' are one and the same. And as Being either is or is not, and cannot not be, IT *IS* (Pv-P, and since -(-P), therefore P).

This confusion between logic and ontology pervades all of Parmenides' system, underlying the identification of his logical principle with the category of being.[18] According to the logic of Pv-P there can be no coming to be or change in the world because this would imply the logical contradiction that something both is and is not (P and -P). But while it is (trivially) true that it makes no sense to say of something both that it is and that it is not, something can be bronze and not marble, or can be a child at one time and a man at another, or both a man and a teacher, or can be wood and burning at the same time. Moreover, the logic of negative predications (that something is not such and such) does not imply the existence of Not-Being as the subject of such predications, as Parmenides seemed to assume; i.e., to deny a property of *something*. In short, the logic of the exclusive alternative (Pv-P) is a static logic, as timeless as Parmenides inferred, merely asserting that something is or is not the case. Parmenides' mistake or confusion was in concluding that because processes in the world do not conform to this restricted logical schema, change and time cannot be real, whereas he should have concluded that the logic itself was too restricted to apply to specific kinds of descriptive judgments about certain natural processes in the world.

Yet Parmenides' approach illustrates a not uncommon tendency among formalistically inclined cosmologists to accept the theoretical implications of a conceptual framework, however disparate its projection of reality may be from that of ordinary experience. Thus Parmenides' reduction of the ordinary empirical world to a qualityless, static, three-dimensional sphere, because of purely formal considerations, is similar in intent to that of the physicist John Wheeler and the philosopher of science John Graves who attempted to reduce physical reality to a "geometrodynamics." Such thinkers are inclined to deny the rich qualitative diversity of the physical world in preference for a formally simpler model. For example, Graves extols the attempt to reduce all physical phenomena, such as electromagnetism, gravitation, electrical charges and physical proportion such as mass, motion, shape, etc., to a single geometrodynamic (GMD) structure. As he says:

> Let us survey just what classical GMD has been able to accomplish. Its basic goal remains the identification of matter with space, which in turn means that all the phenomena traditionally associated with matter and considered conceptually different from space must somehow be incorporated into the natural Riemannian structure of space-time, rather than being imported from the outside as 'foreign' entities. [19]

Thus both Parmenides and Graves advocate a monistic conception of the universe, although their cosmological monism differs in other respects, as Graves himself points out: "Unlike the static, unchanging 'Being' of Parmenides, it [geometrodynamics] is capable of an enormous amount and variety of internal differentiation, while still remaining intelligible."[20] It is this latter contention, of course, that would be denied by Parmenides.

ZENO OF ELEA

Parmenides had attempted to prove that Reality or Being must be one, indivisible, homogeneous, changeless, motionless, timeless, and qualityless (except for its spherical shape) — a rather austere doctrine little suited for general acceptance. As often happens when opponents of a position find it untenable without being able to refute it, they resort to ridicule. Thus, as Plato tells us in his dialogue *Parmenides*, 'there was much mocking' of Parmenides' system. Diogenes, for example, showed his disdain for Parmenides' refutation of motion by getting up and walking around — just as later Dr. Johnson would 'refute' Berkeley's denial of matter by kicking a stone. Both of these remonstrations miss the mark, however, in that neither Parmenides nor Berkeley was denying the *apparent* occurrence of these phenomena, but their ultimate reality (just as pounding on a table would hardly constitute a refutation of the theory that matter is composed of molecules and atoms and hence mainly of empty space).

But Parmenides was more fortunate than Berkeley in having a redoubtable defender of his system in his pupil Zeno. As Plato again tells us, if Parmenides' opponents resort to ridicule, then Zeno 'pays them back in their own coin, and with something to spare.' Zeno's method of attack was to assume his opponent's position, the opposite of that of Parmenides, that Being is a plurality and that change, motion, and time are real, and then show that from these assumptions (as Plato says) 'even more absurd consequences follow.' Thus Zeno invented the powerful, but often specious appearing, form of argument, *reductio ad absurdum*; i.e., the form of argument where one assumes his opponent's thesis and then deduces from it self-refuting, contradictory implications. While this argument usually appeals to the spectator because of its apparently devastating consequences, it is more effective in arousing the wrath of one's opponent than his sense of conviction. Zeno worked out forty such arguments (only a few of which remain) in defense of his master's system, thereby earning from Aristotle the title of "the founder of dialectics."

Unfortunately, limitations of space prevent giving to Zeno's arguments the full consideration they deserve in the light of their influence throughout history. They were taken up by the Sophists and Scholastics as the model of argumentation, and as late as the 20th century philosophers (not to mention mathematicians such as Cantor and mathematical physicists such as Herman Weyl) as diverse as Bergson, Russell, Ryle, Vlastos, and Grünbaum, with entirely different approaches, have attempted to answer Zeno's paradoxes. At most, all we can do is to discuss two of his more typical arguments illustrating their form, presuppositions, and possible weaknesses.

First, in contrast to Parmenides' thesis that Being is one and indivisible, Zeno assumes his opponent's thesis of pluralism, that things are many. He then applies the *reductio* argument by showing that if reality consists of a plurality of units, then this involves the self-refuting consequences that they must be either (1) infinitely small, or (2) infinitely great. For a unit either has magnitude or it does not have magnitude. If a unit has no magnitude, that is, is without size or bulk, then no addition of such units can result in any magnitude. Thus since nothing has any magnitude, everything must be infinitely small. If, however, one makes the contrary assumption that the units have magnitude, then any part of the unit must have magnitude, and a part of that part must have magnitude, *ad infinitum*. Hence any one unit is composed of an infinite number of magnitudes such that it is infinitely large. [21]

In this manner Zeno attempted to prove his contention that if reality consists of a plurality of units they must be either infinitely small or infinitely large, neither possibility being credible. Part of the difficulty in assessing the argument consists in the formulation of the contradictory conclusion: is 'having no magnitude' equivalent to being 'infinitely small' (especially as 'small' suggests some magnitude), and is being composed of 'an infinite number of magnitudes' equivalent to 'being infinitely large'? As mentioned earlier, an unsettling feature of these arguments is that while one does not feel convinced by them, it is difficult to know whether or how they have gone astray. The intended implication of Zeno's argument is clear enough, however, since if it were true he would have refuted the possible existence both of an *unextended* unit and of a *plurality* of extended units. According to this conclusion only an *indivisible extended unit* could be real — namely, the Being of Parmenides. But as Zeno's argument applies to *extended* units, and since Parmenides' Being is extended, the question has been raised as to whether Zeno's second argument would not boomerang and strike down Parmenides' position as well. It would seem, however, that since Parmenides ruled out any division or distinction regarding the continuous and homogeneous nature of his extended Being, Zeno could have assumed that his second argument, which pertains to *divisible parts or units*, would not apply to Parmenides' indivisible Being. Even though extended, if Parmenides' Being is indivisible, then one could hardly argue consistently that it was composed of an infinite number of units.

However, Zeno's argument does lead to questions regarding the differences between geometrical, spatial, and material extension. Was Zeno attacking the now familiar (and paradoxical) mathematical definition of a line as being composed of extensionless points? What, for example, is the significance and meaning in his first argument of a unit 'having no magnitude'? Clearly he means spatial magnitude (for numbers have a magnitude in that '5' is larger than '4'). But then his argument reduces to the obvious conclusion that without beginning with a spatial magnitude you cannot engender spatial magnitudes. Though obvious, the argument does take on significance when applied to the Pythagorean conception of spatial extension being generated by the first unit from the Unlimit. If the unit itself has no magnitude, how does it generate magnitude by separating into further units (analogous to the problem as to how dimensionless points can constitute a line)? The Pythagoreans tried to answer this by saying that the first unit acquired magnitude from the Unlimit: the unit 'inhaled' the void into being from the Unlimit. But Zeno could well have found this answer less than satisfying. In fact, he is reported to have said that 'if anyone would give him an account of the exact nature of the One, he would be able to describe the Many.'[22]

As regards the second argument, does the term 'units with magnitude' signify geometrical figures,[23] pure spatial extension, or material objects — or does it make any difference? Zeno's division of the unit into 'parts' suggests that he had *material* objects in mind, and that his argument was directed against the Pythagoreans, as all commentators maintain. He presumably had material objects in mind because, while it makes sense to talk of dividing bulky objects into units having extension, it is not clear what it would mean to assert that geometrical figures or pure three-dimensional space could be divided into "parts." Moreover, in contrast to Parmenides who denied corporeality as well as divisibility to his Being, the Pythagoreans equated spatial substances with numbers. As Aristotle says, number for the Pythagoreans 'in not separate but sensible substances are formed out of it.' If Zeno's arguments were directed against quasi-material units, this would also explain why he believed that they would be effective against the substantial-mathematical units of the Pythagoreans, without being applicable to Parmenides' Being which admitted neither of spatial distinctions nor of bulk.

This point must have been clear to Parmenides' other follower, Melissus, for he is quoted by Simplicius as saying: 'If Being is, it must be one; and being one, it must have no body. If it were to have bulk, it would have parts and be no longer one.[24] Melissus apparently recognized the destructive effect of Zeno's argument for the Pythagorean conception of substantial unit-points and took pains to deny both divisibility and bulk to Parmenides' Being. This is conjecture, but it seems unlikely that one devoted to the defense of his master's system would deliberately forge a double-edged sword which, though effective in slicing through the arguments of the Pythagoreans, could also be used to undercut Parmenides' position.

The second group of dialectical arguments advanced by Zeno generally have as their common purpose the refutation of motion and time. As the argument of Achilles and the tortoise is well known and typical, we shall take that as exemplifying Zeno's approach. According to Zeno's argument, given any head start by the tortoise, regardless of how swiftly Achilles runs, he can never overtake and pass the tortoise. For, given the lead of the tortoise, it will take Achilles a certain length of time to overtake that point, but during that time the tortoise will have advanced to a new position. Again, it will take Achilles a certain length of time to overtake this newly advanced position, but during that time the tortoise will have advanced again, *ad infinitum*. So while Achilles continues to gain on the tortoise by reducing the percentage of the lead, he can never reduce the percentage to zero and thereby catch up with the tortoise. Thus Zeno achieves his purpose of showing that certain experiences, such as motion and the passage of time, when subjected to (his type of) logical analysis, prove to be self-contradictory and hence impossible — in agreement with Parmenides' system. If one takes the argument seriously, then one is faced with having to choose between the validity of ordinary experience or the validity of the logical critique. The conclusion is analogous to that posed by certain contemporary philosophers of science, such as Adolf Grünbaum, who argue that since the time variable in the equations of physics is reversible, our ordinary experience of time flow or becoming must be an illusion: i.e., the subjective passage of time has no counterpart in physical reality. [25]

Of course very few people, in spite of Zeno's arguments, would deny motion and time, accepting the fact that in races, horses, cars, or runners do overtake and pass one another. Why, therefore, does Zeno's analysis preclude this? Not only do we experience the contrary of Zeno's thesis, but by applying a different logical or mathematical analysis one can easily calculate when Achilles would overtake the tortoise: it would be a function of the distance of the head start given the tortoise divided by the ratio of their speeds. For example, for the sake of simplicity let us assume that the tortoise and Achilles move at constant speeds (to avoid the more complex case of acceleration) or ½ mile per hour and 1 mile per hour, respectively. Let us suppose the tortoise has a head start of ½ of a mile. At the end of one hour Achilles will be exactly one mile from the starting line. And where will the tortoise be? In one hour he will have traveled ½ a mile which added to his ½ mile head start will make his position 1 mile from the starting line, the same position as Achilles. At the next moment he will begin to be passed by Achilles. According to Zeno, however, Achilles can never catch up with the tortoise, much less pass him.

The reason Zeno arrives at his paradoxical conclusion is that instead of setting up the problem in terms of the distances covered per rate of speed, he sets it up in such a way that while Achilles is always diminishing the percentage lead of the tortoise, he never exhausts the infinite progression of diminishing fractional amounts in the series. To use our above example, as Zeno sets up the problem, since Achilles' speed is 1 mile per hour, it will take him ½ hour to reach the initial position of the tortoise; but during that ½ hour the tortoise will have advanced ½ of ½ of a mile (his speed being ½ mile per hour), or ¼ of a mile. So Achilles will have dimished his lead by half. To overtake the new position of the tortoise it will take Achilles ¼ of an hour, but during this quarter hour the tortoise will have advanced $^{1}/_{8}$ th of a mile; it will then take Achilles $^{1}/_{8}$ th of an hour to reach that point during which time the tortoise will have advanced to $^{1}/_{16}$th of a mile, *ad infinitum*. So while Achilles continues to reduce the percentage of the lead by ½ increments, he nevertheless can never reduce the lead to zero and overtake the tortoise, for theoretically space, like numbers, is infinitely divisible.

Thus we have two ways of construing the problem. In the former way, not only does the solution conform to experience, but one can predict exactly when Achilles will overtake the tortoise. In contrast, the second way of construing the problem precludes what actually occurs. What, then, is the answer to the paradox? Superficially, at least, the obvious answer is that Zeno has misconstrued the problem and applied the wrong mathematical-logical analysis to it. It cannot be maintained that *any* mathematical-logical analysis shows the impossibility of Achilles overtaking the tortoise, because according to the first way of construing the problem, it is just as mathematically *necessary* that Achilles both overtake and pass the tortoise, as it is logically *impossible* according to Zeno's analysis. As any school child knows who has taken a course in mathematics, there are correct and incorrect approaches to a problem. Either inadvertently or by design, Zeno set up the wrong procedure (or one might say that it was the *correct* procedure for confounding Parmenides' critics, but the *wrong* procedure for giving a correct analysis of the problem consistent with experience). Thus Zeno's paradox does not seem so paradoxical after all.

Moreover, this way of regarding the paradox is consistent with what was said earlier regarding Parmenides' arguments: namely, that his application of the logical principle of excluded middle was inadequate to the kinds of problems he was analyzing (e.g., change and becoming). More generally, the answer is that one cannot arbitrarily apply a formal system to the analysis of experience, denying *a priori* certain features to the world on the basis of limitations within the formal system itself. In utilizing a formal system, one must ascertain that the system itself is adequate to the data being analyzed. It is logically possible to select any theoretical criterion and deny reality to a class of entities on the basis of this criterion. Berkeley's *"esse est percipi"* was such a criterion, in terms of which he denied the reality of physical objects and matter. There is in fact the constant danger that one's theoretical system will distort one's analysis of empirical phenomena — but the distinction between the purely formal structure of a theory and its application to experience was not as apparent at that time. Thus it was not clear as to what import one should give to Parmenides' or Zeno's analyses, though one could always fall back on ridicule.

Furthermore, I believe this way of interpreting the problem conforms to Aristotle's solution. According to Aristotle, the difficulty with Zeno's analysis is that he confused finite magnitudes with infinite divisibility. Achilles could not overtake the tortoise because he had to pass through an infinite number of spatial segments which is impossible in a finite time. But this conclusion, according to Aristotle, depends upon an equivocal use of 'finite' and 'infinite.' While Zeno assumed that a finite interval of space could be divided into or 'contain' an infinite progression of divisions, he ignored this as regards time. Restricting an interval of time to a finite duration while granting a limited spatial magnitude an infinite series of gradations, he concluded that one could not pass through an infinite number of spatial magnitudes (however infinitesimal) in a finite time; i.e., the intervals of a finite temporal duration could not be put into a one-to-one correspondence with an infinite spatial series. But if, as Aristotle points out, one construes a finite interval of time as also being divisible into an infinite number of intervals, then for every spatial division there will be a corresponding division of time, such that in any finite duration of time there will be a lapse of an infinite series of moments corresponding to the infinitely divisible segments of space. Hence, one can traverse a spatial interval in a certain duration of time (depending upon the speed) contrary to Zeno's contention. Therefore, Aristotle also seems to affirm that the concepts Zeno used to analyze the problem led him to misconstrue it and thereby arrive at a paradoxical conclusion.

In addition to these criticisms, Grünbaum maintains that this particular argument of Zeno involves two additional fallacies. First, the fact that we cannot conceive of an infinite series, in the sense that the minimal duration of a single thought precludes our running through an infinite series in our minds, does not prevent a finite temporal duration from having an infinite number of temporal subintervals. More simply, no finite lapse of consciousness can include an infinitely denumerable metrical series, but this does not prevent a unit of time from containing an infinite number of subintervals.

> Human awareness of time exhibits a positive threshold or minimum. This fact can now be seen to have a consequence of fundamental relevance to the appraisal of Zeno's argument. For it entails that *none* of the infinitely many temporal subintervals in the progression whose magnitude is less than the human *minimum perceptibilium* can be *individually experienced as elapsing* in a way that does metrical justice to its actual lesser duration. To succeed, the attempted individual *contemplation* of all the subintervals would require a denumerable infinity of mental acts, each of which requires or exceeds a positive minimum duration [. . .]. And the resulting compelling feeling that an infinite time is actually needed to accomplish the traversal in turn insinuates the deducibility of this paradoxical result from the theory of motion. [26]

Secondly, Zeno's argument assumes that while Achilles' successive reduction of the time intervals converges to zero, the series never includes the terminating instant. Thus while Achilles can reduce the tortoise's lead by infinitely progressive smaller intervals, this progression never includes the terminal moment, hence he never can overtake the tortoise.

> Zeno illicitly exploits the fact that it is logically impossible for the terminal instant of the motion to belong to any of the subintervals of the unending progression, since it is *later* than all of them. Zeno appeals to this fact to infer wrongly that *there cannot be* any terminal instant at which the runner reaches his destination. [27]

MELISSUS

Zeno's defense of Parmenides' system consisted of a dialectical attack on the assumptions of Parmenides' critics, therefore his contributions were mainly destructive, though important in raising certain mathematical paradoxes. Melissus, on the other hand, another follower of Parmenides (though there is no evidence that he ever directly associated with him) not only defended his position, but made several significant emendations as well. A crucial weakness in Parmenides' cosmology was the fact that while he denied all qualities and distinctions to Being, he still maintained that it was spherical in shape, 'equally balanced from its centre in every direction.' Thus it could be said — and there were those only too anxious to say it — that his system involved inherent contradictions; indeed, it was with respect to this latter point that his views were particularly ridiculed. To eliminate these objections, Melissus accepted the critical implications of Parmenides' detractors by rejecting the spherical shape of Being, maintaining its infinite extension. Reversing Parmenides' argument that if Being *were not limited* it would lack something and hence must be bounded in the shape of a sphere, Melissus argued that if it *were limited* it would have to be limited by Not-Being, which is impossible, hence Being is unlimited or infinite. In contrast to Parmenides, therefore, he concluded that Being was not only eternal in time but infinite in extension: '[. . .] as it Is always, so also its size must always be infinite.'[28]

Except for the characteristic of infinity, Melissus retained the other properties of Parmenides' Being: continuity, homogeneity, lack of change and motion, and incorporeality. But the addition of infinity was an important extension because it marked the first explicit avowal of infinite space as ultimately real and thus edged further the conception of a purely abstract existence.

> Another step has been taken towards the apprehension of the abstract; but it is still only a step in that direction [. . .]. If he had been capable of imagining something that was not only incorporeal

but non-spatial as well, the outcome of his thought would have been different; but the only safe deduction to be drawn [. . .] is that he was not capable. [29]

Still, what a tremendous step it was to have followed the logic of pure reason beyond the horizon where the world of corporeality was left behind in the vision of an infinite abyss. And yet, as if to remind us that a kind of primitive animism lies just beneath the surface of the seemingly most sophisticated forms of thought, he added that Being neither feels 'pain' nor 'grief,' thereby implying that it is in some sense alive and sentient.

The second contribution of Melissus consists of his clearer formulation of the problem of change and identity in his criticism of those who would rely on the testimony of their senses in accepting change and plurality over the unchanging unity of Being. Somewhat in the manner of Descartes who asserted that, though our senses attest to the changing qualities of melting wax, our reason judges it to be the same piece of wax, Melissus argued that if we believe that our senses disclose reality as it truly is, then what are we to conclude when the qualities of things change? Can what is real be changed into something else, or do the apparent changes signify an unchanging reality lying behind the appearance of things?

(3) and it seems to us that the hot becomes cold and the cold hot, and the hard soft and the soft hard [. . .] and that all things change, and that what was and what now is are not at all the same, but iron which is hard is worn away by contact with the finger [. . .]. So that it comes about that we neither see nor know existing things.

(4) [. . .] For although we say that there are many things, everlasting [. . .] having forms and strength, it seems to us that they all alter and change from what is seen on each occasion.

(5) It is clear therefore that we have not been seeing correctly, and that those things do not correctly seem to us to be Many; for they would not change if they were real, but each would Be as it seemed to be. For nothing is stronger than that which is real.

(6) And if it changed, Being would have been destroyed, and Not-Being would have come into Being [. . .] therefore, if Things are Many, they must be such as the One is. [30]

Thus Melissus raises in its modern form the horrendous problem that has afflicted all subsequent philosophy, the problem of appearance as opposed to reality. Are the observable changing qualities and properties of things ultimately real, or are they only apparent manifestations or representations of a more fundamental, unchanging reality? It is primarily on this problem that the following Presocratics, the Pluralists and the Atomists, will focus their attention.

CONCLUSION

While the logical thinking of Parmenides and Zeno manifests syncretic and diffuse aspects, in that differentiations of various uses of the verb 'to be,' as well as distinctions between logic and ontology which are clear to us, were fused in their primitive conception of logic, Parmenides still can be acclaimed for having discovered logical implication and Zeno for having invented dialectics. Moreover, to have abstracted from the empirical contents of knowledge various logical principles, and then adopted them as criteria of reality, was a feat of abstraction indicative of the highest mental capacity. Though limited by the level of knowledge at that time, the Parmenideans (along with the Pythagoreans) establish beyond any doubt that some thinkers of an early age are capable of the same high quality of abstract thought as those of a later age. That contemporary philosophers and mathematicians are still puzzling over the paradoxes of Zeno attests to this fact.

In their mode of philosophizing, the Parmenideans manifest the intellectual trait of what is now called rationalism; i.e., a type of mentality that disdains empirical evidence in favor of principles that appear to be self-evident and indubitable. More generally, a rationalist is one who accords greater credence to the intellectual claims of a formal system than to the empirical evidence of the senses. Thus we find Parmenides denying reality to empirical change, motion, time, and qualities because they do not conform to the logical principle of excluded middle, that something either is or it is not. As interpreted by Parmenides, this principle obviates change in that *it allows of no intermediate state between being and not being*: no change of states or properties, no coming to be or ceasing to be, no motion, etc. Only the changeless, timeless reality of pure Being can be.

Strangely enough, considering the stark implausibility of Parmenides' system, it had a strong impact on subsequent Greek philosophy. As Kirk and Raven claim: "Parmenides seemed, to his contemporaries and immediate successors, to have established once and for all certain canons with which [. . .] all future cosmologies must somehow comply."[31] As interpreted by his successors, these canons were: (1) that Not-Being (the Void) cannot exist; (2) that Being cannot come from Not-Being; (3) that whatever is claimed to be real must therefore be ultimate; (4) that plurality cannot be derived from an original unity; therefore (5) if plurality is admitted it must be ultimate; (6) that change cannot be a coming to be or perishing; and (7) that if motion is injected into the system it must be explained rather than assumed.[32] Still, one is left with the puzzling question as to why Parmenides' critics should have acknowledged the importance of his logical canons, while rejecting his cosmology.

The answer lies, perhaps, in the following considerations. Ever since Kant we have been aware that all knowledge depends upon two basic factors: (1) the empirical content derived from sensory observations and experiments, and (2) the formal element, the basic concepts and theoretical framework, used in the interpretation of experience. While it is obvious that knowledge to be significant must have some empirical content, it is equally true that all knowledge is based upon certain theoretical assumptions and principles. Usually these principles have been justified on the grounds that they are self-evident, indubitable, and thus absolutely necessary. Kant himself believed that he had uncovered the *a priori* conditions of *all* knowledge; i.e., the "universal and necessary" preconditions of all cognition.

As we shall find in the course of this study, the truly revolutionary changes in knowledge occur when the theoretical principles formerly believed to be self-evident, indubitable, and necessary are rejected on the basis of new discoveries and conceptual developments. Originally, the fundamental presuppositions and principles of the theoretical system may not have been considered self-evident and indubitable, but over a period of time a conditioning process often results in their attaining that status. When this happens the time is ripe for a revolution. Einstein's rejection of the invariance of space, time, and mass, considered to be self-evident in Newtonian mechanics, is a recent example of such a revolution.

Although Parmenides' successors were too empirically oriented to accept his rejection of the reality of the empirical world, his logical canons appeared to be incontrovertible and therefore could not be ignored. Was it not self-evident that Not-Being could not exist and that Being could not come from Not-Being? If those succeeding him were to offer a cosmology satisfying the intellectual criteria of the day, the criteria laid down by

Parmenides, then they were compelled to philosophize within his logical strictures. This is an excellent historical example of the controlling effect certain theoretical principles or assumptions can have on the thought of a certain epoch. While the systems of his successors manifest other influences, none was greater than the logical canons of Parmenides.

1   Immanuel Kant, *Critique of Pure Reason*, trans. by Norman Kempt Smith (New York: The Humanities Press [1929] 1950), A798, B825.

2   Plato, *Parmenides*, 127A.

3   The distinction between "logical form" and linguistic structure was first introduced by Russell and was the point of departure of Wittgenstein's philosophy. Cf. Bertrand Russell, *Introduction to Mathematical Philosophy*, 2nd ed. (London: Allen and Unwin, 1920), pp. 170-171, and Ludwig Wittgenstein, *Tractatus Logico-Philosophicus*, new trans. by Pears & Guinness (London: Routledge & Kegan Paul, 1961), 2.18.

4   For an interpretation of the epistemic meaning of each of the allegorical symbols in the "Prologue" see Kathleen Freeman, *The Pre-Socratic Philosophers* (*op. cit.*, ch. IV), pp. 146-147.

5   Kathleen Freeman, *Ancilla To The Pre-Socratic Philosophers* (*op. cit.*, ch. IV), pp. 42-43. The order of the fragments follows Diels-Kranz.

6   Interpreting the alternative (v) as an exclusive alternate; i.e., that P and -P cannot both be true and cannot both be false.

7   Freeman, *Ancilla To The Pre-Socratic Philosophers, op. cit.,* pp. 43-44.

8   *Ibid.*, p. 43.

9   *Ibid.*, p. 44.

10   *Ibid.*

11   Cf. F. H. Bradley, *Appearance and Reality* (Oxford: At the Clarendon Press, 1893, 2nd ed. 1897).

12   Georgio de Santillana, *The Origins of Scientific Thought* (*op. cit.*, ch. IV), p. 104.

13   *Ibid.*, p. 95.

14   Cf. Alexandre Koyré, *From the Closed to the Infinite Universe* (New York: Harper and Brothers, Pub., 1958), pp. 147-154.

15   There are of course important differences between the two views, one of which is that Einstein's universe is four-dimensional (time being the fourth dimension) while that of Parmenides is three-dimensional (time having been excluded).

16   Freeman, *Ancilla To The Pre-Socratic Philosophers, op. cit.*, p. 44.

17   *Ibid.*

18   This is a temptation to which even modern logicians can fall prey. The distinguished American logician Willard Quine has stated, "To be assumed as an entity is, purely and simply, to be reckoned as the value of a variable." Cf. *From A Logical Point of View* (*op. cit.*, ch. I), p. 13.

19   John Cowperthwaite Graves, *The Conceptual Foundations of Relativity Theory* (*op. cit.*, Introduction), p. 312.

20   *Ibid.*, p. 314. (Brackets added.)

21   Cf. Freeman, *The Pre-Socratic Philosophers, op. cit.*, pp. 153-164, for an excellent full discussion of these arguments.

22   *Ibid.*, p. 157.

23   Following Aristotle and others, Freeman suggests (*Ibid.*, pp. 158-9) that Zeno was attacking the conception of geometrical points and lines. But as Zeno himself, in the fragments preserved for us, does not refer to points or lines, it seems safer to confine one's conjectures to the available evidence.

24   Simplicius, *Phys.*, 109, 34. Quoted from Kirk and Raven, *The Presocratic Philosophers* (*op. cit.*, ch IV), p. 302, f.n. 391.

25   Cf. Adolf Grünbaum, *Philosophical Problems of Space and Time*, second, enlarged edition (Boston: D. Reidel Pub. Co., 1973).

26   *Ibid.*, p. 633.

27 *Ibid.*, pp. 634-635.
28 Freeman, *Ancilla To The Pre-Socratic Philosophers, op. cit.*, p. 48.
29 Kirk and Raven, *op. cit.*, p. 304.
30 Cf. Freeman, *Ancilla To The Pre-Socratic Philosophers, op. cit.*, p. 50.
31 Kirk and Raven, *op. cit.*, p. 319.
32 Cf. *Ibid.*

CHAPTER VIII

EMPEDOCLES AND THE RETURN TO THE ELEMENTS

*[. . .] his exuberant genius combined the temperament
of a prophet with a really scientific turn of mind [. . .].*[1]

Cornford

Born of a noble family in Acragas, Sicily, Empedocles reached his prime about the middle of the 5th century B.C. It is said that his grandfather Empedocles won an Olympic victory in 496; beyond that, little is known about the external details of his life, although he, like Pythagoras, was the subject of many apocryphal anecdotes and legends. Of the latter, the most famous is the durable story invented by his adversaries that he lept into the crater of Mt. Etna to eliminate any trace of a mortal ending to substantiate his claim (in the *Purifications*) that he was an immortal god having completed his cycle of purifications.

A founder of the Sicilian school of medicine, his fame as a medical doctor was so widespread and pronounced that miraculous cures were attributed to him, such as bringing to life a woman who had been dead for thirty days. His knowledge of biology, botany and physiology is clearly evident in his thinking and philosophical system. Traces of the cosmological views of Anaximander and Anaximenes also are found in his system, though the influence of the Pythagoreans and of Parmenides is most striking. Moreover, he combined with his extensive empirical interests and broad scientific background the temperament of a prophet and a poet. While somewhat critical of his style, Aristotle called him the "inventor of rhetoric," as he had called Zeno the "inventor of dialectics."

But as pronounced as these various influences were on his thought, they were not more important than the strong religious and mystical tendencies of his varied personality. Apparently following the example of Parmenides, he wrote two poems (as well as other works) entitled *On Nature* and *Purifications*, of which the first presents his cosmological and scientific views, while the latter shows the influence of the mystical Orphic and Pythagorean belief in the transmigration of souls. Earlier commentators (Zeller and Burnet) held that the empiricistic scientific conceptions presented in *On Nature* were incompatible with the religious and mystical doctrines expressed in the *Purifications*. However, more recent scholars have argued that the concept of the cyclical transmigration of souls finds its counterpart in the cyclical cosmological processes of *On Nature*, so that rather than being incompatible, the two works are complementary. In any case, Empedocles combined in his richly diversified personality the scientific capacity of the Ionians, the metaphysical and prophetic orientation of Heraclitus, the respect for rationalistic principles of the Parmenideans, along with the mystical tendencies of the Pythagoreans and the Orphic religion.

THE ELEMENTS

As one would expect from someone with a medical background and a broad interest in natural phenomena in general, Empedocles declares himself in favor of empiricism over rationalism. In direct contrast to Parmenides, who had denied the validity of sensory experience, he begins his poem *On Nature* by emphasizing the importance of sensory observation and the necessity of using 'whatever way of perception makes each thing clear.' Although he later adds that 'eyes are more accurate witnesses than ears,' initially he warns that sight should not be accorded a higher credibility than the other senses, but that the claims of each should be tested by those of the others. Having begun by rejecting Parmenides' rationalism, he continues by rejecting his monism — since it is hardly consistent with his empirical orientation and scientific background to relegate the empirical world to appearance in contrast to an unchanging, qualityless reality. Consequently, he is faced with deciding which of the fundamental substances variously chosen by his predecessors can best account for the rich diversity of empirical phenomena.

Rather than selecting one among the previously chosen substances as ultimately real, he decides in favor of four, fire, earth, air, and water (also the primary opposites in Anaximander's system originating from the Unbounded), introducing them first under the mythical guise of gods: 'Hear, first, the four roots of things: bright Zeus, and life-bearing Hera, and Aidoneus, and Nestis who causes a mortal spring of moisture to flow with her tears.' While it seems obvious that 'bright Zeus' represents fire and that Nestis must be the source of moisture and water, it is not certain as to whether Hera or Aidoneus represents earth or air. The expression 'life-bearing Hera' would suggest the identification of Hera with earth, but since Hera is a goddess of the heavens and Aidoneus a god of the underworld associated with Demeter and Persephone, it has been concluded that Hera represents air and Aidoneus earth. But these mythical names apparently were not of great significance to Empedocles, as he elsewhere refers to fire as Hephaestus. The selection of these four elements as the ultimate components of reality is important, however, for it will remain, with certain important exceptions, the prevalent view until the 18th century revolution in chemistry.

In spite of rejecting his monism, Empedocles indicates his intention of complying with the logical canons of Parmenides described at the end of the last chapter: namely, that what is real cannot come from what does not exist, that change cannot involve an absolute coming to be or perishing of anything, that reality is full and thus admits of no degrees or gaps, and that if a plurality of real elements is admitted they must be ultimate. In words that are hardly more than a paraphrase of Parmenides, he reaffirms these basic canons: [2]

11. Fools! — for they have no long-sighted thoughts, since they imagine that what previously did not exist comes into being, or that a thing dies and is utterly destroyed.

12. From what in no wise exists, it is impossible for anything to come into being; and for Being to perish completely is incapable of fulfillment and unthinkable [. . .].

13. Nor is there any part of the Whole that is empty or overfull.

14. No part of the Whole is empty; so whence could anything additional come?

By selecting four, distinct, imperishable substances that fill the whole of space leaving no room for a void, and explaining all changes in the world as a temporary combination or mixture of these eternal elements, Empedocles deftly preserves the predominant empirical features of the world, those of qualitative diversity, change, and process, while also conforming to Parmenides' logical strictures. For though the world undergoes continuous cyclical changes, the elements which enter into these changes are themselves unchanging: 'Nay, there are these things alone, and running through one another they become now this and now that and yet remain ever as they are.' From one point of view everything is undergoing change as in the system of Heraclitus, while from another there is no fundamental coming to be or destruction, no real change or becoming.

Accordingly, Empedocles has so far satisfied all but one of Parmenides' canons: as the four elements are ultimately real they do not come from what does not exist; as change consists of the mixing of these indestructible elements there is no absolute coming to be or ceasing to be; as the elements (along with the two additional forces to be discussed) completely fill the world no void is possible; and finally, as the qualitative variety of the world is due to the inherent diversity of the four distinct 'roots of things,' plurality is irreducible and ultimate.

> As he is at pains to point out, his elements do not, either in nature or behaviour, break any of the Parmenidean canons. He has already effectively restored, by a mere revision of pre-Parmenidean ideas, plurality and diversity; he is about to restore, by the introduction of his two motive forces, motion, change, and time. In fact the only Parmenidean tenet that he has sacrificed is that which Parmenides himself had valued most of all, his monism. That gone, the rest follows without further infringement. [3]

The only canon of Parmenides that has yet to be satisfied is the demand for an explanation of the cause of change and motion. Having accounted for what Aristotle will call the "material cause" of changes in the world with the four elements, Empedocles has still to explain the "efficient" or motive cause. For this explanation he falls back on two concepts familiar since Hesiod's *Theogony,* Love and Strife. While it is difficult to know exactly what meaning these terms could have had for him, like his predecessors, he seemed to see the changes in the world as dependent upon two contrasting forces, the force of attraction which draws things together, and the force of strife which pulls them apart.

Here again one sees reappearing in the system of Empedocles the primitive motif of political or totemic strife and retributive justice, in which forces contend to preserve their particular domains. Almost in the same words as those of Anaximander, he says that each element 'presides over its own office and each has its own character, and they prevail in the course of Time.' But while he borrows the older concepts of his predecessors, they are considerably refined in his system. Love, for example, is not merely the '*Eros*' or sexual attraction of the *Theogony*, but '*Philotes*,' a term which, in addition to denoting physical attraction, signifies the loftier sentiments of 'Affection,' 'Joy,' and 'Friendship:' 'She it is who is believed to be implanted in mortal limbs also; through her they think friendly thoughts and perform harmonious actions, calling her Joy and Aphrodite.' Thus the same power that courses through human limbs causing sexual attraction, joy, and affection flows through the universe drawing together the elements — further evidence of the strong mystical unity that pervades Empedocles' system, as well as the close analogy in his thinking between anthropology and cosmology.

But the meaning of 'Love' and 'Strife' is not exhausted in their political or affective connotations. Along with these, they also have physical properties, Hate being 'of equal weight in all directions,' and Love being 'their equal in length and breadth.' By 'equal' he means that Love and Hate are coextensive (as well as coexisting) with the four elements or roots of things, thereby instituting six ultimate components of his cosmology. There is no explanation as to why he assigns 'weight' to Strife and 'breadth' to Love, though one can conjecture that there is a kind of affective association of Hate with heaviness and Love with extensiveness. The reason for attributing these physical properties to essentially psychical forces is still the inability to conceive of a form of existence that is non-spatial or intangible.

In addition to these physical properties there is a strong suggestion of the primitive notions of *mana* or soul substance implied in the concepts of Love and Hate. When describing the cycle of cosmic changes brought about by the alternating influences of Love and Strife, Empedocles says that when Strife was on the wane it 'went on quietly streaming out' while 'a soft, immortal stream of blameless Love kept running in.' Here the analogy with male semen may be the predominant factor, but the conception of fluid forces also recalls Codrington's definition of *mana* as "a force altogether distinct from physical power, which acts in all kinds of ways for good and evil [. . .]."[4] This suggestion is reinforced by Aristotle's contention that Empedocles' forces of Love and Strife are *moral forces*, in that 'Love is the cause of good things and Strife of bad.'[5]

While one can conjecture, it is of course impossible to know the exact origin and meaning of these concepts in a mind as complex as that of Empedocles. Was the essential influence that of predecessors such as Hesiod, Anaximander, Heraclitus, etc., or the archaic model of political strife (so apparent in the conflicts among the Greeks and between the Greeks and the Persians), reconciliation, and apportionment, or the physiological and psychological experiences of love, joy, and hate, or the moist sperm of male sexual generation, or the mystical experiences of a rapturous cosmic union alternating with periods of estrangement and separation, or the attenuated soul-substance of *mana* of primitive thought? Or was it the case that, like the elements in his cosmic system, these were all intermingled in his thought?

THE COSMIC CYCLE

Given the four roots of things, fire, earth, air, and water, plus the two motive forces of Love and Strife, there is no beginning or ending in time, nor a coming to be or perishing of the elements in Empedocles' system, but an eternal cyclical process of combining and separating according to whether the effects of Love or Strife are predominant. The cyclical process itself manifests four phases, two polar stages when either Love or Strife is dominant, and two transitional stages during their respective waxing and waning.[6] When Love reigns, all the elements tend to be blended in a silent, unmoving, indistinct mixture since 'there is no strife nor unseemly conflict in its limbs.' It is at this stage that Empedocles describes the shape of the universe as a Sphere, although it apparently retains this spherical shape throughout its various transitions. Once again the paramount influence of Parmenides is seen in Empedocles' description of the Sphere as 'equal in all directions:' 'he (God) is equal in all directions to himself and altogether eternal, a rounded Sphere enjoying a circular solitude.' But the fact that the Sphere is referred to as a kind of personal being, a 'he' or 'god' with 'limbs,' also suggests the influence of Melissus.[7]

Once Love reigns, Hate begins to reassert its influence, motion again starts, and the process of separation recommences. As Simplicius says, quoting a fragment of Empedocles: "But when Strife began once more to prevail, then there is again motion in the Sphere; 'for all the god's limbs in turn began to quake.' "[8] As this occurs the silent harmony that existed under Love is shattered, the One dividing into Many with like elements segregating together. Moreover, he seems to have endowed the Sphere with the characteristic of a fluid vortex in which the alternating influences of Love and Strife 'stream' back and forth from the 'lowest depth of the whirl,' or center of the Sphere, to 'the circumference.' It is as if the elements were in a state of suspension within a spherical vortex through which the streams of Love and Strife flow back and forth causing the elements to unite into an indistinct mixture or to separate into four homogeneous groups. Again as described by Simplicius:

> When Strife had reached to the *lowest depth of the whirl,* and Love was in the *middle of the eddy,* under her do all these things come together so as to be one, not all at once, but congregating each from different directions at their will. And as they came together Strife began to move outwards to the circumference [. . .]. And in proportion as it was ever *running forth outwards,* so *a gentle immortal stream of blameless Love was ever coming in.* And straightway what before had attained to immortality became mortal, what had been unmixed before was now mixed, each exchanging its path [. . .].[9] (Italics added)

At this point the elements are described as 'congregating [. . .] at their will,' though at other times he implies that the process occurs by necessity: 'Strife waxed great in the limbs, and sprang to his prerogatives as the time was fulfilled which is fixed for them in turn by a broad oath.' There is also the suggestion of design, as when he draws an analogy between the blending of colors by temple painters and the harmonious mixture brought about by Love:

> [. . .] all those things which are more suitable for mixture are made like one another and united in affection by Aphrodite [. . .]. As when painters decorate temple offerings with colours [. . .] they take many-coloured pigments in their hands, and have mixed them in a harmony [. . .] [creating] from them forms like to all things, making trees and men and women and animals and birds and fish nurtured in water, and even long-lived gods [. . .].[10] (Brackets added)

This divergence has raised the question as to whether chance, necessity or design rules his system. Aristotle, who was the source of the previous fragment, apparently believed that Empedocles had not resolved the problem, sometimes speaking as if the 'segregating' of the elements was due to chance and sometimes to necessity:

> For though Strife was segregating the elements, it was not by Strife that either was borne upwards; on the contrary, he sometimes speaks as if it happened by mere chance — 'for so, at the time, it chanced to be running, though often otherwise' — while sometimes he says it is the nature of fire to be borne upwards.[11]

Freeman also believes that Empedocles "contradicts himself on this score," speaking sometimes as if the arrangement of the elements were due to chance, and sometimes as if it were the result of design due to the harmony brought about by Love.[12] De Santillana, on the other hand, finds that each has its appropriate domain, Love functioning with design, the elements under the influence of Strife moving with necessity, "but the fearful intrications of their interaction give the effect of chance."[13] In a system as complex as that of Empedocles, it is not surprising to find such unresolved problems.

Returning to the cyclical cosmological processes, between the two polar phases when either Love or Hate reigns there are transitional stages consisting of various combinations of the elements, and it is during these periods that the formation of the physical universe takes place, as well as the generation of organic and plant life. In his description of the origin of celestial phenomena the influence of Anaximander and Anaximenes replaces that of Parmenides. Using Anaximander's term 'separating off,' when the inflow of

Strife begins and the One starts to separate into the Many, the first element to 'separate off' is air, then fire, followed by the earth which 'came together' in the center of the Sphere. The air first forms a spherical layer at the circumference of the Sphere, its outer edge solidifying to form a 'crystalline' or 'frozen' periphery around the sphere. This would seem to be the origin of the notion of "crystalline spheres."

When fire 'separates off,' it moves upward displacing and enveloping the air under the 'solidified periphery.' Then, apparently, some of the fire mixes with the air and sinks to the lower half of the sphere thus forming two hemispheres around the earth, the upper hemisphere consisting of pure fire as the source of light, the lower hemisphere composed of a mixture of fire and air which he 'supposes to be night.' The imbalance due to the mixing sets the hemispheres in circular motion, the alternation of which causes day and night. [14]

In a further adaptation of Anaximander's view that the earth is kept in the center of the universe by equilibrium, Empedocles, according to Aristotle, reasoned that the earth is held in the center of the Sphere by a (centripetal) force caused by the rotation of the hemispheres just described. That he clearly had in mind centripetal force (without using the term 'centripetal') is apparent from Aristotle's statement, and is another striking example of Empedocles' frequent use of excellent analogies:

> [...] all those who try to generate the heavens to explain why the earth came together at the centre [...] then seek a reason for its staying there; and some say, in the manner explained, that the reason is its size and flatness, others, with Empedocles, that the motion of the heavens, moving about it at a higher speed, prevents movement of the earth, as the water in a cup, when the cup is given a circular motion [...] is for this same reason prevented from moving with the downward movement which is natural to it [...]. [15]

Here we have the beginning of that set of assumptions and concepts (e.g., a spherical universe with the earth in the center surrounded by circular crystalline orbs) which, owing to the influence of Aristotle's cosmology, was to become the accepted astronomical view during the later Middle Ages and early Renaissance, until the 17th century revolution wrought by Copernicus.

The pressure of the centripetal force causing the earth to remain in the center of the Sphere also explains the origin of water which is 'pressed out' of the earth: 'From the earth, as it was exceedingly constricted by the force of the rotation, sprang water.' The sea is referred to as 'the sweat of the earth,' with air coming from water by evaporation. The further details of his astronomical view are rather sketchy. The fixed stars are made of fire and attached to the solidified periphery (somewhat as Anaximenes described the fixed stars as like nail heads on the vault of the heavens) while the planets are unattached. Unlike the fixed stars, the sun and moon are not fiery in nature (the latter composed of air), the sun deriving its light from the reflection of the fire of the upper hemisphere on the earth itself or on the water of the earth. The moon in turn 'gets its light from the sun' and causes eclipses when it comes between the sun and the earth, casting 'a shadow over as much of the earth as was the breadth of the pale-faced moon.'

THE GENERATION OF LIVING ORGANISMS

Like the cosmic cycle, the generation of plants, animals, and 'mortal things' involves a 'double process,' 'the coming together of all things' under the prevalence of Love and the 'division' and 'separation' of things under the influence of Strife. As one would expect of a learned physician, Empedocles' account of the origin, reproduction, growth, and physiological functioning of living organisms is much more detailed than that of the preceding Presocratics. Rather than a complete description, however, an outline of his system will be presented, as well as some suggestion of the scope and complexity of his more important observations and theories.

Along with a 'double process' of generation and destruction coinciding with the two transitional periods of the cosmic cycles, this double process in turn is divided, according to Aëtius, into four stages of generation: (1) the emergence of grotesquely disjointed limbs, (2) the uniting of the disjointed limbs into monstrous creatures, (3) the creation of 'whole-nature forms,' and (4) the usual sexual generation of organisms. [16] There has been considerable discussion as to which stages of generation correspond to which phase in the cosmic cycle. Following the reconstruction of Kirk and Raven, the first two stages occur when Love is gaining ascendancy, the latter two when Strife is gaining upon Love, the fourth stage coinciding with the present state of the world.

The first stage, when the 'gentle immortal stream of blameless Love' begins to penetrate the Sphere, is bizarrely surrealistic in that various and sundry bodily parts wander about disconnected: 'Here sprang up many faces without necks, arms wandered without shoulders, unattached, and eyes strayed alone, in need of foreheads.' As the unifying influence of Love gains ascendancy during the next phase, these disconnected bodily organs are arbitrarily united in a nightmarish composition of organisms: 'Many creatures were born with faces and breasts on both sides, man-faced ox-progeny, while others sprang forth as ox-headed offspring of man, creatures compounded partly of male, partly of the nature of female, and fitted with shadowy parts.' In this state, Empedocles describes the survival of these forms in a way that suggests Darwin's evolutionary "survival of the fittest" (as Anaximander had anticipated the origin of animals and humans from aquatic-creatures). In the words of Aristotle:

> Wherever then all the parts came about just what they would have been if they *had come to be for an end,* such things survived, being organized spontaneously *in a fitting way;* whereas those which grew otherwise *perished* and continue to perish, as Empedocles says his 'man-faced ox-progeny' did. [18] (Italics added)

During the third stage when Strife reenters the uniform mixture of elements and the process of 'separating off' begins, the separation of fire causes 'whole-natured forms' or relatively undifferentiated but completely integrated organisms to 'spring from the earth.' In the fourth and final phase, when the further penetration of Strife causes increased differentiation, we have the present stage of the world where the first living things to appear were egg-bearing plants from which arose sexually differentiated organisms, followed by bisexual reproduction as a result of female attraction. As an indication of his deep interest in embryology, Empedocles maintained that the substance of the child is derived from both the mother and the father, the sex determined by whether the male seed falls into the warmer or colder part of the womb, the colder resulting in females and the warmer males. The features of the child are determined by whichever of the parent's seed is the hotter: if they both are equally hot the child resembles the father, if equally cold, the mother; and if the father's seed is hotter, a boy with his mother's features is born; if the mother's seed is the hotter, a girl would resemble her father. [19]

He goes on to explain sterility, the birth of twins and triplets, and of monsters. He observed that the fetus is enclosed in a 'sheepskin' (or amnion) and noted the differentiation of various organs during the embryological development, claiming that the articulation of the heart began on the 36th day. Analogous to the Pythagoreans, he explained the nature of bones, flesh, and blood in terms of the proportions of the elements combining to form each: blood and flesh contain about an 'equal proportion' of the elements while white bones contain four parts of fire and two of earth and water.[20] Although this is only a cursory list of some of his biological observations and theories, it suggests the wealth of his investigations and knowledge.

RESPIRATION AND PERCEPTION

Empedocles also presents a detailed description of respiration and perception, illustrating again his adroit use of analogies. As regards respiration, he noted that all living things inhale and exhale through 'tubes stretched under the bodies' surface' and terminating in the 'pores' of the skin. There is even a suggestion that these terminal ducts are different from arteries and veins, being 'bloodless channels' or capillary tubes, as we would now call them. The blood ebbs and flows, inhalation occurring when the blood recedes from the surface drawing the air in while exhalation occurs when the blood returns forcing the air out, as in the process of filling a porous vessel with water or wine. As described by Aristotle:

> So do all things inhale and exhale: there are bloodless channels in the flesh of them all, stretched over their bodies' surface, and at the mouths of these channels the outermost surface of skin is pierced right through with many a pore, so that the blood is kept in but an easy path is cut for the air to pass through. Then, when the fluid blood rushes away thence, the bubbling air rushes in with violent surge; and when the blood leaps up, the air is breathed out again, just as when a girl plays with a klepsydra [a vessel perforated at the bottom] of gleaming brass. When she puts the mouth of the pipe against her shapely hand and dips it into the fluid mass of shining water, no liquid enters the vessel, but the bulk of the air within, pressing upon the frequent perforations, holds it back until she uncovers the dense stream; but then, as the air yields, an equal bulk of water enters [. . .].[21] (Brackets added)

This reference to the *klepsydra* has generated a controversy as to whether it represents one of the first examples of scientific experimentation. Burnet stated that "the first recorded experiment of a modern type is that of Empedokles with the klepsydra,"[22] but Cornford offers his own experiment to refute this.

> Now this theory could have been tested by anyone who would sit in a bath up to his neck in water and observe whether any air bubbles passed through the water into, or out of, his chest as he breathed. Why did no one try this simple experiment before dogmatically stating that this is how we breathe? It seems certain that it was never tried, either to prove or to disprove the doctrine.[23]

As Cornford argues, no one familiar with the use of experiments today would claim that Empedocles was implying a test of his theory. From Aristotle's description, it is clear that Empedocles is reasoning by analogy, using the everyday employment of the *klepsydra* in drawing water, to illustrate the mechanism of the respiratory process which cannot be directly observed. This is an appropriate use of analogy, but in no sense does it suggest setting up an experimental situation to confirm or disconfirm the theory. The essential purpose of an experiment is to test a theory, while the intent of analogical reasoning is to use a familiar process to illustrate the mechanism of an unknown process based on similarities between the two. Cornford's reference to the bath clearly is experimental and quite different from Empedocles' example.

A further ingenious use of analogy occurs in Empedocles' explanation of perception.

Even at this early stage the Greeks were aware that while perception appears to be a *direct disclosure* of the external world, an intervening causal process must mediate this disclosure.[24] Though we just seem to see and hear objects at a distance from us, the actual experience of seeing and hearing is a result of a complex chain of processes from the object to the appropriate sense organ. Recognizing this fact, Empedocles asserts that all objects send out 'effluences,' though the primary effluences are those of the four elements. Perception occurs when the effluences from objects pass through the 'right sized' passage of the appropriate sense organ and meet the same element or mixture of elements in the body, or as in the case of sight, the elements come together either inside or outside the eye. As Theophrastus says, 'Empedocles has the same theory about all the senses, maintaining that perception arises when something fits into the passage of the senses.' Aristotle adds that 'with earth do we see earth, with water water, with air bright air, with fire consuming fire; with Love do we see Love, Strife with dread Strife.'

But while Theophrastus claims that Empedocles has the same general theory about all the senses, vision requires a somewhat more complex account since it is primarily by sight that we are aware of things outside of us in space. In this explanation he uses the analogy of a lantern to explain visual perception.

> As when a man, thinking to make an excursion through a stormy night, prepares a lantern, a flame of burning fire, fitting lantern-plates to keep out every sort of winds, and these plates disperse the breath of the blowing winds; but the light leaps out through them, in so far as it is finer, and shines across the threshold with unwearying beams: so at that time did the aboriginal Fire, confined in membranes and in fine tissues, hide itself in the round pupils; and these (*tissues*) were pierced throughout with marvellous passages. They kept out the deep reservoir of water surrounding the pupil, but let the Fire through (*from within*) outwards, since it was so much finer.[25]

So just as a lantern by means of the fire contained within it illuminates the surroundings, the eye sees by means of the fire enclosed behind the pupil (which accounts for the shiny property of the pupil). Because of their small size and the fact that fire is finer than water, the delicate passages in the pupil allow the fire in the eye to escape, while containing the water that surrounds the pupil. The fire of the pupil has a slight admixture of earth, enabling one to perceive material things, since like perceives like (in contrast to Heraclitus and Anaxagoras who claim that perception occurs between like and unlike elements). However, on the principle that like perceives like it is not clear how we perceive substances other than fire and earth. In any case, vision is a result of the fire from the eyes meeting the fiery effluences from outside. Whether this occurs within the eye, or outside as the analogy with the lantern would suggest, is not clarified. Aristotle claimed that "Empedocles speaks sometimes as if sight were due to the fire in the eye shining out, and sometimes as if emanations from objects entered the pores of the eyes."[26] Empedocles adds that 'One vision is produced by both eyes.'

Smell occurs with respiration when effluences are taken in as we breathe and thus is adversely affected by colds. The ear functions like a bell when the turbulence of the outer air strikes it like a bell clapper. All of these perceptual processes involve consciousness which has yet to be explained. As all existing phenomena are a result of the temporary combination of the four elements pervaded by the forces of Love and Strife, consciousness too is a function of the mixing of these six elements. While he says that 'all things [. . .] have intelligence and a portion of Thought' — a further example of the mystical unity pervading his system — in man thought and consciousness are located in the blood: 'for the blood round the heart is Thought in mankind.' (This is a view often attributed to Aristotle, although I have found no reference in Aristotle to support it.) The fact that the blood is composed of nearly equal portions of the four elements, plus Love and Hate, enables man to perceive and be conscious of the various combinations of these elements (or their effluences) in the outer world. Differences in intelligence depend upon the composition of the blood, but all things have some intelligence. Dreams are due to changes in the body: 'In so far as their natures changed (*during the day*), so does it befall men to think changed thoughts (*in their dreams*).'[27]

This physiological account of consciousness raises the question, mentioned earlier, whether the book *On Nature* is incompatible with the religious doctrine of the immortality of the soul, and Empedocles' belief in his own reincarnation, described in the *Purifications*. On the basis of a passage from Plutarch, Kirk and Raven conjecture that Empedocles held two concepts of the soul, the "conscious soul" that has its seat in the blood and perishes with the body, and the "life-soul" which survives the death of the body.[28] The striking similarities between *On Nature* and the *Purifications*, particularly the statement in the former that 'all things have a portion of Thought,' and the assertion in the latter that God 'is Mind, holy and ineffable, and only Mind, which darts through the whole universe with its swift thoughts,' would suggest that the two works are not entirely irreconcilable. As often happens in minds given to mystical insight, what appears on the empirical or logical level to be distinct and incompatible is united on the level of mystical unity.

CONCLUSION

In a system that exhibits as many different influences and facets as that of Empedocles, it is difficult to draw any hard and fast conclusion regarding the nature or level of his thought. On the one hand, his abstraction of the four elements, fire, earth, air, and water, as the eternal, immutable substances involved in all natural changes, represents a laudable attempt to satisfy the logical canons of Parmenides, and thereby reconcile monism and pluralism, permanence and change, appearance and reality. Moreover, this doctrine of the four elements is a forerunner of one of the most fruitful scientific theories ever developed, atomism.

On the other hand, his use of the Hesiodic concepts of Love and Strife to account for natural changes, anthropomorphic concepts retaining associations with the ancient notions of retributive justice, *mana,* and soul-substance, is a reversion to a more primitive level of explanation. In addition, there seems to be a strong element of diffuse or syncretic thought in his identification of the quasi-physical properties of extension and heaviness with the psychological forces of Love and Hate, all mingled with the pervasive presence of thought, mind, and god. And yet his experience as a physician — for the medical art requires one to observe accurately the symptoms and physiological changes in the patient — led him to be an astute observer of organic phenomena.

What we seem to find displayed by Empedocles, therefore, is a clear example of the thesis pointed out in Chapter II, that the thought of creative individuals often manifests markedly different levels, depending upon whether the phenomena or problems are empirical and close at hand, or more speculative and remote. As Piaget has found with respect to conceptual developments and formal inferences in the child, when individuals can closely observe and interact with objects their concepts and logical principles tend to be more objective and appropriate to the situation.[29] Thus we find Empedocles presenting more exact, detailed observations and more refined, sophisticated explanations of organic life than the previous Presocratics because they relate to his training and development as a physician. His embryological conjectures and physiological investigations in particular exhibit a highly developed critical awareness.

131

However, as regards more remote astronomical phenomena and cosmological speculations, he not only relies on the theories of his predecessors, especially the Milesians, but his thinking displays considerable anthropomorphic tendencies, regressing to the mythopoetic notions of Love and Strife found in Hesiod. Nonetheless, his conception of a spherical cosmos with the earth in the center surrounded by crystalline spheres is the first clear representation of the two-sphere, geocentric universe developed by Aristotle and adopted in the later Middle Ages. Also, his conjecture that the earth is held in the center of the cosmic sphere by vortical or centripetal motions generated by the revolving heavens presages the vortical theory of Descartes and was remarkable for that time.

In addition to these achievements, Empedocles' main contribution to the development of scientific rationalism lies in his astute use of empirical analogies to explain physiological processes. For the development of scientific knowledge depends not only on new observations, discoveries, and experimental investigations, it also depends upon explaining strange or unobservable phenomena in terms of more familiar structures and occurrences. The most striking example of this is the representation of the interior structure of the atom as analogous to the solar system. But much earlier Descartes relied on analogy when he described the functioning of the nervous network as similar to a hydraulic system, the nerves serving as conduits for the animal spirits and the heart performing as a furnace and a pump. Modern neurophysiologists do the same when they compare the functioning of the brain and nervous system to the complex processing activities of a sophisticated electronic computer.

Although both Anaximander and Anaximenes attempted to account for certain astronomical and meteorological phenomena in terms of terrestrial analogues, Empedocles' use of analogy is in many respects more deliberate and sophisticated, especially his reference to the *klepsydra* to depict the reciprocal flow and exchange of blood and air during the respiratory process and his use of the lantern to illustrate the processes of vision. Here we see the emergence of an empirical, systematic investigation of certain physiological phenomena that can serve as the basis or background for a developing medical tradition. Moreover, his comparison of the artistic creation of figures by the painter and the harmonious organization of the elements by Love is the first glimmer of the argument from design. When one combines these scientific accomplishments with his religious doctrines and mystical tendencies, one realizes that he confronts a personality and intellect extraordinary in their diversity and versatility.

NOTES

CHAPTER VIII

[1] Francis M. Cornford, *From Religion to Philosophy* (*op. cit.*, ch. IV), p. 150.

[2] Kathleen Freeman, *Ancilla To The Pre-Socratic Philosophers* (*op. cit.*, ch. IV), pp. 52-53.

[3] G. S. Kirk and J. E. Raven, *The Presocratic Philosophers* (*op. cit.*, ch. IV), p. 329.

[4] Cf. ch. III.

[5] Aristotle, *Met.*, 985a4. (Ross trans.)

[6] Cf. Kirk and Raven, *op. cit.*, p. 327.

[7] While Empedocles decries anthropomorphism in others (god has ' no feet, no swift knees, no organs of reproduction'), in both *On Nature* and the *Purifications* (the fragments are almost identical in both), his reference to Love and Hate working through the ' limbs' of the god or sphere suggests that he was unable to avoid anthropomorphism in his own thinking.

[8] Simplicius, *Phys.*, 1184, 2. Quoted from Kirk and Raven, *op. cit.*, p. 331, f.n. 430.

[9] *Ibid., De Caelo*, 529, 1 and *Phys.*, 32, 13. Quoted from Kirk and Raven, *op. cit.*, pp. 346-347, f.n. 464.

[10] Freeman, *op. cit.*, p. 55.

[11] Aristotle, *De Gen. et Corr.*, 334a1. Quoted from Kirk and Raven, *op. cit.*, p. 346, f.n. 462.

[12] Cf. Freeman, *The Pre-Socratic Philosophers* (*op. cit.*, ch. IV), p. 187.

[13] Giorgio de Santillana, *The Origins of Scientific Thought* (*op. cit.*, ch. IV), p. 116ff.

[14] Cf. the original description by Plutarch, in Kirk and Raven, *op. cit.*, p. 332, f.n. 434.

[15] Aristotle, *De Caelo*, 295a13. (Stocks trans.)

[16] Cf. Kirk and Raven, *op. cit.*, p. 000, f.n. 442.

[17] Cf. *Ibid.*

[18] Aristotle, *Phys.*, 198b29. (Hardie and Gaye trans.)

[19] Cf. Freeman, *The Pre-Socratic Philosophers*, *op. cit.*, p. 194.

[20] De Santillana gives the proportions as eight of earth, two of water, and four of fire, *op. cit.*, p. 120. This, however, does not seem consistent with the expression ' parts of,' implying that eight is the total number.

[21] Aristotle, *De Respiratione*, 473b9. Quoted from Kirk and Raven, *op. cit.*, pp. 341-342, f.n. 453.

[22] John Burnet, *Early Greek Philosophers* (*op. cit.*, ch. IV), p. 27.

[23] Cornford, *Principium Sapientiae* (*op. cit.*, ch. IV), p. 6.

[24] The Pythagorean physician Alcmaeon presented a detailed analysis of sense perception which was the basis for the accounts given by Empedocles and Democritus.

[25] Freeman, *Ancilla To The Pre-Socratic Philosophers*, *op. cit.*, pp. 60-61.

[26] Freeman, *The Pre-Socratic Philosophers*, *op. cit.*, p. 198.

[27] Freeman, *Ancilla To The Pre-Socratic Philosophers*, *op. cit.*, p. 63.

[28] Cf. Kirk and Raven, *op. cit.*, pp. 360-361.

[29] Cf. Bärbel Inhelder and Jean Piaget, *The Growth of Logical Thinking*, trans. by Anne Parsons and Stanley Milgram (New York: Basic Books, Inc., Pub., 1958).

CHAPTER IX

OF SEEDS AND THINGS

[. . .] the breach made by Anaxagoras in the fortress of mythology
[. . .] had few parallels in the long history of the ancient world.[1]

Sambursky

Like the Milesians by whom he was strongly influenced, Anaxagoras was from Ionia, born in the city of Clazomenae which was not too distant from Miletus and Ephesus. The date of his birth is given as 500 B.C., which would have made him older than Empedocles, although Aristotle states that he was 'later in his philosophical activity' (however, this is disputed by Kahn[2]). Perhaps as a result of Persian suppression of Clazomenae, he left for Athens at the age of twenty and remained there for the greater part of his life. He began his philosophical pursuits in Athens and was the teacher of such prominent figures as Pericles, Euripides, and Archelaus, who was himself a teacher of Socrates, who taught Plato, who was a teacher of Aristotle, who taught Theophrastus, etc. Thus Anaxagoras headed a long and distinguished philosophical dynasty, undoubtedly the most distinguished of all time. Toward the end of his life he was prosecuted, condemned, and exiled from Athens, whereupon he settled in Lampsacus where he died.

The accounts of the circumstance of the prosecution and the trial are varied, but the specific charge brought against him was impiety: the accusation that he maintained that the sun and the moon were material bodies rather than deities. When Socrates was brought to trial on a similar charge (only in his case an obvious pretense), he asked his accuser Miletus whether he was not confusing him with Anaxagoras who said that 'the sun is a stone, the moon earth,' for Socrates was not interested in such physical theories but in questions of morality and ethics. However, the situation of Anaxagoras does recall the unfortunate prosecution of Galileo. Anaxagoras was better received at Lampsacus, for when he was near death he was asked by the rulers how they might honor him, whereupon he requested that the children be given an annual holiday in the month of his death. This shows something of his character in that his bequest surely brought joy to more individuals than is normally the case.

Anaxagoras apparently wrote one rather short book as it is said that it sold for merely one drachma. While on first reading it may appear that Anaxagoras sets forth his doctrine very clearly, the more often and the more carefully one reads him the more difficult it becomes to find a single, consistent interpretation of his views. This had led Kirk and Raven to assert:

No Presocratic philosopher has given rise to more dispute, or been more variously interpreted, than has Anaxagoras [. . .]. It is actually very doubtful whether any critic, ancient or modern, has ever fully understood Anaxagoras, and there are some points on which certainty is now unattainable. [3]

One point, at least, is clear, that Anaxagoras, like his contemporary Empedocles, although rejecting the static monism of Parmenides, maintained his thesis that being cannot come from not-being, and therefore concluded that what exists must be ultimate and indestructible.

> The Greeks are wrong to recognize coming into being and perishing; for nothing comes into being nor perishes, but is rather compounded or dissolved from things that are. So they would be right to call coming into being composition and perishing dissolution [4]

Although this statement agrees with Empedocles' theory that coming to be and perishing should be understood as a process of composition and dissolution, Anaxagoras denied that the tremendous diversity of empirical phenomena and variety of changes could be accounted for in terms of Empedocles' four elements. Could substances such as blood, bones, hair, etc. be explained as compounds formed from such unlike elements as air, earth, fire, and water? As Anaxagoras asks, 'How could hair come from what is not hair or flesh from what is not flesh?' In claiming that combinations of the four elements can give rise to such different substances, was not Empedocles inadvertently violating Parmenides' canon that what exists cannot come from what does not exist?

To avoid these difficulties, Anaxagoras begins his book by stating that the original mixture from which everything arises must contain an infinite number of things, but so infinitely small that their individual properties and qualities were not discernible except for the prevalence of air and aither (fire).

> All things were together, infinite in respect of both number and smallness; for the small too was infintite. And while all things were together, none of them were plain because of their smallness; for air and aither covered all things, both of them being infinite; for these are the greatest ingredients in the mixture of all things, both in number and size. [5]

In asserting that the original mixture contained elements that were infinitely small, as well as infinite in number, Anaxagoras contradicts not only Parmenides' monism, but the first of Zeno's arguments in support of it. In his attack on the pluralistic units of the Pythagoreans, Zeno had advanced the *reductio ad absurdum* argument that if there were such units they would have had to be both infinitely small and infinitely great, which he concluded was self-contradictory and hence impossible.

The argument was based on the premise that the units must either have a magnitude or not have a magnitude: if they have no magnitude then the units must be infinitely small (since they arc without any size), but if they have a magnitude, then any part of the unit must have a magnitude such that the units must be infinitely large (since any unit can be divided into an infinite number of parts and hence contains an infinite number of components). Thus Zeno concluded: 'So, if there is a plurality, things must be both small and great; so small as to have no size at all, so great as to be infinite [. . .].'[6] He then added: 'If there is a plurality, the things that are are infinite [. . .].'[7] But though Zeno rejected this conclusion, Anaxagoras embraces it. So what Zeno assumed to be an impossible conclusion becomes the very cornerstone of Anaxagoras' philosophy; namely, that if there is a plurality then things must be infinite, *including the paradoxical conclusion* that things are both large and small:

> Neither is there a smallest part of what is small, but there is always a smaller (for it is impossible that what is should cease to be). Likewise there is always something larger than what is large. And it is equal in respect of number to what is small, each thing, in relation to itself, being both large and small. [8]

Thus Anaxagoras incorporates the conclusion of Zeno's *reductio* argument into his system: insofar as things are infinitely divisible the units must be infinitely small, but insofar as any entity contains an infinite number of components, it is infinitely large. Thus depending upon which aspect one emphasizes, its divisibility or its composition, both of which are characteristics of things, anything is 'in relation to itself [. . .] both large and small.' Such, it would seem, is the meaning of this paradoxical statement.

There is an additional problem, however, regarding the meaning of the statement that air and aither (fire) being infinite 'covered all things.' His explanation is that they are the 'greatest ingredients' in the mixture, but what can this mean? In a mixture containing an infinite number of things, and in which each thing itself is infinitely divisible, does it make any sense to talk about a 'greatest number of ingredients'? Can there be greater or lesser infinite quantities? Also, what is the nature of the air and the 'aither'? Are they, as the simplest interpretation would suggest, merely different aggregates of the same elements, air and 'aither'? Or are they collective names, as Aristotle implies, not for aggregates of the *same kinds* of things, 'homoeomerous' substances as Aristotle calls them (although Anaxagoras does not use this term), but for mixtures of things?

> His elements are the homoeomerous things, viz. flesh, bone and the like. Earth and fire are mixtures, composed of them and all the other seeds, each consisting of a collection of all the homoeomerous bodies, separately invisible [. . .]. [9]

But if earth and fire are not themselves simple elements, then what properties of the aggregation of the other elements give them their distinctive characteristics as earth and fire?

Having maintained that the original mixture contains an infinite number of infinitely small entities, Anaxagoras goes on to add diversity of quality and kind to the composition of the mixture, even though originally its components were so finely blended that it did not admit of color or of any distinctions:

> [. . .] before [. . .] things were separated off, while all things were together, there was not even any colour plain; for the mixture of all things prevented it, of the moist and the dry, the hot and the cold, the bright and the dark, and of much earth in the mixture and of seeds countless in number and in no respect like one another [. . .]. And since this is so, we must suppose that all things are in the whole. [10]

Thus the original mixture contains things infinite in variety as well as in number and size. But what significance is to be given to the distinction between the pairs of opposite qualities (moist-dry, hot-cold, bright-dark), the earth, and the countless number of different seeds? Do the qualities have an existence independent of the seeds and things, or are they qualities of them? Why is earth mentioned separately?

Equally difficult is the question of the composition of the seeds themselves. Are the seeds *'homoeomeries'* (Aristotle's term meaning 'things with like parts') or do they contain a 'mixture' or 'portion' of everything? In the only passage in which he uses the term 'seeds,' Anaxagoras says: '[. . .] we must suppose that there are many things of all sorts in everything that is being aggregated, seeds of all things with all sorts of shapes and colours and tastes [. . .].' But while this tells us that there are many kinds of seeds in *ag-*

gregated or composite groups, and that they have all sorts of shapes, colors, and tastes, it tells us nothing about the composition of the seeds themselves. He repeats innumerable times that 'all things have a portion of everything,' but he does so in reference to the things that exist *after* the process of separating off has begun, so that this in itself does not tell us whether the seeds in the original mixture *before* the separating off began are composite or simple. The neatest interpretation would be that they are simple or '*homoeomerous*,' but several passages suggest the contrary, as we shall see.

THE PROCESS OF SEPARATING OFF

Anaxagoras' description of the 'separating off' process that results in the existence of ordinary things seems to be a composite of Anaximander's and Empedocles' theories. The conception of 'separating off' itself of course is due to Anaximander (though used by Anaximenes and Empedocles), and just as the first opposites to separate off from Anaximander's Unbounded were air-mist and fire, so air and aither (fire) are the first to separate off from Anaxagoras' original mixture: 'For air and aither are being separated off from the surrounding mass, which is infinite in number.' It is as if Anaxagoras filled Anaximander's Unbounded (which, like his mixture, was unlimited, imperishable, and so blended as not to be characterized by any specific quality) with an infinite number of infinitely diversified elements from which things come to be by a process of 'separating off.'

One can also conjecture that the 'steering principle' that guided the separating off process in Anaximander's system has now become 'Mind' (literally '*Nous*') in that of Anaxagoras. For Mind is defined as 'self ruled,' having 'knowledge about everything and the greatest power,' as well as controlling all things 'that have life.' Unlike 'all other things that have a portion of everything,' Mind is unmixed, 'the same throughout,' being 'the finest of all things and the purest.' Here again one can discern an attempt to differentiate a form of existence beyond that of the corporeal and the spatial — yet insofar as Mind is only the finest and purest among existing 'things,' the distinction was still unrealized.

When Anaxagoras describes the *motion* of the mixture that results in the elements being separated off, he seems to turn from Anaximander to Empedocles, for his description of the motion as 'rotary' is similar to the 'whirl' of Empedocles' sphere. Mind apparently initiates the rotary motion and then exerts a controlling influence over it: 'Mind controlled also the whole rotation, so that it began to rotate in the beginning.' This is a very abstract rendering of the animistic notion that only psychic powers can be a source of motion. No explanation is given as to how the Mind began the rotation, introducing the problem, to which Aristotle gave the ingenious answer of 'unmoved movers,' as to how non-mechanical or psychic forces can give rise to mechanical motions (the Atomists will solve the problem by reducing all causes to mechanical ones).

The rotation begun in a small area spreads throughout the mixture, always being controlled and known by Mind which extends throughout the mixture without being mixed with it. Once the rotation starts it increases rapidly in speed thereby creating, in contrast to Empedocles' centripetal force, a centrifugal force which is the mechanical cause of further separating off: '[. . .] these things rotated thus and were separated off by the force and speed (of their rotation). And the speed creates the force. Their speed is like the speed of nothing that now exists among men, but it is altogether many times as fast.'

Empedocles and Anaxagoras thus were aware of the two contrasting forces (centripetal and centrifugal) generated by circular motion, but they still followed the older tradition in assuming that only some kind of psychic force was capable of *originating* motion. In fact, Anaxagoras' system contains the most explicit statement thus far of the most troubling of all philosophical dualisms, mind and matter. In the opinion of Theophrastus:

> [. . .] Anaxagoras would appear to make his material principles infinite, but the cause of motion and coming into being one only, namely Mind. But if we were to suppose that the mixture of all things was a single substance, indefinite both in form and in extent, then it follows that *he is really affirming two first principles only,* namely *the substance of the infinite and Mind.* [11] (Italics added)

Having described Mind and the nature of the rotation, Anaxagoras then goes on to give a fuller description of the process of 'separating off.' In addition to the pairs of opposite qualities already mentioned, he now includes the dense and the rare of Anaximenes:

> [. . .] Mind arranged [. . .] all, including this rotation [. . .]. And this rotation caused the separating off. And the dense is separated off from the rare, the hot from the cold, the bright from the dark and the dry from the moist. But there are many portions of many things, and nothing is altogether separated off nor divided one from the other except Mind. [12]

He then continues with a statement which could provide an answer to the question raised earlier as to whether the seeds and things in the original mixture were *homoeomerous* or contained portions of everything else. In the statement he makes two important points: (1) he distinguishes Mind from all other things in that Mind is 'all alike' *in the largest and the smallest portions,* while *nothing else* is, and (2) he maintains that each thing is and 'was' a mixture characterized by what it contains the most of. The use of the past tense is particularly significant because it explicitly implies that not only *after* the separating off process began, but *originally,* all things contain portions of everything.

> Mind is all alike, both the greater and the smaller quantities of it, while nothing else is like anything else, but each single body is and *was* most plainly those things of which it contains most. [13] (Italics added)

In addition, the phrase 'nothing else is like anything else' is similar to the one in the earlier second fragment in which he stated that seeds are 'in no respect like one another' and that 'none of the other things either are like to the other.' The fact that the phrase in the just preceding fragment is coupled with the assertion that things are determined by whatever they contain the most of is at least circumstantial evidence that he had the same state of affairs in mind in both cases. Moreover, the following quotation would add additional, though not unequivocal, support to this interpretation:

> And since the portions of the great and of the small are equal in number, so too all things would be in everything. Nor is it possible that they should exist apart, but all things have a portion of everything. Since it is not possible that there should be a smallest part, nothing can be put apart nor come-to-be all by itself, but *as things were originally,* so they must be now too, all together. In all things there are many ingredients, equal in number in the greater and in the smaller of the things that are being separated off. [14] (Italics added)

While at first glance it might seem that this fragment clearly settles the issue, it actually is less conclusive than the one analyzed previously. For on the one hand, while the phrase 'as things were originally, so they must be now too, all together' clearly refers

138

to the original mixture, it does not state that *each thing* was a mixture, but merely that originally things were 'all together.' On the other hand, while the last sentence clearly states that 'in all things there are many ingredients,' the reference to 'the things that are being separated off' indicates that he is not referring to things as they were in the original mixture, but to those which are being separated off. This appears to be true also of the phrase 'all things have a portion of everything,' since it seems to refer to things after the separating process began — or at least it is not unambiguous as to which state it refers. Nonetheless, taken all together, the evidence strongly indicates that both *before* and *after* the separating off process began everything contains 'a portion of everything except Mind.' If there are portions of everything in everything, things are distinguished and given names, as Aristotle says, according to which portions predominate:

> So they assert that everything has been mixed in everything, because they saw everything arising out of everything. But things, as they say, appear different from one another and receive different names according to the nature of the particles which are numerically predominant among the innumerable constituents of the mixture. For nothing, they say, is purely and entirely white or black or sweet, bone or flesh, but the nature of a thing is held to be that of which it contains the most. [15]

Although this interpretation that all things contain a portion of everything seems to be the one most consistent with what Anaxagoras says, the problem does not end there because, if this interpretation *is* correct, then Anaxagoras' position is actually untenable. For if everything is what it is because it contains a predominant amount of a certain kind of thing, but each of these things in turn are what they are because *they* contain a predominant kind of thing, and if this is true of *all things*, then the position is self-refuting, for there are no fundamental, irreducible, simple kinds of things which, in a sufficient portion, could give things their particular characteristics — or as Anaxagoras says, 'make them plain.' If he had held that the seeds or things in the original mixture were *homoeomerous*, then he would have avoided this *reductio ad absurdum* consequence; but, as we have tried to show above, his own statements indicate that he did not avoid this. One could contend that by introducing the pairs of opposite qualities he obviated the infinite regress, but since the qualities, too, seem to be merely part of the innumerable portion of things, this would not be a way out of the difficulty.

Most interpreters have refused to believe that Anaxagoras could have been guilty of this contradiction or ambivalence and therefore have supported one or the other of the alternatives. As Cornford maintains:

> Anaxagoras' theory of matter [. . .] rests on two propositions which seem flatly to contradict one another. One is the principle of Homoeomereity [. . .]. The other is: 'There is a portion of everything in everything' [. . .]. Unless Anaxagoras was extremely muddleheaded, he cannot have propounded a theory which simply *consists* of this contradiction. [16]

But this judgment assumes that Anaxagoras would have had as clear an understanding of this inherent problem as those who came after him and who had the benefit of successive critical examinations of his views. That a consistent interpretation of his system, either by ancient or by modern commentators, has been so illusive, suggests that at the time of its original formulation the implications might have been so complex, obscure, and deeply embedded in his thought as not to have been "simply" (or obviously) contradictory to him.

If the general orientation of this study is correct, then this is another example of diffuse thought, though on a very high level of abstraction, such that we should not be surprised that distinctions and implications that appear evident to us, and to his successors, were not as obvious to him. Unless this were true we should have to charge many of our greatest thinkers with being "muddleheaded." Even as prominent a logician as Willard Quine could miss a contradiction in his earliest logic text which was discovered by others.[17] In addition, there is the example of Einstein's relativity theory, the theoretical implications of which were not at all immediately obvious. For that reason, a controversy arose as to whether time dilation and space contraction were merely apparent, owing to the "perspective of velocity," or whether they were actual physical effects. Only after considerable correspondence and several international congresses was it finally determined that the effects were apparent in the special theory, but actual for the general theory. [18]

Along with the previous problems, there is a further question as to whether the opposite qualities that he has described as the first to separate off had a superior status to that of the seeds, and whether they constituted the basic or largest portion of things. It would seem, however, since he immediately adds, after describing the separating off of the opposite qualities, 'But there are many portions of many things,' that he is anxious to emphasize the fact that the seeds and things have an equal status with that of the qualities, in that they all comprise the original mixture.

But why, then, was it necessary to distinguish between the seeds, earth, and the opposite qualities? Having said that originally 'all things were together,' why did he not let it go at that? As indicated earlier, the answer seems to lie in the fact that he did not believe Empedocles had adequately accounted for the original diversity and changes in things in terms of the four elements, and thus had not really satisfied Parmenides' canons.

To insure himself against the vulnerability of Empedocles' system, therefore, he packed his own original mixture with all kinds, sizes, and categories of things. At least Aristotle saw this as his intention:

> The theory of Anaxagoras that the principles are infinite in multitude was probably due to his acceptance of the common opinion of the physicists that nothing comes into being from not-being. For this is the reason why they use the phrase 'all things were together' and the coming into being of such and such a kind of thing *is reduced to change of quality,* while some spoke *of combination and separation.* Moreover, the fact that the contraries proceed from each other led them to the conclusion. The one, they reasoned, must have already existed in the other; [. . .] namely that things come into being out of existent things, i.e., out of things already present, but imperceptible to our senses because of the smallness of their bulk. So they assert that everything has been mixed in everything, because they saw everything arising out of everything [. . .].[19] (Italics added)

Aristotle's interpretation provides a further reason for distinguishing between seeds and opposite qualities, namely, the difference between 'change of quality' and change due to 'combination and separation.' Modern commentators seem to have overlooked the importance of this distinction in Anaxagoras' thought. As Aristotle's summary implies, there is a fundamental difference between accounting for the formation of hair, blood, and bronze, and explaining the change of something from warm to cold or from bright to dark, etc. While the generation and alteration of *substances* might be explained as a combination and separation of elements, the changes of *quality* would seem to involve just that — a change of *quality* rather than of elements. In other words, his distinction between seeds and qualities seemed to imply a categorical distinction between substances and qualities, the kind of distinction that would be so important later to Plato and especially to Aristotle (which probably explains why Aristotle was aware of the implied distinction in Anaxagoras' system).

Furthermore, as Aristotle's summary also suggests, in positing pairs of opposite qualities and deriving one from the other (hot coming from the cold and the cold from the hot, etc.), he avoids the implication that the origination of the qualities is due either (1) to the combination of the elements (involving the problems discussed previously) or (2) to an absolute coming to be or ceasing to be of the qualities themselves. For if the hot separates from the cold and the cold from the hot (and so with the dense and the rare, bright and dark, dry and wet), then their occurrence or cessation is a result of separating from, or being dominated by, the opposite quality, not a coming to be from nothing or perishing into nothing.

If that provides a probable rationale as to why he distinguished between the opposite qualities and the seeds, it still leaves unanswered the question why he added seeds to the original mixture, as contrasted with things or elements. Again one can conjecture that the explanation lies in his awareness of the difference between three kinds of changes: (1) change due to the 'separating off' of the opposite qualities, (2) change due to the combination and separation of things, and (3) change due to growth. By endowing his mixture with seeds (which in the original Greek means 'sperm') he could more readily explain generation (as compared to mere change of composition): 'Anaxagoras [. . .] says the air contains the seeds of all things and that it is these seeds which, when carried down with the rain, give rise to plants [. . .].' Additionally, the notion of seeds is perhaps a more suitable means for explaining nourishment and growth than the concepts of 'thing' or 'quality.'

As the seeds contain portions of all things, when one drinks wine and eats meat one is taking in various portions of the elements which can supply the body with the substances it needs, meat contributing to the flesh, wine to the fluids and the blood, and elements of hair and nails restoring those that have been lost. As Aëtius states, 'we take in nourishment [. . .] such as bread and water, and by this are nourished hair, veins, arteries, flesh, sinews, bones and all the other parts of the body.' Aëtius goes on to add that though we do not perceive such elements by our senses, *we can apprehend by reason* that such food must contain 'parts' that nourish the body: 'For there is no need to refer the fact that bread and water produce all these things to sense-perception; rather, there are in bread and water parts which only reason can apprehend.'[20] This is the first explicit statement of what probably has been the most important principle in scientific procedure, the inference that unobservable entities are the probable constituents and causes of observable phenomena.

COSMOLOGY AND ASTRONOMY

Returning to Anaxagoras' description of the generation of the world, the rotation begun by Mind progressively spreads throughout the mixture increasing rapidly in speed, causing air and aither (fire) to separate off, then, as the rotary motion increases the denser elements are forced to the center while the lighter move outward: 'The dense and the moist and the cold and the dark came together here, where the earth now is, while the rare and the hot and the dry went outwards to the further part of the aither.' It thus appears that air is a collective name for one set of the contrasting paired qualities (the dense, moist, cold, and dark) and aither for the other (the rare, dry, hot, and bright). Air is the more important of the two as it is from the qualities separating off from the air that most of the common substances arise, while aither serves mainly to carry round the stars and the planets once they are formed.

The pressure of the rotary motion along with the consequent solidification cause further separation into distinct substances. The dense, moist, and cold having separated from the air and congregated at the center, the moist forms clouds, water condenses from the clouds, and earth solidifies from the water.[21] The water which remains in the 'hollow earth' feeds the rivers and gives rise to the sea. The earth is flat, suspended in the center of the rotation and supported by the air 'because of its size,' there being no void so that the air, 'which is very strong, keeps the earth afloat on it.' In denying the void, Anaxagoras has complied with another of Parmenides' canons.

From the earth stones are solidified by the cold and apparently 'move outwards' to form the planets and the stars which, as 'red-hot stones,' could not be formed from the qualities in the aither. Although no explanation is given as to how these celestial stones become hot, it is reasonable to assume that they acquire their heat from the aither. He does state that we do not feel the heat of the stars 'because they are far from the earth' and 'occupy a colder region' than the sun (why this should be so since they are both in the aither (fire) is not explained). Having maintained that the moon is a red-hot stone, he nevertheless follows Parmenides and Empedocles in holding that it 'has not any light of its own but derives it from the sun' which 'exceeds the Peloponnese in size.' Probably because he knew of the large meteorite which fell in Aegospotami in 476 B.C., he held that 'the moon was made of earth, and had plains and ravines on it.' It was not until Galileo's telescopic observations of the moon that the latter inference acquired additional empirical evidence, while the theory that the moon is a fragment of the earth has only recently been disconfirmed by the lunar landing from which it was learned that the composition of the moon's surface is unlike that of the earth. Though today it may seem commonplace to extrapolate from the appearance of a meteorite to the composition of the planets and the stars, at the time of Anaxagoras it was a remarkable feat in the face of the persistent tendency to characterize the planets and the stars as "celestial," "heavenly," or "divine" bodies. As Sambursky justly remarks:

> Anaxagoras' astronomical hypotheses are throughout dominated by a "terrestrial" approach which makes no distinction between phenomena "there" in the sky and those "here" on the earth, and gives a purely physical evaluation of astronomical data and their possible causes. The heavenly bodies are nothing more than flaming stones; "the sun is larger than the Peloponnesus" — what uninhibited freedom of thought is revealed by this comparison of the mightiest of the celestial bodies, apotheosized by the deep-rooted irrational beliefs of mythology, with a geographical object, a part of the inhabited earth! [22]

Such is the superstitious folly and ignorance of mankind that Anaxagoras was rewarded for his brilliant suppositions by being prosecuted and exiled from Athens. But even as late as the 16th century, most astronomers refused to admit that the bright apparition or "nova" that was observed in 1572 in the constellation Cassiopeia, and then disappeared after shining for eighteen months, was actually a new star. For if it were a new celestial body, then this would challenge the time-honored conception of the immutability of the heavens and with it the qualitative distinction between the celestial and the terrestrial realms. Old superstitions die hard as the continued belief in astrology indicates.

Anaxagoras also held correct views regarding the relative positions of the sun and the moon to the earth ('the moon is nearer to us'), the phases of the moon (caused by the sun from which it derives its light), and of eclipses: 'Eclipses of the moon are due to its being screened by the earth, or, sometimes, by the bodies beneath the moon; those of the sun to screening by the moon when it is new [. . .].' He held that the stars are carried from east to west by the aither, moving under the earth, and that 'Beneath the stars are certain bodies, invisible to us, which are carried round with the sun and moon' — but whether he meant by these the inferior planets is unknown.

The Milky Way is due to the effect of the light of the stars, normally hidden by the sun's rays, shining through the shadow cast on the sky by the earth when the sun passes below it. "Comets are a concatenation of planets" and "shooting stars are broken off from the aether like sparks, and are immediately quenched."[23] Like Anaximander, he apparently speculated as to whether there are innumerable worlds, although it is a matter of conjecture again as to whether he meant successive worlds or various contemporary worlds. In any case, his astronomical theories were devoid of all mythical and anthropomorphic representation. As Kirk and Raven state: "Clearly Anaxagoras' astronomy is much more rational than most of his predecessors' [. . .]."[24] Indeed, so farsighted were his speculations that he would have looked upon the recent explorations of the moon by the astronauts as merely confirming most of his own theories.

He followed the Milesians in his interest in meteorological phenomena, explaining clouds, snow, and hail in a way similar to Anaximenes. Hail is caused when moisture is squeezed out of dense clouds, the drops becoming frozen and rounded in their descent. Lightning and thunder have their origin in the aither, lightning being a flash of fire bursting through the clouds, the sound of thunder due to its being quenched. Rainbows are a result of the sun's rays reflecting from a thick cloud. [25]

There are a few fragments to indicate that he had an interest in organic life, but nothing like the wealth of investigations of Empedocles. As mentioned earlier, he believed that 'the air contains seeds of all things and that it is these seeds which, when carried down with the rain, give rise to plants [. . .].' Again he followed Anaximander in maintaining that 'Animals originally arose in the moisture, but later from one another.' The white of the egg he identified as the nourishment for the embryo, calling it 'bird's milk.'

All living things, he concluded with Empedocles, have a portion of mind. The 'soul' he equated with life, saying that it "causes motion in the organism just as Mind did in the Whole."[26] He antedated modern thinkers in recognizing the significance of man's use of his hands in the development of intelligence: 'Anaxagoras says that it is his possession of hands that makes man the wisest of living things.' Contrary to Empedocles, he asserted that we perceive unlike by unlike or by contrasts:

> Anaxagoras thinks that perception is by opposites, for like is not affected by like [. . .]. A thing that is as warm or as cold as we are does not either warm us or cool us by its approach, nor can we recognize sweetness or bitterness by their like; rather we know cold by warm, fresh by salt and sweet by bitter in proportion to our deficiency in each. [27]

CONCLUSION

Slightly more than a hundred years separate Anaxagoras from Anaximander, a century that marked tremendous progress in the analysis and clarification of the fundamental problems posed by the Milesians. Yet the original questions, as well as the basic schema underlying the various cosmological systems, remain essentially unchanged. Although the theory of a primal mixture of infinitely divisible and diversified elements reflects a considerable advance over the mythical conception of Chaos or the vague notion of the Boundless, it still represents an answer to the original inquiry of Hesiod: what was the first state of things out of which everything now existing originated? Moreover, the imprint of the traditional mode of explanation in terms of a conflict of opposites is faintly discernible in Anaxagoras' conception of contrasting qualities separating off from one another, even though the rationale is now determined by the more rigorous canons of Parmenidean logic (i.e., that since what exists cannot come from Not-Being, the different qualities must originate in their opposites). Nevertheless, the subsequent explanations of the origin and transformation of phenomena have undergone considerable refinement, particularly because of the Parmenidean attack on the general notions of plurality and change. As a result, we find Anaxagoras beginning to distinguish different kinds of changes as well as various types of entities or states, such as qualities, elements, compounds, seeds, etc.

Throughout the century, Anaximander's concept of 'separating off' remained the typical and favorite mode of explaining the formation of the empirical world from the original state of things, but this concept also was made more precise owing to the identification by Empedocles and Anaxagoras of the two basic forces, centripetal and centrifugal, generated by circular motion. And though Anaxagoras still followed the ancient tradition in evoking a psychical force (*Nous* or Mind) as the initiator and controller of the mechanical rotation causing the separating off action, the concept of 'Mind' was shorn of much of the primitive, anthropomorphic connotations associated with the archaic concepts of Love and Strife. In addition, while still resembling the 'Steering principle' of Anaximander, 'Mind' now has acquired more of the sense of a regulative force than of retributive justice. In fact, it could be argued that though still animistic, Heraclitus' concept of the Logos and Anaxagoras' notion of *Nous* were forerunners of the later idea that natural law "governs" physical processes. Furthermore, in separating Mind from the other elements as 'purer' and 'finer,' Anaxagoras came within a razor's edge of recognizing a reality beyond that of the corporeal and the spatial, thereby anticipating the dualism of mind and matter.

But it is in his astronomical theories that one finds the most striking examples of scientific rationalism in his thought. His contention that the sun and the stars were made of stones derived from the earth, and that the moon with its plains and ravines resembled the earth, is extraordinary for the time. While considerable differences in knowledge, technology, and scientific method separate Anaxagoras from contemporary scientists, the nature of his thinking remains fundamentally the same. Also, he clearly recognized the need to supplement ordinary observations with inferences to unobservable entities to explain natural phenomena. Thus he attempted to preserve the qualitiative diversity and natural changes in the world, while satisfying the logical canons of Parmenides, by positing an infinite number of unobservable elements as the essential reality behind and cause of natural processes. However, though his attempt to solve the problems of plurality and change by positing *ab initio* an endless variety of things was ingenious, the infinite regress implied in the infinite divisibility and composition of these elements was incompatible with his doctrine that things are characterized by that of which they contain the most and his explanation of the generation of things.

Given the premise that everything is infinitely divisible and composed of an infinite number of things, there is no way of avoiding the *reductio* argument. To prevent this consequence it would be necessary to posit, as the Atomists will, an indivisible element as the irreducible ingredient in all change. Yet, insofar as all subsequent attempts to discover the basic, irreducible elements of the world have proven futile, as recent investigations in quantum mechanics affirm, the theory of Anaxagoras, in spite of its apparent self-contradiction, seems to have been a correct forecast of the present state in sub-atomic physics. Borrowing his terminology, one could say that all attempts at discovering the fundamental particles of physical reality have resulted in further particles being 'separated off' — however extensively and deeply one probes, it seems that there are 'infinite dimensions' of reality lying beyond.

CHAPTER IX

[1]   S. Sambursky, *The Physical World of the Greeks* (*op. cit.*, ch. IV), p. 40.

[2]   Cf. Charles Kahn, *Anaximander And The Origins Of Greek Cosmology* (*op. cit.*, ch. IV), pp. 163-165. According to Kahn, Aristotle's remark means "Anaxagoras comes before Empedocles in time, but after him in his philosophical achievements," p. 164.

[3]   G. S. Kirk and J. E. Raven, *The Presocratic Philosophers* (*op. cit.*, ch. IV), p. 367.

[4]   Simplicius, *Phys.*, 163, 20. Quoted from Kirk and Raven, *op. cit.*, p. 369, f.n. 497. Unless otherwise indicated, the fragments quoted in this chapter are from this latter book.

[5]   Simplicius, *Phys.*, 155, 26. *Ibid.*, p. 268, f.n. 495.

[6]   Simplicius, *Phys.*, 141, 1. *Ibid.*, p. 288, f.n. 365.

[7]   *Ibid.*

[8]   Simplicius, *Phys.*, 164, 17. *Ibid.*, p. 370, f.n. 499.

[9]   Aristotle, *De Caelo*, 302a31-302b4. (Stocks trans.)

[10]   Simplicius, *Phys.*, 34, 21. *Ibid.*, p. 368, f.n. 496.

[11]   Theophrastus, *Phys., Op.* fr. 4, *ap.* Simplicius, *Phys.*, 27, 17. *Ibid.*, p. 375, f.n. 507.

[12]   Simplicius, *Phys.*, 164, 24 and 156, 13. *Ibid.*, pp. 372-373, f.n. 503.

[13]   *Ibid.*

[14]   Simplicius, *Phys.*, 164, 26. *Ibid.*, pp. 375-376, f.n. 508.

[15]   Aristotle, *Phys.*, 187b2-6. (Hardie and Gaye trans.) This statement of Aristotle contradicts another already quoted (f.n. 9) in which he asserted that flesh and bone were *homoeomerous* substances: 'His elements are the homoeomerous things, viz. flesh, bone and the like [. . .].'

[16]   F. M. Cornford, 'Anaxagoras' theory of matter,' in *Classical Quarterly*, 24(1930), p. 14ff. and p. 83ff. Also, see the discussion by Kathleen Freeman, *The Pre-Socratic Philosophers* (*op. cit.*, ch. IV), p. 266. She maintains that if everything contains portions of everything, then "This would reduce Anaxagoras' metaphysics to absurdity, for he would merely have postulated in the sphere of the elements the same diversity as is found in our world, without explaining either."

[17]   Cf. Willard Van Quine, *Mathematical Logic*, revised edition (Cambridge: Harvard Univ. Press, 1951), p. ix.

[18]   Cf. Milič Čapek, *The Philosophical Impact of Contemporary Physics* (*op. cit.*, Introduction), pp. 191-205.

[19]   Aristotle, *Phys.*, 187a28-187b2. (Hardie and Gaye trans.)

[20]   Aëtius, 1, 3, 5. Quoted from Kirk and Raven, *op. cit.*, p. 385, f.n. 523.

[21]   Cf. Simplicius, *Phys.*, 179, 8 and 155, 21. *Ibid.*, p. 382, f.n. 517.

[22]   Sambursky, *op. cit.*, pp. 39-40.

[23]   Freeman, *op. cit.*, p. 270.

[24]   Kirk and Raven, *op. cit.*, p. 392.

[25]   These meteorological descriptions are from Freeman, *op. cit.*, p. 271.

[26]   *Ibid.*, p. 274.

[27]   Theophrastus, *De Sensu*, 27ff. Quoted from Kirk and Raven, *op. cit.*, p. 394, f.n. 538.

CHAPTER X

ANCIENT ATOMISM

The fertility of the Greek atomic philosophy proves the power of speculative reason. [1]

Whyte

In another ironic twist of history, two of the most brilliant physical theorists of all time, Leucippus and Democritus, remain relatively unkown. While the works of Plato and Aristotle are regarded as the major fonts of Western intellectual culture, very few people outside the halls of academe have even heard of Leucippus and Democritus. And yet, measured by the contemporaneity of their scientific theories, they are in most respects superior to either Plato or Aristotle.

In part, the eclipse of the founders of Atomism is due to the calamitous loss of their writings — one cannot read "the works" of Democritus as he can those of Plato and Aristotle. A man of vast interests and encyclopaedic learning, Democritus was one of the most prolific of all ancient authors having written as many as fifty-two separate works on such diverse subjects as art, astronomy, ethics, mathematics, medicine, music, politics, psychology, etc., as well as treatises on the lore of Egypt and Babylon. Of these extensive writings, the little that survives consists mainly of ethical aphorisms. As for Leucippus, one fragment is preserved. Seldom has history been more unjust.

But the cause of their eclipse lies much deeper, raising profound questions regarding the intellectual versus the symbolic needs of man, the conflict between wish-fulfilling fantasy and the reality principle, or the capacity of mankind to face up to a truly naturalistic, albeit alien, account of the world and its deep-rooted need to feel "at home" in the universe. For though Atomism in Greece flourished for a period of about 150 years, from 430 to 280 B.C., owing partially to its adoption by Epicurus, and later became widely known in Rome because of its poetic expression by Lucretius, it was, as an outstanding scholar on Greek Atomism states, "even so regarded rather as the eccentric adjunct of a popular theory of morals than as a scientific explanation of the world." [2] Because of the incompatibility between the materialism of Atomism and the supernaturalism of Christianity, Atomism was totally abandoned during the Middle Ages, then revived again by Gassendi, Boyle, and Newton in the 17th century as "the metaphysical foundation of modern science." [3] As Bailey exclaims:

> To the modern mind few facts in the history of human thought are probably more astonishing than that the cardinal theories, on which two great branches of science, astronomy and chemistry, are now respectively built, should have been propounded by Greek philosophers only to be abandoned and lie dormant for centuries [. . .]. [4]

Perhaps even more disconcerting is the disparaging, if not contemptuous, manner in which some contemporary physicists refer to the necessarily cruder concepts of the early Atomists. And yet, considering the unhistorical orientation of the scientific discipline, it is not surprising that scientists, comparing the solid, indivisible, 'hooked' atoms of Leucippus and Democritus which became 'entangled' in compounds, with the more sophisticated Rutherford-Bohr atomic model of modern physics (not to mention the more recent conceptions of the atom in quantum mechanics), find the earlier concepts exceedingly primitive. Still, one should remember that the notion of the atom as an indivisible solid particle was held by Newton ("[. . .] God in the Beginning form'd Matter in solid, massy, hard, impenetrable, moveable Particles, of such Sizes and Figures [. . .] as most conduced to the End for which he formed them [. . .]."[5]) and remained the prevalent conception until the end of the 19th century. It was this image of the atom that the Viennese physicist, Ernst Mach, at the turn of the century, disparagingly referred to as "a convenient fiction."

Furthermore, though ancient Atomism was formulated independently of any experimentation, the basis of most theory construction in science today, one ought to recognize that progress in science also depends upon creative imagination, the source of new concepts and theories. The reflective abstraction of the atomic framework from the concrete world of sensory experience, a framework which, in terms of its essential schema (the reduction of the complex qualitative entities and processes of the world to the motion of irreducible particles interacting in the void), was fundamentally the same as that of modern physics and chemistry for over 200 years, is one of the major theoretical scientific achievements of mankind. As Whyte has stated, "Atomism has proved the power of the intellectual imagination to identify aspects of an objective truth deeply rooted in the nature of things."[6]

As brilliant as are the overall philosophical reflections of Plato and Aristotle, judged from the contemporary point of view their physical and cosmological theories are vastly inferior to those of the Atomists. There is no doubt that Platonism and Aristotelianism played a much more important role throughout most of Western intellectual history, especially before the 17th century, partially accounting for their predominant position; but that a theoretical framework in the long run proves truer and more fruitful should be at least as important in evaluating its significance as its historical influence. In terms of these criteria, there is little question regarding the superiority of Atomism over Platonism and Aristotelianism. As Čapek says:

> Certainly the early Greek atomists came as close as was possible to the modern kinetic and mechanistic conception of nature; and the laws of inertia as well as the law of the conservation of energy were, if not explicitly anticipated, then at least foreshadowed by them.[7]

Had Atomism rather than the qualitative, geocentric, organismic cosmology of Aristotle been the accepted theory at the end of the Middle Ages and the beginning of the Renaissance, the emergence of modern science would have occurred earlier with less traumatic consequences intellectually and culturally. For much of the history of science from Copernicus to Newton is the story of the struggle to break away from the presuppositions and tenets of Aristotelianism that had become enshrined as dogma through the influence of the Scholastics.

LEUCIPPUS

Leucippus, as Bailey states, "remains wrapped in considerable obscurity" for practically nothing is known about his personal history and very little about his writings. It is generally agreed, at least, that he originated Atomism, which he expounded in a book entitled the *Great World-system*, and that Democritus took over the theory, developing its implications and working it out in greater detail. According to Cicero: 'Leucippus postulated atoms and void, and in this respect Democritus resembled him, though in other respects he was more productive.' It is not certain as to where or when Leucippus was born or when Democritus became associated with him. Simplicius says that 'Leucippus of Elea or Miletus (both accounts are current) had associated with Parmenides in philosophy;' however, since his writings reveal somewhat more of the doctrine of Melissus of Ionia than of Parmenides, as well as the strong influence of the Milesians, it is now considered more likely that he was from Ionia, but that he may have visited Elea.

In any case, Leucippus, like Empedocles and Anaxagoras, attempted to synthesize the *a posteriori* scientific speculations of the Milesians and the *a priori* logical rationalism of the Eleatics. As Greek Atomism was based less on inductive inference than on deduction from Eleatic principles, it was essentially *a priori* in origin, although offered as an explanation of empirical phenomena. In that respect, however, it is not basically different from theories in modern physics which are derived from, or modifications of, previous explanations — the difference lying in the lack of experimentation, quantification, and prediction and verification in Greek Atomism. But the speculations of the Atomists, like those of Plato and Aristotle, were concerned to give a true account of reality, not to provide useful empirical knowledge. Apart from the medical tradition and a few biological inquiries of Aristotle, the scientific investigations of the Hellenic period were wholly unexperimental and unpragmatic. The aim was to acquire truth, insofar as this is ever possible, not to manipulate or control the physical world.

DEMOCRITUS

In contrast to Leucippus, Bailey asserts that "Democritus is a far less shadowy figure." While Diogenes Laertius states that he was 'a citizen of Abdera or, as some say, Miletus,' he usually is known simply as 'Democritus of Abdera.' Abdera was close to Stagera, the birthplace of Aristotle, both being in Thrace. Born in a wealthy family, when his father died Democritus asked for his patrimony in money so that he could be free to travel, whereupon he is reputed to have gone to Egypt to 'visit the priests and learn geometry,' as well as to Persia, the Red Sea, and perhaps even India and Ethiopia. As was true also of Thales, these extensive travels provided him with the broad background evident in his prolific writings and encyclopaedic knowledge which earned for him the nickname of "Wisdom" among his contemporaries.

The date of his birth can be reliably established from his own statement, in the *Little World-system* (after the title of Leucippus), that he was forty years younger than Anaxagoras, which would make it approximately 460 B.C., in which case he would have been ten years younger than Socrates and (according to Bailey) Leucippus. During his travels he visited Athens and met Socrates, but whether out of modesty (as Demetrius of Magnesia suggests) or reticence, he did not make himself known to him. According to his own statement, 'I came to Athens, but no one knew me.' He seems to have been in many ways a rather strange personality although generally cheerful in disposition, thus acquiring the famous title of 'the laughing philosopher' owing to "his good-natured amusement at 'the vain efforts' of men."[8] Although critical of his system, Aristotle says of Democritus that he alone of his predecessors made a thorough examination of the problems of coming-to-be and passing away.

A similar criticism [to that applied to Plato] applies to all our predecessors with the single exception of Democritus. Not one of them penetrated below the surface or made a thorough examination of a single one of the problems. Democritus, however, does seem not only to have thought carefully about all the problems, but also to be distinguished from the outset by his method.[9] (Brackets added)

Coming from Aristotle, that is high praise indeed.

ATOMS AND THE VOID

The cardinal factor in understanding Greek Atomism (as well as the new science of the 17th and 18th centuries) is that it developed almost exclusively to solve the problem of motion. While Aristotle's system reflects his main interest in growth and qualitative changes, ancient Atomism, like the investigations of Copernicus, Kepler, Galileo, and Newton, was concerned to describe and explain various kinds of motions, reducing qualitative change itself to a form of motion. From this standpoint, the basic concepts of atomism, whether they be those of the 5th century B.C. or of the 17th century A.D., follow deductively from certain theoretical considerations regarding the motions of solid particles in empty space. Though lacking the *explicit* formulation of such later concepts as force, momentum, mass, and inertia, as well as the formulation of general laws, Greek Atomism did accomplish the necessary first step of *abstracting from experience and clearly formulating the basic concepts needed to conceive and describe the motions of unobservable particles.* As such, it provided the initial theoretical framework in terms of which more sophisticated conceptual refinements and precise investigations could be carried out later.

Like Empedocles and Anaxagoras, Leucippus and Democritus rejected the static monism of the Eleatics and thus were faced with the same theoretical problem of reconciling change and motion with the Parmenidean canons. Specifically, they had to show that the coming to be and ceasing to be of phenomena did not involve an origination from nothing or a perishing — the one canon of Parmenides which no one dared reject. Apparently realizing that without some irreducible elements Anaxagoras' solution that 'everything contains an infinite portion of everything' leads to a *reductio ad absurdum,* Leucippus posited an infinite number of *indivisible* elements, *homogeneous* as regards quality though infinitely variable as regards shape and arrangement. Presumably, the infinite differences in the shapes of the atoms, as well as the possibility of limitless variations in their spatial positions and arrangements, could account for the ceaseless changes and motions in the world, thereby avoiding the self-refuting concepts of infinite divisibility and portions. Leucippus thus satisfied Parmenides' main canon by positing, as he had, an indivisible *plenum* as the ultimate Being, but whereas Parmenides' Being was a continuous, immobile, homogeneous Unity, that of Leucippus consisted of an infinite number of discrete, eternally moving, homogeneous atoms. As described by Simplicius:

> They (sc. Leucippus, Democritus, Epicurus) said that the first principles were infinite in number, and they thought they were indivisible atoms and impassible owing to their compactness, and without void in them; divisibility comes about because of the void in compound bodies [. . .].[10]

For Leucippus it was the smallness of the atoms that prevented their being divisible, while Democritus based their permanence on their hardness.

Of the Eleatics, it seems to have been Melissus who exerted the strongest influence on the Atomists, for it was he who had declared that 'if things are Many, they must be such as the One is,' namely, homogeneous. It was also Melissus who replaced Parmenides' limited Sphere with an infinite extension, stating 'as it is always, so also it's size must be infinite.' The Atomists similarly posited an eternal, infinitely extended universe, along with the infinite number of indestructible atoms.

Having satisfied the Eleatic criterion as to the existence of Being, the Atomists then take the further giant step of affirming the existence of a form of not-Being, the void. Neither Empedocles nor Anaxagoras had dared admit reality to the void, maintaining, as Parmenides and Melissus had, that all Being must be continuous. But Leucippus recognized that the concepts of discreteness (separateness) and of motion (displacement from one position to another) presuppose a *spatial separation* of the atoms, as well as a space through which they can move and to which they can change their position. As Aristotle reasons:

> The void, they argue, 'is not': but unless there is a void with a separate being of its own, 'what is' cannot be moved — nor again can it be 'many', since there is nothing to keep things apart [. . .]. The "many" move in the void (for there is a void): and by coming together they produce "coming-to-be", while by separating they produce "passing-away." [11]

With his usual, remarkably clear grasp of the problems, Aristotle presents the rationale for the final admission of a form of not-Being, the void. While the system of both the Pythagoreans and the Eleatics had implied the reality of pure spatial existence, neither school was capable of attributing independent ontological status to that which was non corporeal, leaving Leucippus to conceive of a new form of existence. As Bailey asserts:

> In effect, Leucippus had introduced a new conception of reality: in the old sense empty space is not real, for it has not the most elementary attribute of matter, it cannot touch or be touched. But it none the less exists: we must form a new idea of existence, something non-corporeal, whose sole function is to be where the fuller reality is not, an existence in which the full reality, matter, can move and have its being. [12]

However, Leucippus and Democritus were very careful to differentiate between the *existence* of the atoms and the *form of non-existence* of the void. Leucippus did this by distinguishing what is real from what exists; both atoms and the void exist, but only atoms are truly real: "though empty space is not in a sense a real thing, it none the less exists: 'the real exists not a whit more than the not real, empty space no less than body.' " [13] On his part, Democritus attempted to distinguish between the two with a play on words: "*Mēdén*, Naught, exists no less than *dén*, Aught." [14] Or, according to another version:

> Democritus [. . .] calls space by these names — 'the void', 'nothing', and 'the infinite', while each individual atom he calls 'hing' [i.e. 'nothing' *without* 'not'], the 'compact' and 'being'. [15]

While the atoms are *real* being, the void is *un*real, but not non-existent.

One must beware of reading into this conception of the void the same explicit sense of an empty coordinate system ontologically independent of, and logically prior to, matter, as in Newtonian science. According to Bailey, while Democritus may have held the notion of space as "the whole extent of the universe some parts of which were occupied by matter," Leucippus tended to think of it merely "as the 'empty' parts, the intervals between body." [16] As one would expect of its first formulation, rather than a self-contained entity, space tended to be thought of as merely the sum total of all those domains which at any moment are not occupied by matter: "the atomists had no conception of bodies *occupying space*, and for them the void only exists where atoms are not, that is, it forms gaps between them."[17] No attempt is made to explain the origin of the atoms or the void; as in Melissus, the infinitude in dimension seemed to be associated with infinitude in time.

As regards the infinite shapes (and sizes) of the atoms, Leucippus appears to resort to the negative application of the principle of sufficient reason: 'Leucippus [...] held that the number of their shapes was infinite, on the ground that there was no reason why any atom should be of one shape rather than another [...].'[18] Furthermore, unless the size of the atoms also varied infinitely, as Bailey noted, it would be difficult to understand how a finite limitation on size would allow for an infinite variation in shapes. While this problem may not have been obvious to Leucippus, it seems to have been evident to Democritus who posited large atoms as well as small.[19] In any case, the two main features of the atomic view have been defined: (1) an infinite number of indivisible atoms infinitely variable in shape (and probably size) and (2) the admission, finally, of another form of existence, that of space or the void. In terms of this latter concept alone, which represented a major breakthrough at the time, and which has played such an important role throughout the history of modern science, Leucippus deserves to be accorded a foremost position in the early history of science, at least.

COLLISION AND ENTANGLEMENT

Assuming that physical reality is reducible to these two fundamental existents, how is one to explain the ordinary appearances and changes in the world? Since the atoms are homogeneous, indivisible, and unalterable, possessing the sole primary qualities of shape, size, and solidity, explanation of the composition of entities, as well as of the changes of phenomena, has to be in terms of the different positions and arrangements of the mobile atoms according to their diverse shapes and sizes. Using the model of motes flitting in a sunbeam (analogous to Brownian motion), the Atomists pictured a random movement of the atoms 'colliding,' 'rebounding,' and 'scattering' in space, or sometimes becoming 'entangled' or 'intertwined' because of their various shapes, and thus forming gross perceptible objects. As Aristotle states, while the atoms

> are so small as to elude our senses [...] they have all sorts of forms and shapes and differences in size. So he [Democritus] is already enabled from them [...] to create by aggregation bulks that are perceptible to sight and the other senses.[20] (Brackets added)

Such is the mode of explanation of the most powerful scientific theory ever developed: namely, that the observable qualities and changes of ordinary perceptual objects are dependent upon the properties and motions of imperceptible particles. The interaction of the atoms due to their motion is described in various but similar ways: 'They struggle and move in the void because of the dissimilarities among them [...] and as they move they collide and become entangled;' 'all things are generated by the intertwining and scattering around of these primary magnitudes;' 'For they say that the atoms move by mutual collisions and blows.' But the most explicit description is given by Simplicius:

> [...] these atoms move in the infinite void [...] overtaking each other they collide, and some are shaken away in any chance direction, while others, becoming intertwined one with another according to the congruity of their shapes, sizes, positions and arrangements, stay together and so effect the coming into being of compound bodies.[21]

Contrary to modern physics and chemistry, the atoms do not (chemically) combine in such a way as to lose their identity in new substances, their interaction always being a form of mechanical combination. This combination follows from the collision of the atoms causing some to become 'intertwined' or 'entangled' owing to their 'hooks' or because they fit together as a result of such 'congruent shapes' as 'concavity' and 'convexity.' As such, the account is far removed from the Rutherford-Bohr conception of the atom modeled after the solar system with protons and neutrons making up the mass of the atom in the nucleus, while the number of electrons in the outer orbits comprise most of its volume and give the element its electric charge and chemical properties. Nonetheless, in spite of the significant differences in detail, and between a mechanical as compared to a chemical-electrical explanation of interactions, the Atomists deserve credit for constructing the essential schema — it is more difficult to create a new theoretical framework than to refine its concepts and work out its implications later.

Aristotle relates how the Atomists, using letters of the alphabet, illustrated the possible atomic permutations, graphically adding that 'Tragedy and Comedy are composed of the same letters:'

Leucippus and his associate Democritus [. . .] say the differences in the elements are the causes of all other qualities. These differences [. . .] are three — shape and order and position. For they say the real is differentiated only by 'rhythm' and 'inter-contact' and 'turning'; and of these rhythm is shape, inter-contact is order, and turning is position; for A differs from N in shape, AN from NA in order, ⊥ from H in position. [22]

As de Santillana states, "This is excellent use of analogy in constructing physical theory, and it is natural to a Greek for whom *stoicheia* meant both letters and elements in a general sense."[23] Just as letters can be conjoined to make up endless combinations of words and sentences, so atoms can combine in innumerable arrangements to make up the diversified entities of the observable physical world.

THE ORIGIN AND NATURE OF MOTION

According to the previous description the atoms collide, then scatter or combine to form compounds, as a result of their continuous motion. But how did they acquire this original motion? What was its origin and nature? In Aristotle's judgment, neither Leucippus nor Democritus, 'who say that the primary bodies are in perpetual movement in the void,' adequately explained 'the manner of their motion and the kind of movement which is natural to them.'[24] The form of Aristotle's criticism, that the Atomists did not specify the motion 'natural' to the atoms, reveals his own presuppositions or biases, for he believed that different elements had their own 'natural motions.' In addition, he complained that the Atomists had not provided an explanation of the original source of motion: 'If there is no ultimate natural cause of movement and each [. . .] is moved by constraint, we shall have an infinite regress.'[25] For Aristotle, such an explanation was unacceptable because an infinite series of movers precludes an initial, original, first cause of the motion. Without a 'prime mover' that 'causes motion not by constraint but naturally,' there would not be anything to set the whole system in motion.

From the contemporary standpoint, however, the preeminent superiority of the scientific reasoning of Leucippus and Democritus over that of their predecessors, as well as Plato and Aristotle, consists in the fact that *they did not* require an additional, external source or cause of motion. Nor did they attribute different kinds of motions to the atoms.

Instead, in a solution whose simplicity belies its brilliance, they claimed that from eternity the atoms have possessed the same intrinsic random motion. An original source of motion was unnecessary because there never was a time when the atoms were at rest. As Democritus himself asserted, 'Of that which ever is and has been there is no reason to enquire for the cause.' Though the cause of all else, the eternal motion of the atoms itself is uncaused. Indeed, motion was such an inherent property of the atoms that it constituted part of their meaning: "We are told that the technical name for the atom in the School was *rhysmos*, 'onrush,' so little could it be thought at rest."[26] Herein lies the kinetic theory of motion and the branch of physics known as kinematics.

Thus in a flash of genius too brilliant and out of time to endure, awaiting the lapse of two millennia before its eventual acceptance, the Atomists' conception of motion dispelled the animistic specters that had haunted previous systems and that would reappear again in the cosmologies of Plato and Aristotle. As Bailey says, the world

> did not require the intervention of Empedocles' 'Love and Strife' or Anaxagoras' 'Mind' to bring it into being or keep it going: it is a wholly physical existence whose action is purely mechanical, controlled by the law of its own being and nothing more.[27]

In all theories one of the most difficult and crucial decisions pertains to what can or should be accounted for, in contrast to what must be assumed as given or inexplicable. Without this judgment all explanations would be infinitely regressive. In attributing an eternal motion to the atoms, thereby obviating an additional explanation, the Atomists demonstrated the meaning of elegant simplicity in scientific theories. Should there be any doubt as to the superiority of their mode of reasoning, one has only to compare to it Aristotle's account of the source of the motion of the celestial bodies in terms of empathetic love or desire — the famous "first mover" argument. Apart from the archetypical notion that only living beings can be an adequate source of motion, it is difficult to understand how Aristotle could have believed that positing 'prime movers' or 'intelligences' as the original causes of motion was superior to simply attributing motion to the atoms themselves.

WEIGHT

A related question to the nature of the motion of the atoms pertains to their weight. Since the collision of atoms results in compounds being 'shaken' and 'scattered apart,' the atoms manifest a force of impact, but do they also have weight? The Ancient Greeks had a word for weight, but no notion of gravitational forces, therefore their conception of weight was necessarily different from ours. We understand by weight the attraction toward the earth of an object as a result of its mass and the gravitational force of the earth — thus outside the earth's gravity an object is weightless, or heavier or lighter depending upon the strength of the gravitational field (as on the moon).

With no notion of gravitational forces, the Greek conception of weight resembled what ours would be without any understanding of the influence of gravity: namely, equivalent to mass and density or size and bulk. Thus weight was thought of either as a relational property dependent upon the comparative size or density of an object (i.e., an object is heavier or lighter than another if it is either larger or smaller or more or less compact), or as an inherent quality belonging to the object owing to its material nature. Prior to Aristotle, it was generally believed that weight is merely a relative or comparative property.[28]

In the latter's view, however, 'weight' is the 'quality' of heaviness or lightness of certain basic elements correlated with whether they naturally rise or fall: thus air and fire possess the 'quality of lightness,' in virtue of which they rise, while water and earth possess the 'quality of heaviness,' accounting for the fact that they fall (ether, the fifth element, is weightless). For Aristotle, comparative weights depend both on the kinds of substances and the amounts involved.

Since there is no mention of Leucippus' ever discussing the problem of weight, the question pertains to whether Democritus held that the atoms have weight and what he intended by the conception. There are various accounts as to what he said and meant, two of them seemingly contradictory. Both Aristotle and his student Theophrastus maintained that Democritus held that the atoms have weight in proportion to their size or bulk. According to Aristotle, 'Democritus says that each of the indivisible bodies is heavier in proportion to its excess (sc. of bulk),' and Theophrastus states that 'Democritus distinguishes heavy and light by size [. . .].' If the atoms have weight, then *variations* in weight would *have* to depend upon their shapes or sizes, for since each atom is completely solid and homogeneous (made of the same material), both density and differences in atomic material are ruled out as determinants of their weight. In contrast, the weight of *compound bodies* would depend upon their density or bulk, as Theophrastus goes on to add: 'in compound bodies the lighter is that which contains more void, the heavier that which contains less [. . .].'

According to their account, then, weight is not an inherent, primary quality of the atoms like their sizes and shapes, but a derived, comparative property relative to their sizes. Yet, if Democritus did hold the view that weight depends upon the size of the atoms, this would imply that the solidity (the mass) of the atom gives it its weight, variations in size (or shape) merely accounting for the *differences* in weight — for size alone could not explain the weight of atoms apart from the solid material of which they are composed. That is, if the atoms initially had no weight, then an increase in size would make no difference as regards their weight: e.g., in outer space objects are still weightless regardless of their size or mass.

On the other hand, if he held that weight was merely a *relative* property depending on the comparative sizes of the atoms, then the conception would have no meaning in isolation. While it would make sense to say that a particular atom was lighter or heavier than another atom (because of its size), it would be meaningless to refer to the weight of an atom in isolation. This view is analogous to Einstein's relational conception of space and time in comparison to Newton's absolutistic theory.

In contrast to the interpretation of Aristotle and Theophrastus, Aëtius maintains that Democritus did not hold that weight was a property of the atoms, even if derived, the latter conception having been introduced later by Epicurus who held that weight caused a downward movement of the atoms.

> Democritus named two (sc. properties of atoms), size and shape; but Epicurus added a third to these, namely weight [. . .] — Democritus says that the primary bodies (i.e., the solid atoms) do not possess weight but move in the infinite as the result of striking one another. [29]

According to Aëtius, atoms in the 'infinite void' do not possess weight but move as a result of collision and deflection. In contemporary terms, they have inertial mass but no weight: i.e., they tend to continue in motion indefinitely (and presumably in a straight line for there would be no reason for them to move in any particular direction) until deflected from their course by striking another atom. This is a further example of Democritus' remarkable capacity to deduce by reflective abstraction the essential concepts of the atomic theory. As de Santillana justly concludes:

> "Atoms have no weight, but they move by mutual impact in infinite space."
>
> Such is the earliest statement of the inertial principle, a principle so abstract and unfamiliar to ordinary thought that the grasp of it was almost immediately lost [. . .]. That Democritus had a clear intuition of the difference between weight and inertial mass, but could not express it for lack of proper terms, is shown by the other, seemingly contradictory statement: "The more an indivisible exceeds, the heavier it is." Mass, in other words, is there and goes with volume [. . .]. [30]

Confronted with similar problems in the 17th century, Galileo would arrive at a *restricted* formulation of the principle of inertia, maintaining that on a frictionless table a rolling ball would tend to continue indefinitely. His principle was inferior to that of Democritus, however, in that he believed that the inertial motion of an object in outer space would be circular rather than linear. It was Descartes who arrived at the full and correct expression of the principle of inertia, as Democritus had, by considering the possible motion of the atoms in infinite space.

As far as this latter motion is concerned, Aëtius' interpretation would be the consistent one. Since directional motion "downward" is the only kind that would be evidence of an atom's having weight, in contrast to random inertial motion, and since "directional motion downward" is inconsistent with the conception of an infinite isomorphic space in which atoms continue their motion until deflected by other atoms, the only consistent interpretation would be that atoms in the void 'do not possess weight' but do possess inertial mass. In the system of Epicurus atoms do fall downward and hence possess weight, but this could not be true of the system of Leucippus and Democritus, as far as *the motion of atoms in the void* is concerned.

However, as we shall find shortly, within the confines of the formation of individual worlds, as compared to atoms moving in the void, the distinction between downward and upward movements does have significance and hence presupposes that the atoms have acquired weight. According to Kirk and Raven:

> 'Weight' only operates in a vortex, in a developed world, and is an expression of the tendency of bulky objects towards the centre of a whirl. Before becoming involved in a vortex an atom is not activated by weight at all. [31]

Thus if Aristotle and Theophrastus were referring to atoms within the context of the formation of worlds, then there would be no conflict between their view and that of Aëtius'. But the only evidence to support this conjecture is a statement by Bailey:

> [. . .] there is no doubt that Democritus did speak, as Aristotle reports him, of 'heavier' and 'lighter' atoms, as a derivative quality, immediately dependent on size, and called into being, as it were, in the vortex, as a counteraction of the rotary motion of the whirl. [32]

Kirk and Raven, in contrast, appear to accept as contradictory the accounts by Aristotle and Theophrastus on the one hand, and Aëtius on the other. [33]

In conclusion, as far as motion in the void is concerned, since there is no reason for the atoms to fall or rise, they have no weight but do possess inertial mass along with inertial kinetic motion. Although it is not certain that Aristotle and Theophrastus intended their assertion that Democritus' atoms possess weight to apply only within the context of created worlds, this is the case. As Bailey says:

> It may then safely be inferred [. . .] that though Democritus did indeed speak of 'heavier' and

'lighter' atoms, he did not attribute absolute weight to them, still less did he regard weight as the initial cause of perpendicular motion downwards in the void, but considered it rather as a derivative property from size, acting, not when the atoms were free in the void, but only in the cosmic whirl [. . .].[34]

According to this interpretation, Democritus held as we do that mass is an inherent property of matter and that weight is a derived property, but while we derive it from, and make it relative to, the strength of a gravitational field, he derived it from, and made it relative to, the size of the atoms within the domain of formed worlds.

WORLDS IN FORMATION

Nowhere as in their cosmological views, particularly those of Democritus, do the sheer originality and contemporaneity of the Atomists stand forth so strikingly. There is not the slightest trace of mythical thought in their system. If one were not aware of the true authors, one could easily imagine he were reading a very general contemporary account of the nature of the universe and of the formation of particular galaxies according to the "steady-state theory" and the "nebular hypothesis."[35] *Perhaps nowhere in the whole history of thought is the sheer power of reflective reasoning to arrive at a true speculative account of the universe, as supported by later and more sophisticated developments in science, more clearly evident.*

Originally, the universe consisted of moving atoms separated and scattered throughout the infinite universe. Given an endless number of atoms in an eternity of time, an occasional chance collision and grouping of a few atoms would result in a concentration of more of them in a particular area of the void. These atoms in turn would form a 'whirl' attracting more atoms, thus initiating the formation of individual worlds. No explanation is given as to how the random motion of the aggregated atoms became transformed into a vortical motion. Here, as in the description of the formation of worlds, the Atomists often relied on the theories of their predecessors, particularly Anaxagoras, with the familiar image of the winnower's sieve again evident. Once the 'whirl' originates and 'separates off' from the rest of the void, the vortical motion causes the atoms to collide, thus coalescing or separating. Like Anaxagoras, Democritus held that like is attracted to like, whether of animate or inanimate things.

> For creatures (he says) flock together with their kind, doves with doves, cranes with cranes and so on. And the same happens even with inanimate things, as can be seen with seeds in a sieve and pebbles on the sea-shore [. . .].[36]

As the 'whirl' continues, the larger atoms are separated and sifted toward the center while the smaller, 'round and smooth,' fiery atoms are 'squeezed' to the periphery, thus illustrating how, within individual worlds, the atoms exhibit weight: "[. . .] Democritus' school thinks that everything possesses weight, but that because it possesses less weight fire is squeezed out by things that possess more, moves upwards and consequently appears light."[37] The shape of these whirls or worlds is spherical, although the infinite universe is without shape or boundary. A clear summary of the process is provided by Diogenes Laertius.

> Leucippus holds that the whole is infinite [. . .] part of it is full and part void [. . .]. Hence arise innumerable worlds, and are resolved again into these elements. The worlds come into being

as follows: many bodies of all sorts of shapes move 'by abscission from the infinite' into a great void; they come together there and produce a single whirl, in which, colliding with one another and revolving in all manner of ways, they begin to separate apart, like to like. But when their multitude prevents them from rotating any longer in equilibrium, those that are fine go out towards the surrounding void as if sifted, while the rest 'abide together' and, becoming entangled, unite their motions and make a first spherical structure. [38]

Each of these 'spherical structures' is enveloped by a 'containing membrane' formed of 'hooked atoms' which 'is itself increased, owing to the attraction of bodies outside.' The larger atoms in the center of the whirl form the earth with some breaking away later to form the planets and the stars. In this manner the solar system is formed.

While the description of the formation of the earth and the solar system follows the account of Anaxagoras, the original, accidental grouping of the atoms is analogous in explanation to the "nebular hypothesis" of the universe; namely, that a chance aggregation of atomic particles brings about a concentration of matter which gradually condenses into a solar system. Moreover, in their conception of innumerable worlds in various stages of formation and disintegration, differing as regards their planetary structure as well as the possibility of various forms of life, the Atomists anticipated one of the foremost contemporary theories of the universe, the "steady-state theory." While there were hints of innumerable worlds in the theories of Anaxagoras and others, the Atomists were the first explicitly to assert this view:

> [...] there are innumerable worlds, which differ in size. In some worlds there is no sun and moon, in others they are larger than in our world, and in others more numerous [...]. The intervals between the worlds are unequal; in some parts there are more worlds, in others fewer; some are increasing, some at their height, some decreasing; in some parts they are arising, in others failing. They are destroyed by collision one with another. There are some worlds devoid of living creatures or plants or any moisture. [39]

It would be difficult to find a more precise statement of the contemporary "steady-state" theory. When one realizes that this conception was formulated about twenty-four centuries ago, more than four centuries before the Christian era, it is evident that the previous assertion regarding the extraordinary originality of the cosmological speculations of the Atomists was not at all exaggerated. It certainly attests to the Greek genius that within two centuries they were able to move from an essentially mythological framework to the formulation of a cosmological system comparable to one of the leading contemporary theories.

Just as one asked how the original motion of the atoms came about, so the question was raised as to whether the 'separation' of the various 'whirls' came about by chance, necessity, or design. As one would expect, there is no indication that the universe came about by design which would have run counter to the whole mechanistic temper of the Atomists' thought. There are differences of opinion, however, as to whether it originated by accident or by necessity. In the only fragment remaining by Leucippus, he states that 'Nothing occurs at random, but everything for a reason and by necessity.' It is clear from other reports that by 'reason' he did not mean purpose or intention, as would be the case if the universe were created for a reason or for a purpose, but that he meant a necessary, causal, antecedent condition. At least this was the view attributed to Democritus by Diogenes Laertius: 'Everything happens according to necessity; for the cause of the coming-into-being of all things is the whirl, which he calls necessity.'

By 'necessity' Democritus meant that everything is produced by antecedent mechanical causes which are theoretically, if not actually, determinable. The fact that there may

Democritus explains sight by the visual image, which he describes in a peculiar way; the visual image does not arise directly in the pupil, but the air between the eye and the object of sight is contracted and stamped by the object and the seer; for from everything there is always a sort of effluence proceeding. So this air, which is solid and variously coloured, appears in the eye [. . .].[46]

According to this account, color would be a property of the air (rather than a subjective sensation) caused jointly by the effluences from the object fusing with those from the eyes of the seer. As such, it would be neither purely subjective nor objective, but acquire the status of an objective quality when a seer is present. As for the adequacy of the explanation itself, Theophrastus raises a number of difficulties. For example, if we perceive an image of the object imprinted on the air, how can we perceive several objects in the same place? If the image is due to the imprint of the object on the air in front of it, why do we not see the object in reverse?[47] Also, what can be meant by the expression that the visual image 'appears in the eye'? It seems obvious, however, that Democritus was attempting to explain the fact that colors, unlike tastes and sensations, appear to qualify objects outside of us in space.

Sounds are produced by the motion of "sound-particles"[48] striking the air causing it to break up into similarly shaped atoms which are pushed along by the sound-atoms: 'Democritus says that air is broken up into bodies of like shape and is rolled along together with the fragments of the voice.' These sound-atoms enter throughout the whole body (apparently because the entire body is responsive to sounds), but especially through the ears which serve as channels. It is not clear what is meant by "sound-particles." The atoms themselves do not possess sound qualities, so presumably these arise when the appropriate atoms in the body are set in motion.

All sensation, perception, and thought occur as a result of the impact of the effluences or images on the body which set in motion the soul-atoms: '[...] for they represent all perception as being by touch.' The Atomists rejected a dualistic conception of the mind and the body, attributing to the soul an atomic composition consisting of spherical atoms distributed throughout the body: 'Democritus says that the spherical is the most mobile of shapes; and such is mind and fire.' When one breathes one inhales spherical atoms from the atmosphere; therefore the loss of consciousness and death are due to the lack of a sufficient number of soul-atoms. The mind consists of unmixed soul-atoms located in the breast, thought being analogous to sensation or perception in that it occurs when the soul-atoms are set in motion by the impact of effluences or images from outside.

Thought is not a "mental process" independent of sensation, but a more subtle form of sensation which in turn is a motion of the soul-atoms. Thus while thought is a more 'genuine' form of knowledge than perception, Democritus remains committed to empiricism: '[...] Wretched mind, do you, who get your evidence from us [the senses], yet try to overthrow us? Our overthrow will be your downfall.' Both in his theory of knowledge as well as in his reduction of mental processes to the motion of atoms, Democritus antedated the theories developed by Thomas Hobbes in the 17th century.

Even his account of religion is consistent with atomism, for he apparently believed in a corporeal divinity composed of fire-atoms like the soul-atoms found in ourselves.[49] The origin of the belief in gods he attributed to visions or dreams which can be either beneficial or harmful. Like Feuerbach, he claimed that these beliefs are supported by fear of the overwhelming forces of nature and the ardent hope that they can be propitiated by gifts to the gods. Democritus himself, however, believed that prayer is useless since there is no divine creator, all events being the result of natural causes.

be no way of ascertaining the causes at any one time is the reason, apparently, for the alternative description that they occur by 'chance.' By this it is not meant that they occur without a cause or a reason, but that the cause is too complex to be known — as we use the word 'chance' or 'accident' today when we say that something happened by "accident" or that a particular run of cards occurred by "chance." Thus Simplicius states: "When Democritus says that 'a whirl was separated off from the whole, of all sorts of shapes' (and he does not say how or through what cause), he seems to generate it by accident or chance." In contrast, Aristotle used the terms 'chance' or 'accident' to refer to events which do not fulfill any purpose or final cause. As he says, 'There are some who make chance the cause both of these heavens and of all the worlds: for from chance arose the whirl and the movement which, by separation, brought the universe into its present order.'

The Atomists, consequently, were the first to present a consistently physical account of the universe whose interaction is purely mechanical, based solely on natural causes. Like the contemporary scientist, they were aware that the ultimate causes are at any time too complex to be completely ascertainable and therefore are described as occurring by chance. According to Bailey:

The Atomic conception of 'chance' then is, as we may say, the purely subjective conception which is proper to a scientific view of nature. 'Chance' is no external force which comes in to upset the workings of 'necessity' by producing a causeless result; it is but a perfectly normal manifestation of that 'necessity', but the limits of the human understanding make it impossible for us to determine what the cause is.[40]

It is only recently that atomic physicists, because of particular experimental results in quantum mechanics requiring the use of probability formalisms, have been willing to consider certain natural events as partially uncaused and therefore somewhat indeterminate. But many physicists, such as Einstein and Bohm, have been dissatisfied with these developments, arguing that they indicate that quantum mechanics is incomplete.

Returning to the process of world formation, the vortical motion of the whirl causes the larger heavier atoms to aggregate at the center thus forming the earth. Later, other earthly substances break away from the earth to form the moon, sun, and fixed stars (for some reason Leucippus thought the orbit of the fixed stars was between that of the sun and the moon, but this view was corrected by Democritus). As the planets and the stars were originally 'stones' from the earth, they lacked heat and brightness which they acquired by their rapid movement through the heavens (or perhaps by contact with the outer layer of fiery atoms). Thus, as in the system of Anaxagoras, the sun and stars are a molten mass of fiery stones and the surface of the moon consists of mountains, glens, and valleys. The Atomists followed Parmenides, Empedocles, and Anaxagoras in maintaining that the moon acquired its light from the sun.

Democritus anticipated the 1st century Roman architect Vitruvius in holding that the velocity of the planets and the stars increases with their distance from the earth, explaining the fact that although the moon seems to move more rapidly than the sun, and the sun than the fixed stars, this merely appears to be so because they are closer to the earth and have smaller orbits. The earth also moved when it was small and light, but as it became larger and heavier by the aggregation of more atoms it finally came to rest. It is tilted towards the south because of less support at that end which accounts for the inclination of its axis. The Atomists believed that the earth was 'oblong,' the length being one and a half times the breadth, perhaps holding this odd view because of the elongated shadow cast by the earth on the stars accounting for the appearance of the Milky Way.

Like Anaxagoras, Democritus believed that as the sun goes below the earth at night the earth casts a shadow on the sky revealing stars which normally cannot be seen because of the brightness of the sun's rays. The Milky Way consists of the concentrated light of these stars since 'owing to the distance they seem to be one, as when grains of salt are thickly sprinkled' (an excellent illustration of analogical reasoning). Two thousand years thereafter, Galileo similarly will describe the Milky Way as "a congeries of very minute stars."

THEORY OF PERCEPTION AND KNOWLEDGE

Reality, according to the Atomists, consists of atoms and the void. Since the atoms are insensible and homogeneous, differing only as regards shape, size, position or motion, and arrangement, the Atomists were faced with the same epistemological problems as their 17th century successors: e.g., Galileo, Gassendi, Locke, Boyle, and Newton. If the atoms consist of the same material, however infinitely varied they might be as regards shape and size, as well as arrangements, then how can one explain the kaleidoscopic diversity of sensory qualities actually manifested by the world? The atoms, though originally imperceptible because of their smallness, can, by combining or aggregating, form compound objects large enough to be perceived. Moreover, the primary qualities of these objects, their solidity, shapes, sizes, positions, etc., would be derived from the similar properties of the atoms. But how does one account for the other qualities, the colors, sounds, odors, and tastes, that are now generally referred to as secondary qualities?

While the answer is not always consistent, either because of the imprecision or variations in the interpretations of their successors or owing to the complexity of the problem itself (which still has not been resolved), the prevailing evidence suggests a solution similar to the later atomists. First, 'two forms of knowledge, one genuine, one obscure,' were distinguished. 'To the obscure belong all the following: sight, hearing, smell, taste, touch. The other is genuine, and is quite distinct from this [. . .].' In contrast to sensory experience, then, 'genuine knowledge' pertains to what is real, the atoms and the void, implying that the secondary qualities have an epistemic and ontological status less real than the primary qualities. In fact, in one of the most frequently quoted passages of the Atomists, they describe the origin and nature of secondary qualities as we would today.

> By convention are sweet and bitter, hot and cold, by convention is colour; in truth are atoms and the void [. . .]. In reality we apprehend nothing exactly, but only as it changes according to the condition of our body and of the things that impinge on or offers resistence to it. [41]

As in the modern conception of physical reality, atomic particles are not characterized in terms of colors, sounds, smells, or tastes; similarly, hot and cold are viewed as sensations, caused by the impact of the atoms on the body, rather than as belonging to the atoms themselves. Like sensations of pain or tickling, which are recognized as bodily effects of external stimuli, the secondary qualities, although experienced as if they belonged to objects existing independently of us in space, are really dependent upon the 'condition' or sense organs of the body itself. This position is such a natural consequence of the atomic theory that it was expressed in practically identical words by Galileo two millenia later:

> [. . .] I think that these tastes, odours, colours, etc., on the side of the object in which they seem to exist, are nothing else than mere names, but hold their residence solely in the sensitive body; so that if the animal were removed, every such quality would be absorbed and annihilated. [42]

This subjectivistic interpretation is partially supported by another fragment:

> Bitter taste is caused by small, smooth, rounded atoms, whose circumference is actually sinuous; therefore it is both sticky and viscous. Salt taste is caused by large, not rounded atoms, but in some cases jagged ones [. . .]. [43]

Here it is evident that different tastes are experiences explained in terms of the effects of specific primary qualities of the atoms on our senses. But the fragment also implies that 'sticky' and 'viscous' are properties of the circumference of the atoms, whereas in consistency, they too, should be tactual qualities dependent upon the impact of the atoms on our senses. In another passage it is also unclear as to whether the secondary qualities of the natural elements are merely correlated with their inherent primary qualities or are reducible to them.

> They (sc. Leucippus and Democritus) did not further define what particular shape belonged to each of the elements but merely attributed the sphere to fire; air, water and the rest they distinguish by magnitude and smallness [. . .]. [44]

Consistent with their overall position, this means that the different elements are distinguished in terms of such primary properties as shape and size, but it does not explain the origin of such qualities as hotness, wetness, dryness, etc. Do these qualities arise owing to the sizes, shapes, and motion of the atoms or are they secondary qualities, like tastes, caused in the body by the effects of the atoms on our senses?

Freeman presents a very interesting passage indicating that Democritus made a careful study of colors, dividing them into primary and mixed, explaining the differences among the primary colors in terms of the various shapes, sizes, positions, and arrangements of the atoms.

> There are four primary colours: white, black, red, yellow; all others are made by combining these. White is identified with the smooth, that is, with whatever is not shadow-forming. Bright white is made of shapes 'like the inside surface of shells'; powdery white of circular atoms arranged in groups of two set slantingly with regard to each other. Black is made up of rough uneven particles which cast shadows and are difficult to penetrate. Red is made up of the same kind of atoms as fire, but larger; things grow red when heated. Yellow is made from solid and emptiness combined, the colour arising out of position and arrangement. All other colours are made by admixture [. . .]. [45]

While this description implies that colors are objective qualities of aggregated objects themselves, it could well be that Democritus was just explaining in objective language the origin of mixed colors from primary ones, ignoring for the moment the dependence ultimately of all colors on the sense of sight. This is precisely what we do when accounting for the same phenomena today.

Democritus, in his theory of vision, appears to have held a directly representational view according to which 'images' (eidola) conveyed to the eyes are copies of the ordinary objects around us: 'They attributed sight to certain images, of the same shape as the object, which were continually streaming off from the objects of sight and impinging on the eye.' More specifically, sight occurs when the 'effluences' (Empedocles' term) of the object and the 'seer' impress or 'stamp' an image of the object on the air which then is communicated to the eye where it enters and is transmitted to the rest of the body. As described by Theophrastus:

CONCLUSION

In this chapter an attempt has been made to redress the injustice done to Leucippus and Democritus owing to the loss of their works, with the consequence that their contributions to the growth of Western thought have been entirely eclipsed by the reputations of Plato and Aristotle — in spite of the remarkable originality and modernity of their thinking, as well as the superiority of their cosmological theories. Also, an effort has been made to balance the stress put on the discovery of scientific laws and mathematical correlations, experimentation, and prediction and verification in scientific inquiry, by modern philosophers of science, with a complementary emphasis on the importance of the creation of new concepts and novel theoretical frameworks for the advance of science. Precisely because it is so fundamental, it is easy to forget that all of our thinking about the world, including that which guides experimentation, prediction, and the discovery of scientific laws, depends first upon conceptual interpretations. It was Leucippus and Democritus who created the atomic theory on which so much of modern scientific inquiry, explanation, and progress has depended.

The concept of the atom, which is what the *archē* or *physis* has now become, was formed by abstracting from the multifarious qualities of the world the tactual-visual properties of impenetrability or solidity, shape, and size, and then reconstructing the atom in terms of these primary qualities. In addition, since all change was attributed to the mobility of these atoms, they were also given the properties of kinetic motion, inertial mass, and relative weights. Then, realizing that the separateness and motion of the atoms presuppose spatial intervals, the Atomists made the conceptual leap of positing an even more abstract mode of existence, the void, conceived as a series of unoccupied spatial intervals corresponding to the gaps between the atoms. In this way they conceptually replaced the ordinary macroscopic world, the appearance of which they attributed to the impact of the atoms on the sense organs of the body, with their atomic framework, exactly as the contemporary scientist does. The schema of interpretation is the same; only the specific concepts and ways of justification have been refined.

Furthermore, reflecting on the possible cosmological implications of their theoretical model, they went on to imagine a simplified form of the "nebular hypothesis" to explain the formation of particular worlds (or galaxies) and a "steady-state" description of the universe itself, thereby approximating contemporary cosmological speculation. In accomplishing this, they eliminated from their conceptual framework all traces of animism and myth, along with such extraneous notions as a creator, prime mover, or mind causing and controlling the motions in the world. If today such theorizing is abetted by experimentation with very sophisticated apparatus and mathematical deduction, it should be with more, not with less, amazement and admiration that we acknowledge the extraordinary theoretical attainments of Leucippus and Democritus based on reflective reasoning alone.

There were some respects, however, in which the Atomists' views were inferior to those of Plato and Aristotle. Unlike the astronomical theories of the latter, which included a spherical earth and a description of the positions and motions of the inferior and superior planets, as far as we know the Atomists only described the motions of the sun and the moon and retained the erroneous conception of an elongated earth. In addition, there is no evidence to indicate that they were influenced, as Plato so profoundly was, by the mathematical speculations of the Pythagoreans, and therefore lacked (as Aristotle would) an appreciation of the possibility of discovering mathematical proportions or harmonies underlying accoustics, the interactions of the elements, and the motions of the planets.

Nor were their biological inquiries as extensive as those of Aristotle, though Democritus held theories regarding procreation and embryological development, and there is evidence indicating that he practiced dissection. Though critical of his work on biology, Aristotle used it in composing his own treatise on the subject. It is said that either Democritus or one of his pupils was a teacher of the great physician Hippocrates which, if true, would mean that Democritus was not only a creator of the most powerful scientific *theory* of all time, he also was a source of the most important *empirical* scientific tradition in antiquity.

But it was in their attempt to explain all phenomena, physical, psychological, biological, and religious in terms of one "unified theory," that the greatness of the Atomists lies. If it remained a mystery as to how the mechanical combination of qualityless atoms and their impact on the body could give rise to such secondary qualities as colors, sounds, and tastes, or how the motion of rounded, fiery soul-atoms could be the occasion of sensations, perceptions, and thoughts, we are hardly nearer to a solution of these problems today. What modern chemist, for example, can explain how the combination of atoms of sodium and chlorine, none of which are white or salty, gives rise to these qualities in ordinary salt? Are they, as they appear to be, inherent properties of macroscopic quantities of salt, or do the white color and salty taste, though experienced as independent properties of salt *within the context of ordinary experience*, nevertheless depend upon the sense organs and nervous system of the perceiver, and therefore, as Galileo and most scientists today conclude, do not exist apart from the organism? Like the Atomists, modern scientists regard the atomic domain as being different from the ordinary macroscopic world which raises the same questions as to the origin and status of the qualities of the world as directly experienced.

Furthermore, with the recent progress in neurophysiology and the rise of cybernetics, the model of the human nervous system as nothing more than an extremely complex machine has gained considerable support — another instance of the affinity of modern science with Greek Atomism. But can the contemporary neurologist explain, any better than Democritus could, how purely chemical-electrical nerve impulses eventuate in the experience of sensory qualities, sensations, images, and thoughts? The fact that the schema of the Ancient Atomists raises the same kinds of profound questions is further testimony to their extraordinary foresight. As Bailey states:

> In reading or writing the history of Greek Atomism it is impossible not to have an eye on modern science. Yet the temptation to see closer parallels than actually exist and to read modern ideas into ancient speculation is very real and must be resisted [. . .]. But even a layman may assert without contradiction that it was the Greeks who put the questions which modern science is still endeavouring to answer: had their problems been different, the whole course of European scientific thought might never have existed. [50]

We owe to the Ancient Atomists, therefore, the essential concepts and basic theoretical framework, as well as the major epistemological problems, characteristic of modern classical science from the 17th through the 19th centuries. Since the beginning of the 20th century, however, revolutionary developments in physics, such as relativity theory and quantum mechanics, have led to drastic revisions in the underlying presuppositions and central concepts of classical mechanics and atomic physics. Because of this, the physicalistic, mechanistic, and deterministic framework of Ancient Atomism finally is being superseded by a newer and more adequate conception of physical reality. Nonetheless, atomic physics still provides the paradigm for research in astronomy, chemistry, biophysics, and genetics, while particle physics continues to be one of the most fruitful and exciting areas of experimentation in quantum mechanics. As long as this is true, our debt to Leucippus and Democritus remains outstanding.

NOTES

CHAPTER X

[1] Lancelot Law Whyte, *Essay on Atomism* (Middletown: Wesleyan Univ. Press, 1961), p. 3.

[2] Cyril Bailey, *The Greek Atomists and Epicurus* (Oxford: At the Clarendon Press, 1928), p. 1. After almost half a century, this still remains an outstanding work on the Greek Atomists.

[3] Cf. E. A. Burtt, *The Metaphysical Foundations of Modern Science*, rev. ed. (Garden City: Doubleday & Co., Inc., [1932] 1954).

[4] Bailey, *op. cit.*, p. 1.

[5] Isaac Newton, *Opticks*, based on the London 4th edition, 1730 (New York: Dover Pub., Inc., 1952), p. 400.

[6] Whyte. *op. cit.*, p. 3.

[7] Milič Čapek, *The Philosophical Impact of Contemporary Physics* (*op. cit.*, Introduction), p. 72.

[8] Bailey, *op. cit.*, p. 112.

[9] Aristotle, *De Gen. et Corr.*, 315b. (Joachim trans.)

[10] Simplicius, *De Caelo*, 242, 18. Quoted from G. S. Kirk and J. E. Raven, *The Presocratic Philosophers* (*op. cit.*, ch. IV), pp. 407-408, f.n. 556.

[11] Aristotle, *op. cit.*, 325a5-8; 31-34.

[12] Bailey, *op. cit.*, p. 75.

[13] *Ibid.*

[14] Giorgio de Santillana, *The Origins of Scientific Thought* (*op. cit.*, ch. IV), p. 144.

[15] Aristotle, *On Democritus ap.*, Simplicius, *De Caelo*, 295, I. Quoted from Kirk and Raven, *op. cit.*, p. 407, f.n. 555.

[16] Bailey, *op. cit.*, p. 77.

[17] Kirk and Raven, *op. cit.*, p. 408.

[18] Simplicius, *Phys.*, 28, 4. Quoted from Kirk and Raven, *op. cit.*, p. 400, f.n. 546.

[19] Cf. Kirk and Raven, *op. cit.*, p. 409, f.n. 560.

[20] This quotation is a continuation of f.n. 15 above.

[21] Simplicius, *De Caelo*, 242, 21. Quoted from Kirk and Raven, *op. cit.*, p. 419, f.n. 582.

[22] Aristotle, *Met.*, 985b4. (Ross trans.)

[23] de Santillana, *op. cit.*, p. 149.

[24] Aristotle, *De Caelo*, 300b8. (Stocks trans.)

[25] *Ibid.*

[26] de Santillana, *op. cit.*, p. 145.

[27] Bailey, *op. cit.*, pp. 121-122.

[28] *Ibid.*, pp. 144-145.

[29] Aëtius, 1, 3, 18. Quoted from Kirk and Raven, *op. cit.*, p. 414, f.n. 574.

[30] de Santillana, *op. cit.*, p. 145.

[31] Kirk and Raven, *op. cit.*, p. 416.

[32] Bailey, *op. cit.*, p. 146.

[33] Cf. Kirk and Raven, *op. cit.*, pp. 415-416.

[34] Bailey, *op. cit.*, p. 132.

[35] There are two leading cosmological models of the universe: (1) the evolutionary or "big bang theory" advanced by George Lemaitre and supported by George Gamow, and (2) the "steady-state theory" developed by Herman Bondi, Thomas Gold, and Fred Hoyle. According to the "big bang theory" the present universe began with the explosion of an incredibly dense atom and has been expanding and evolving ever since, such that the various evolved epochs of the universe show signs of a later development. In contrast, the "steady-state theory," like the theory of the Atomists,

maintains that the universe contains innumerable galaxies scattered throughout the universe in various stages of formation and disintegration, and hence that the total universe shows no sign of a temporal evolution. Of the two theories, the red shift in the spectrum indicating an expanding universe, supports the "big bang theory." The "nebular hypothesis," that individual galaxies were formed from the chance concentration of dust particles, was formulated independently by Kant and by Laplace during the 18th century. Cf. Milton K. Munitz, ed., *Theories of the Universe* (New York: The Free Press, 1957), pp. 390-429. Also, cf. Fred Hoyle, *The Nature of the Universe* (New York: Harper & Bros., 1950).

[36] Sextus, *Adv. Math.*, VII, 117. Quoted from Kirk and Raven, *op. cit.*, p. 413, f.n. 569.

[37] Simplicius, *De Caelo*, 712, 27. *Ibid.*, p. 415, f.n. 575.

[38] Diogenes Laertius, IX, 31. *Ibid.*, pp. 409-410, f.n. 562.

[39] Hippolytus, *Ref.*, 1, 13, 2. *Ibid.*, p. 411, f.n. 564.

[40] Bailey, *op. cit.*, p. 143.

[41] Sextus, *Adv. Math.*, VII, 135. Quoted from Kirk and Raven, *op. cit.*, p. 422, f.n. 589.

[42] Galileo Galilei, *Opere*, IV, 333ff. Quoted from Burtt, *op. cit.*, p. 85.

[43] Theophrastus, *De Sensu*, 66. Quoted from Kirk and Raven, *op. cit.*, p. 423, f.n. 591.

[44] Aristotle, *De Caelo*, 303a12. Quoted from Kirk and Raven, *op. cit.*, p. 421, f.n. 584.

[45] Kathleen Freeman, *The Pre-Socratic Philosophers* (*op. cit.*, ch. IV), pp. 311-312.

[46] Theophrastus, *De Sensu*, 50. Quoted from Kirk and Raven, *op. cit.*, p. 421, f.n. 587.

[47] Cf. Freeman, *op. cit.*, p. 311. Also, see Plato's explanation in the *Sophist*, 266B and the *Timaeus*, 46A-C.

[48] The term "sound particles" is used by Freeman, *op. cit.*, p. 112.

[49] Cf. Freeman, *op. cit.*, p. 315.

[50] Bailey, *op. cit.*, pp. 4-5.

CHAPTER XI

PLATO AND THE WORLD OF FORMS

Wherever I go in my thoughts I find Plato returning.[1]
Buchanan

Whitehead stated that "the safest general characterization of the European philosophical tradition is that it consists of a series of footnotes to Plato."[2] While this statement (though somewhat exaggerated owing to Whitehead's Platonic bias) attests to the enduring influence of Plato on Western philosophy, it neglects the profound impact he had on Western culture in general. Probably no individual has been as extraordinarily gifted as Plato nor has had as deep and lasting an effect on our civilization. Hardly an area of culture does not attest to his influence: art, astronomy, jurisprudence, literature, mysticism, philosophy of mathematics, physics, political science, religion, and so forth. Even when he did not himself make an original contribution to a particular discipline the import of his teachings often had a tremendous effect, as in the case of mathematics. As Heath asserts:

> Whatever original work Plato himself did in mathematics (and it may not have been much), there is no doubt that his enthusiasm for the subject in all branches and the pre-eminent place which he gave it in his system had enormous influence upon its development in his life-time and the period following.[3]

The same could be said of astronomy, although there Plato may have made some original contributions, as we shall find.

The fact that Plato's brilliance irradiated so much of Western civilization makes it difficult to evaluate the quality of his thought and its effect on the development of scientific rationalism. This difficulty is augmented by the rich diversity of his thinking and writings, a blend of rationalism, mysticism, scientific speculation, and poetic fantasy with a superb literary imagination, explaining why he has represented so many different things to such diverse people: a rationalistic philosopher, an exponent of number mysticism, an impeder of the development of empirical science, a precursor of mathematical physics, a source of demonology and astrology, a primary influence on the development of astronomy, etc. One can find in some passage in the extensive writings of Plato, either literally or figuratively, evidence for almost any interpretation so that, as Cornford states, the "temptation to read into Plato's words modern ideas that are in fact foreign to his thought has proved too much for some commentators."[4] In any case, one can agree with Dicks that Plato exhibits "an intellect which has as many facets as there are stars in the heaven."[5]

The problem in assessing Plato's influence on the growth of rational thought is further augmented by the fact that he avoided specialized language, often used terms imprecisely, and usually resorted to ingenious metaphors, allegories, and myths to present and develop his views on such technical subjects as the philosophy of mathematics, astronomy, and cosmology. In addition, he antedated Popper's notion that scientific theories are just conjectures by presenting such theories not as factual truths about reality, but at most as "likely stories." In the *Republic*, for example, after presenting two of the most beautiful and illuminating allegories in the entire history of literature and philosophy, he has Socrates conclude:

> The ascent to see the things in the upper world you may take as standing for the upward journey of the soul into the region of the intelligible [...]. Heaven knows whether it is true; but this, at any rate, is how it appears to me. [6]

Similarly, in the *Timaeus*, Plato's major astronomical and cosmological work, he continually cautions the reader against a dogmatic acceptance of his view.

> If then, Socrates, in many respects concerning many things — the gods and the generation of the universe — we prove unable to render an account at all points entirely consistent with itself and exact, you must not be surprised. If we can furnish accounts *no less likely* than any other, we must be content, remembering that I who speak and you my judges are only human, and consequently it is fitting that we should, in these matters, *accept the likely story and look for nothing further.* [7] (Italics added)

One must constantly remember, when reading the *Timaeus*, that it is essentially a mythical account of creation, not a scientific treatise on astronomy — although there is some serious astronomy presented. Finally, one has the further difficulty of determining when the views presented by the various speakers in the dialogues represent those of Plato himself.

BACKGROUND

In the space available, no more than the briefest sketch can be given of the rich, eventful, and productive life of Plato.[8] He was born in Athens in 428/27 B.C. of an aristocratic family which traced its lineage from the ancient gods. The times were turbulent: the death of Pericles in 429 marked the close of the golden age of Athens, the beginning of the loss of its empire, and the end of its political stability. All of Plato's youth was spent under the cloud of the Peloponnesian War; he was twenty-three when Athens was defeated by Sparta and ceded her empire to her. In the same year members of his family and close friends took part in the revolution that brought about the rule of the Thirty; however, rather than a source of satisfaction, this proved to be bitterly disappointing because of the corrupt rule of the Thirty who themselves were removed after only a year and a half in power. In a letter written near the end of his life, Plato recalled this earlier period in terms which have been sighed throughout the ages: "[...] the whole fabric of law and custom was going from bad to worse at an alarming rate."

Thus Plato's lifelong concern with the "art of statesmanship" and the construction of an ideal republic was not a matter of mere speculation, but an outgrowth of his close acquaintance with, and concern for, the frequent political upheavals of his day. Twice he traveled to Syracuse in Sicily with the intention of converting either the despot Dionysius I, or his brother-in-law Dion, to the life of the philosopher-king, thereby hoping to transform Syracuse into an ideal republic. However, like all attempts to establish a perfect state in this imperfect world, Plato's efforts to reform the court of Dionysius I were met with eventual resentment and hostility, such that he was forced to leave under conditions that either threatened his liberty or endangered his life. This brought an end to his attempt to bring about the political actualization of his belief that mankind "would never see the end of trouble until true lovers of wisdom should come to hold political power, or the holders of political power should [. . .] become true lovers of wisdom."

The model for the "true lovers of wisdom" was the older philosopher Socrates whom Plato immortalized in his dialogues. Plato was only a young man when Socrates was condemned to death in 399 (as if to explain his absence from the death scene described in the *Phaedo*, he has Phaedo say, "I believe that Plato was ill"). But so indelible was the impact of the older philosopher, described by Plato in his "Autobiography" as "a man whom I should not hesitate to call the most righteous man then living," that it is generally believed that it was he who served as the model of the true philosopher, one who turns away from the deceptive lures of the physical world to cultivate his soul by searching for Truth, Wisdom, Virtue, and Goodness.

At the age of forty, after his unsuccessful ventures in Syracuse, Plato founded the Academy which was the most successful achievement of its kind in history. The Academy had the longest duration of any academic institution, lasting for 900 years when it finally was closed by order of the Christian Emperor Justinian on the grounds that it was a "pagan" school. During its early years the Academy attracted some of the most outstanding scholars of Greece, such as Theaetetus, Heraclides of Pontus, and Eudoxus. The latter, the originator of the theory of concentric spheres of planetary motion adoped by Aristotle, taught there, and Aristotle himself began his studies there at the age of seventeen and remained twenty years until the death of Plato (at the age of eighty) in 348/7. Plato spent the last forty years of his life directing the Academy, teaching, and writing the dialogues that are among the most valued legacy of Greek civilization. It would be difficult to overestimate the importance of the Academy in stimulating research in such areas as mathematics and astronomy, as well as in philosophy, and in the enrichment of almost the entire field of learning.

Considering the range of Plato's interests and writings, this chapter will focus on the problem as to what extent Plato contributed to the development of abstract thought in terms of two crucial doctrines: (1) the theory of forms, and (2) Plato's conception of mathematics and its role in astronomy. As we shall find, the latter conception is a complex one such that the conclusion can at best be conjectural and tentative. While the authenticity of Plato's writings is not in question as it was in the discussion of the Presocratics (with the exception of the *Epinomis*), the multiple levels of Plato's thought, as well as the form in which his doctrines are presented, occasionally precludes a definitive interpretation of his position. As Crombie states, in some instances there is a "variety of possible positions which we might ascribe to Plato, and I do not think that his language is clear enough to allow us to choose with confidence between them." [9]

In such cases, we shall try to let Plato's words speak for his thought, rather than interpret his meaning in terms of what we would expect someone today would have meant by his expressions.[10] But even this is difficult because one finds, as regards crucial passages, that the translation will reflect the attitude and interpretational orientation of the translator. For example, the essential passage for the interpretation of Plato's conception of the role of mathematics in astronomy occurs in the *Republic*, VIIa-c, and a comparison of the translations of Cornford, Heath, Dicks, and Shorey reveals strong divergences reflecting their different interpretations.[11] Beyond that, however, the passage just *does not allow* for a definitive interpretation: one would have to know what Plato meant by several of the key terms and the context does not provide this knowledge. Thus one can only grope for an answer to one of the most provocative problems in Plato. Keeping these provisos in mind, we turn to the first of Plato's doctrines.

THE THEORY OF FORMS

The doctrine for which Plato is most famous, his most original and significant contribution, is the theory of Forms; i.e., the doctrine that in addition to the perceptible physical world of change and becoming there is a transcendent supersensory realm of unchanging eternal truths and values. This conception of the separate existence of an intelligible realm of Forms was connected, at least in the early dialogues, with his (and Socrates') belief in the immortality of the soul. As Cornford states, the "two pillars" of Plato's philosophy are "the immortality and divinity of the rational soul, and the real existence of the objects of its knowledge — a world of intelligible 'Forms' separate from the things our senses perceive." [12]

Along with and complementing the above conceptions, Plato also held the doctrine of *'anamnesis'* or recollection, the theory that knowledge is not acquired through our senses or through the conveying of genuinely new information by education, but by 'recollecting' the realities and truths known to the soul in its existence before birth. Plato went so far in the *Phaedo* as to maintain that these three doctrines were "logically" interrelated.

> If all these absolute realites, such as beauty and goodness [...] really exist [...] does it not follow [...] that it is *logically* just as certain that our souls exist before our birth as it is that these realities exist [...]? It is perfectly obvious [...] that the same *logical necessity* applies to both. [13]
> (Italics added)

But it is the *Meno* that contains the well-known example of the doctrine of 'recollection' when Socrates, by questioning the slave boy, elicits the answer that the square on the diagonal of the original square is the correct answer to the question, "What figure would be double the size of the square?" Here again Plato repeats his belief in the immortality of the soul and in *'anamnesis'*:

> Thus the soul, since it is immortal and has been born many times, and has seen all things both here and in the other world, has learned everything that is. So we need not be surprised if it can recall the knowledge of virtue or anything else which, as we see, it once possessed [...] for seeking and learning are in fact nothing but recollection. [14]

The doctrine of the immortality of the soul is a typical Socratic belief, along with the conception that there are absolute values such as Justice, Beauty, Virtue, Piety, Holiness, Courage, etc., the *essential definition* of which is the aim of philosophical inquiry. According to Aristotle, this latter Socratic thesis is one of the two sources of the Platonic conception that these essential definitions have as their referents the independently existing, eternal, intelligible Forms. As Aristotle states, in contrast to the cosmological investigations of the Ionians and the physical inquiries of the Pluralists,

> Socrates [...] was busying himself about ethical matters and neglecting the world of nature [...] but fixed thought for the first time on definitions; Plato accepted his teaching, but held that the problem applied not to sensible things but to entities of another kind — for this reason, that the common definition could not be a definition of any sensible thing, as they were always changing. Things of this other sort, then, he called Ideas, and sensible things, he said, were all named after these, and in virtue of a relation to these [...] . [15]

The second source of the theory of Ideas or Forms, according to Aristotle, was the

Pythagorean belief in the reality of numbers. However, unlike the Pythagoreans who conceived of numbers as quasi-spatial, tangible entities with the power of engendering physical objects, Plato abstracted numbers from the spatial, temporal, material domain, attributing to them a transcendent, supersensory existence either as mathematical objects occupying an intermediate position between sensible things and the Forms, or as having an existence analogous to the Forms themselves. [16]

> Further, besides sensible things and Forms he [Plato] says there are the objects of mathematics, which occupy an intermediate position, differing from sensible things in being eternal and unchangeable, from Forms in that there are many alike, while the Form itself is in each case unique [...]. His divergence from the Pythagoreans [consists] in making the One and the Numbers separate from things [...]. [17] (Brackets added)

Assuming that Aristotle was correct that the theory of Forms had its origin in these two sources, the collective arguments running through all the dialogues in support of the theory are more numerous, as indicated by the following categories.

Semantic: As Aristotle claimed, Socrates' search for the essential definition of such ethical terms as 'goodness,' 'virtue,' 'courage,' 'piety,' 'holiness,' 'justice,' etc., led Plato to postulate a realm of supersensible unchanging entities as the referents of these definitions. Underlying this assumption is what has been called, after Russell and Wittgenstein, the "name theory of meaning;" namely, that if a word is to have a sense it must name or designate 'something,' whether that word refers to a perceptible physical entity (Socrates), a sensory quality (white), a general class (man), a universal (tallness), a relation (larger than), a number (10), an aesthetic value (beauty), or a moral ideal (virtue). This theory requires that in addition to the sensory world which serves as the referent for empirical descriptive statements, there be a supersensible realm of intelligible entities or Forms which are named by, and thus constitute the meaning of, archetypical abstractions, universals, mathematical concepts, classes, value terms, etc.

For example, in such expressions as "this statue of Apollo is beautiful," "the Athenian constitution is just," "Socrates is snub-nosed," "few men are wise," the terms 'beautiful,' 'just,' 'snubness,' and 'wise' are intelligible because they name an abstract supersensible entity which is their meaning. In addition, for Plato this explanation accounts for the fact that words retain the same meaning when used by different individuals on different occasions. As he has Parmenides assert in the dialogue by that name:

> When you use any word, you use it to stand for something. You can use it once or many times, but in either case you are speaking of the thing whose name it is. However many times you utter the same word, you must always mean the same thing. [18]

Thus Plato did not hold the prevalent contemporary view that words acquire their sense by convention and thus have a conceptual or nominal meaning only. Instead, he believed in the independent existence of abstract universal meanings, thereby originating the position now known as "Platonic Realism."

Axiological: Plato also believed in absolute moral values and in the meaning of such absolute value terms as 'The Good,' 'Beauty Itself,' 'Perfect Justice,' etc. And since these terms presumably do not have referents in the world of changing, imperfect objects and events, they too must refer to a supersensible realm of unchanging, ideal archetypes or Forms. As indicated earlier, in the *Phaedo* this knowledge of absolutes was related to the belief in the previous existence of the soul, since the knowledge of absolute standards or of moral values could not be derived from the sensory world.

> Then if we obtained it before our birth, and possessed it when we were born, we had knowledge, both before and at the moment of birth, not only of equality and relative magnitudes, but of all absolute standards [. . .] [such as] beauty, goodness, uprightness, holiness, and, as I maintain, all those characteristics which we designate in our discussions by the term 'absolute.'[19] (Brackets added)

Similarly, in the *Republic*, Plato distinguishes between the plurality of *observable things* we call good or beautiful, and the unchanging, unique, *intelligible Form* of Beauty or Goodness itself.

> Let me remind you of the distinction we drew earlier and have often drawn on other occasions, between the multiplicity of things that we call good or beautiful [. . .] and [. . .] Goodness itself or Beauty itself [. . .]. Corresponding to each of these sets of many things, we postulate a single Form or real essence, as we call it [. . .]. Further, the many things, we say, can be seen, but are not objects of rational thought; whereas the Forms are objects of thought, but invisible.[20]

To further illustrate the transcendent position and irradiating power of Absolute Goodness, Plato creates the celebrated simile between the Good and the Sun in the *Republic* — the image that had such a splendid and lasting effect on intellectual history owing to its influence on such illustrious thinkers as Plotinus, Copernicus, and Kepler. Although refusing to define the Good, Plato refers to it as both the ontological *source* and *illumination* of the Forms.

> This, then, which gives to the objects of knowledge their truth and to him who knows them his power of knowing, is the Form or essential nature of Goodness. It is the cause of knowledge and truth [. . .]. You will agree that the Sun not only makes the things we see visible, but also brings them into existence and gives them growth and nourishment [. . .]. And so with the objects of knowledge: these derive from the Good not only their power of being known, but their very being and reality; Goodness is not the same thing as being, but even beyond being, surpassing it in dignity and power.[21]

In this metaphysical vision of reality, Plato views the Good as lying beyond Being itself. Unlike the cosmology of the previous Atomists and that of modern science, but similar to the world's great religions, he maintains that values are the epitome of all creation.

Epistemic: It is a curious fact of human history that the confidence in attaining absolute truth (and of absolute values) has varied inversely with the progressive acquisition and development of knowledge. Today significant knowledge (as well as value judgments) is generally considered to be empirically derived, conjectural, probable, conditional or relative, approximate, and pragmatic. While this conception of knowledge was held also by the Sophistic tradition (e.g., Protagoras, Gorgias, etc.) of ancient Greece, it was not characteristic of the two most prominent traditions, the Platonic and Aristotelian.

For the latter, knowledge must be *certain*, having the *real* as its object of knowledge. Thus Plato distinguishes between knowledge (*episteme*) which is infallible and has real intelligible objects (the Forms) as its referent, and belief or opinion (*doxa*) that can be either true or false since it refers to the changing world of sensory appearances. For example, the following dialogue occurs in the *Republic*:

> [. . .] a little while ago you agreed that knowledge and belief are not the same thing.
> Yes: there could be no sense in identifying the infallible with the fallible.
> Good. So we are quite clear that knowledge and belief are different things?
> They are.

If so, each of them, having a different power, must have a different field of objects.
Necessarily.
The field of knowledge being the real; and its power, the power of knowing the real as it is.
Yes. [22]

'Knowledge,' he adds, 'has for its natural objects the real - to know the truth about reality,' while belief refers to the changing world of appearances, the domain of existence intermediate between the real and the (theoretically) unreal.

This position is fully developed in the allegory of the "divided line" which is too well-known to be more than briefly summarized here. In the allegory Plato distinguishes between the various faculties of knowledge and their corresponding objects. Starting with the state of cognitive innocence in which one uncritically accepts all impressions as equally real, Plato describes the ascent from this state of belief in the world of appearances to the higher levels of cognition with their more abstract objects existing in the intelligible world. Thinking about abstract mathematical objects with the aid of diagrams and mathematical deduction, one then proceeds by a dialectical examination of the meanings of crucial terms to a rationally intuitive knowledge of the unique, unchanging, eternal essences of all truths and values, the Forms. As he states in his incomparably beautiful way:

> When its gaze is fixed upon an object irradiated by truth and reality, the soul gains understanding and knowledge and is manifestly in possession of intelligence. But when it looks towards that twilight world of things that come into existence and pass away, its sight is dim and it has only opinions and beliefs which shift to and fro, and now it seems like a thing that has no intelligence. [23]

Knowledge of the Forms is described as analogous to perception or vision; just as when we see a real physical object it has an objective existence independent of us, so when we truly know, an object also is disclosed to us, but in this case an intelligible object or Form. Beyond the intuition of the intelligible Forms Plato points to an even higher level of cognition, a kind of mystical apprehension ('*noesis*') of the supreme Form, that of Goodness itself. As mentioned earlier, it is the Form of Goodness which ultimately is the source of the knowledge of the Forms (i.e., their illumination) as well as their existence.

Mathematical: As previously indicated, according to Aristotle the two sources for the belief in the independent existence of the Forms were the Socratic search for the essential meaning of certain words and the Pythagorean assumption of the fundamental reality of numbers. Just as Plato was lured by the question of the ontological status of the referents of the Socratic definitions, so he was enticed by the problem of the ultimate status of mathematical entities. While the Pythagoreans accorded numbers a unique position in their system, their level of intellectualization was such as not to allow them to conceive of numbers as having a status independent of the spatial-physical world. For Plato, however, just as discourse presupposed a realm of unique, timeless, intelligible meanings beyond the imperfect, changing, empirical world, so mathematics presupposed a realm of ideal, supersensible mathematical objects transcending the sensory world of space and time. Thus he was the first to conceive of a purely abstract form of existence, or *subsistence*, independent of the spatial-corporeal world. Had Plato done nothing else, this feat of abstraction alone would have assured his status as one of the world's greatest thinkers.

To this day a number of mathematicians and logicians (e.g., Frege, Russell when he wrote *Principia Mathematica* with Whitehead, and Alonzo Church) have subscribed to the position of Platonic Realism; i.e., to the belief in the independent existence of mathematical entities, relations, and functions as the ground of mathematical discoveries and operations. Just as the disclosure of new elements and their properties in the periodic table presupposed the prior, independent existence of those elements, so for Platonic Realists the discovery of mathematical properties (e.g., that the natural numbers are even or odd, that certain numbers, called irrationals, cannot be expressed as a quotient of two integers, or Goldbach's conjecture that the sum of any two prime numbers is an even number) presupposes the independent existence or subsistence of mathematical units. That is, the mathematician does not appear to create such numerical properties but to disclose or discover them.

Unlike the contemporary mathematicians, however, Plato in his early period not only vigorously disapproved of *applied* mathematics, he did not even advocate the study of *pure* mathematics for its own sake. In the *Republic* he actually chides mathematicians for their concern with mathematical 'operations:'

> [...] no one who has even a slight acquaintance with geometry will deny that the nature of this science is in flat contradiction with the absurd language used by mathematicians [...]. They constantly talk of 'operations' like 'squaring,' 'applying,' 'adding,' and so on, as if the object were to *do* something, whereas the true purpose of the whole subject is knowledge — knowledge, moreover, of what eternally exists, not of anything that comes to be this or that at some time and ceases to be. [24]

As he goes on to state, the real purpose and value of studying mathematics is to lead the mind away from the world of material things, diagrams, and mathematical operations to the apprehension of the eternal Forms, and ultimately, 'towards a comprehension of the essential Form of Goodness.'

Accordingly, while Plato's conception of the status of mathematical entities and their formal properties was superior to that of the Pythagoreans, in that he conceived of numbers as having an abstract subsistence independent of the spatial-temporal world, his understanding of mathematical operations and the application of mathematics in scientific inquiry was considerably less sophisticated and farsighted than that of the Pythagoreans. This is evident, for example, in the contrast between the following statement by the Pythagorean mathematician Archytas, a contemporary of Plato, and that of Plato above:

> Mathematicians seem to me to have excellent discernment, and it is in no way strange that they should think correctly concerning the *nature of particular existences* [...]. Indeed, they have handed on to us a clear judgment on the speed of the constellations and their rising and setting, as well as on geometry and Numbers and solid geometry, and not least on music; for these mathematical studies appear to be related. [25] (Italics added)

In the *Timaeus*, written much later than the *Republic*, Plato's position changed to a more Pythagorean conception of the role of mathematics, applying it to astronomy and even attempting to reduce the primary elements and their interactions to geometrical configurations and mathematical proportions. But even here the primary aim was not to extend knowledge by the use of mathematics, but to show that the universe exhibited a mathematical or intelligible order as evidence of its not having arisen by mere chance or mechanical necessity.

Ontological: Plato's theory of Forms was developed not only to solve certain semantic, moral, and epistemic problems, but also to provide a solution to the ontological questions raised by Parmenides: questions as to the reality of the One and the Many and of Being and Becoming. Parmenides had reduced reality to a motionless, qualityless Being, the One, while Heraclitus, focusing on the aspect of change in experience, had maintained that 'all is flux.' Plato attempted to incorporate both of these ontologies in his conception of knowledge and reality.

The ordinary physical world, consisting of a material domain of individual, imperfect entities in a constant state of perpetual change, represents Heraclitus' world of flux, while the unchanging, supersensible, eternal realm of Forms attests to Parmenides' demand for a timeless, static Being. Unlike Parmenides' Unity, however, there were a plurality of Forms the interrelationship of which constitutes another important aspect of Plato's theory of Forms. In describing Plato's view, Aristotle says,

> to each thing there answers an entity which has the same name and exists apart from the substances, and so also in the case of all other groups, there is *a one over many*, whether these be of this world or eternal.[26] (Italics added)

Aristotle's expression, "a one over many," indicates that the Forms, in addition to being archetypes and the essential meanings of universals and absolute values, also represent the defining characteristics or exemplars of classes of entities. That is, as in Aristotle's system, the Forms represent the various ideal 'kinds' of things, the genera and species, that determine the nature and range of classes of particular entities. There are Forms exemplifying more general or universal classes, such as animal, geometric figure, or number, as well as Forms standing for more restricted groups such as man, triangle, two, etc.

These generic Forms, capable of being combined or 'blended' in various ways (e.g., man blends with animality and rationality, but not with plantness or barking), comprise the interrelated structure of Reality, the pattern emulated by the Demiurge in its attempt to 'persuade' the intractable Receptacle to 'imitate' or 'mirror' this Ideal Realm. The science that investigates the structure or patterning of the Forms is called by Plato 'Dialectic,' or the method of investigation by 'Collection and Division.'

While mathematics consists of deductive reasoning downward from unquestioned assumptions, axioms, or first principles, Dialectic reasons from these assumptions upward, endeavoring to determine the essential meanings or Forms on which the mathematical concepts composing the axioms are based (e.g., point, line, unity, number, etc.). Thus Dialectic, a form of philosophic discourse which seeks the essential definition of terms, also attempts to disclose the hierarchical order of Forms constituting the structure of reality. As Plato asks in the *Sophist*: 'Dividing according to kinds, not taking the same Form for a different one or a different one for the same — is not that the business of the science of Dialectic?'

But he had already given an affirmative answer to this question in the *Republic*:

> [. . .] the summit of the intelligible world is reached in philosophic discussion by one who aspires, through the discourse of reason unaided by any of the senses, to make his way in every case to the essential reality and perseveres until he has grasped by pure intelligence the very nature of Goodness itself. This journey is what we call Dialectic.[27]

This method of Dialectic will aspire to correctly represent, by affirmative and negative judgments, the ideal combination or disjunction of the Forms, culminating in the Form of Goodness, which comprises the transcendent pattern of reality. According to Cornford:

> The expert in Dialectic will guide and control the course of philosophic discussion by his knowledge of how to 'divide by Kinds' not confusing one Form with another. He will discern clearly the hierarchy of Forms which constitutes reality and make out its articulate structure, with which the texture of philosophic discourse must correspond, if it is to express truth.[28]

Plato, accordingly, solved the logical and ontological problems Parmenides posed by affirming two modes of Being, the Forms which satisfied Parmenides' demand for an eternal, unchanging Reality, and the world of Becoming which satisfied Heraclitus' insistence on change and plurality. As the Stranger says in the *Sophist*:

> [. . .] it seems that only one course is open to the philosopher who values knowledge and the rest above all else. He must refuse to accept from the champions either of the one or of the many forms the doctrine that all Reality is changeless, and he must turn a deaf ear to the other party who represent reality as everywhere changing. Like a child begging for 'both', he must declare that Reality or the sum of things is both at once — all that is unchangeable and all that is in change. [29]

Furthermore, the theory of the 'blending of the Forms' enables Plato to avoid the Eleatic conclusion that negative judgments are meaningless because they refer to Not-Being. Contrary to Parmenides' thesis, such judgments do not refer to Not-Being but merely assert (or deny) that certain Forms are combined or related in a certain way. When one asserts that "Ethiopeans are not blond haired" or that "winged men do not exist," one is not asserting the existence of what is not, nor making any claim about Not-Being; instead, one is asserting that certain Forms do not combine in certain ways; i.e., that 'blond-haired' does not combine with 'Ethiopeans,' and that the Form of 'existence' does not combine with that of 'men' and 'being winged.' As Cornford says, the "whole texture of philosophic discourse will consist of affirmative and negative statements about Forms, which should correctly represent their eternal combination or disjunction in the nature of things."[30] While Wittgenstein in the *Tractatus* described statements as "pictures" or "projections" of possible matters of fact, Plato in the *Sophist* conceives of statements as signifying possible combinations of the Forms.

SUMMARY AND CRITICAL EVALUATION

As we have seen, the theory of Forms was introduced as an explanation of the following problems: (1) to account for words having an objective and common meaning for different individuals, (2) to explain how one can refer to absolute values and eternal verities in a world of change and imperfection, (3) to account for the resemblances among objects in terms of their sharing or participating in a common but unique Form, (4) to provide archetypes for the creation of individual copies, (5) to account for mathematical discoveries and the autonomous status of arithmogeometric objects, (6) to satisfy the demand for an unchanging, eternal Being to complement the restless, contingent world of Becoming, and (7) to provide an ideal pattern of possible Formal interconnections as a model for empirical existents and interactions. Thus, as Aristotle said, the Forms were not introduced primarily to solve the kinds of scientific questions raised by the Ionians and the Pluralists, but to provide a metaphysical solution to the questions asked by Socrates and implied in the formalistic inquiries of the Pythagoreans and Parmenideans.

Plato's theory of the Forms has been submitted to such relentless criticism, both in his own time and in the subsequent epochs of intellectual history, that it would be redundant to repeat it here.[31] Suffice it to say that the problems for which the theory of Forms was seen by Plato to be the best possible solution can now be answered in ways that are more precise, more illuminating, and that require fewer metaphysical presuppositions. For example, one can account for words having a common, objective meaning without assuming a direct intuitive knowledge of a supersensory domain of intelligible essences. In fact, the "name theory of meaning," which Wittgenstein also had defended in the *Tractatus*, though in a different version from Plato's, was later abandoned in the *Philosophical Investigations* after an intensive analysis and devastating criticism. [32] Similarly, the argument that a subsisting realm of eternal Ideals must exist to account for absolute value judgments presuppposes that such judgments are possible, a conviction less subscribed to today. Also, there are alternative positions to Platonism to account for mathematical properties and discoveries, such as the logistic, formalistic, and intuitionist schools[33] — Quine, in particular, has attempted to divest mathematics and logic of the ontological presuppositions of Platonism.[34] In addition, as Aristotle caustically remarks, the attempt to account for natural phenomena, as well as their similarities, in terms of empirical objects 'copying' or 'participating in' a realm of transcendent Forms, really explains very little.

> Above all one might discuss the question what in the world the Forms contribute to sensible things, either to those that are eternal or to those that come into being and cease to be; for they cause neither movement nor any change in them [. . .] further, all other things cannot come from the Forms in any of the usual senses of 'from'. And to say that they are patterns and the other things share in them is to use empty words and poetical metaphors.[35]

So rather than a detailed review of the traditional criticisms of the theory of Forms, we shall evaluate the theory as a manifestation of a mode or level of cognitive development, following the main purpose of this study. When one reads the dialogues, one can hardly resist being moved by the imaginative grandeur and moral loftiness of Plato's vision of a transcendent reality of eternal truths and values culminating in the Good. But as soon as we lapse back into the actual world mirrored around us, Plato's view does seem to shrink to a mere "poetic metaphor." With a perspective on the dark side of human history extending twenty-four centuries beyond Plato, with an awareness of the chance, lowly evolutionary origins of man, with an enforced comprehension of the enormous complexity of nature, with the universe appearing so vast as to be beyond any ultimate comprehension, with the realization that human life could be obliterated at any moment — with all this crowding in on one's consciousness it hardly seems credible that reality should be dependent upon some ultimate, absolute good.

Also, while it was a considerable feat of abstraction to free meanings, values, and mathematical objects from their traditional embodiment in spatial, substantial objects, the consequent projection of them onto a separate ontological plane was analogous to the primitve propensity, described earlier, to hypostatize words. That is, initially primitive peoples cannot conceive of abstract qualities, forms or concepts (such as mathematical objects) except as inherent properties of concrete objects or collections of objects. Thus colors, shapes and kinds originally are not thought of as abstractable qualities and forms, but as inseparable from their manifestation in things. Similarly, numbers at first are not conceived as abstract designations for collections of the same units of objects, but as inherent in the collections themselves, as the derived meanings of numbers indicate; e.g., 'digit' referring both to fingers and numerical units, the word for five and hand being the same in many languages.

Gradually, qualities, forms, processes, aspects, etc., were abstracted from their inherence in things and thought of separately, but then it was natural initially to hypostatize or reify these entities, as occurred in the systems of the Presocratics: e.g., love, strife, harmony, logos, *nous*, etc. Similarly, recognizing the importance of meanings, universals, and values in discourse, and abstracting mathematical objects and archetypical forms from concrete existents, Plato hypostatized or reified these abstractions, attributing to them a unique form of existence, that of subsisting beyond the spatial-temporal world.

Thus while the introduction of a unique mode of existence was a tremendous intellectual advancement, Plato's reification of meanings and mathematical objects was not essentially different from the primitive's tendency to hypostatize such aspects of experience as spirit, eros, evil, etc.

The conception of the realm of Forms presented in the *Phaedrus* and the *Sophist* [36] as an interrelated system of 'kinds,' a fixed hierarchy of genera and species, reminds one of the complex classificatory systems of primitives described by Lévi-Strauss. In fact, Plato's conception of the 'method of Collection and Division,' the dialectical representation of the structure of reality, could be considered an abstract rendering of the primitives' attempt to invest reality with a totemic classification. Consider, for example, Wedberg's description of Plato's theory of kinds:

> According to Plato's view, all existing things are eternally distributed into a system of kinds and within this system of kinds he assumes a fixed hierarchy of genera and species. The dialectician who endeavours to become aware of this hierarchical order may start from below, from the particulars, and work his way upwards [. . .] or he may start from above and work downwards, at each stage dividing the genus at hand into its proper species. [37]

Compare this description with Lévi-Strauss's representation of totemism as a complex rational schema of classification.

> As medial classifier [. . .] the species level can widen its net upwards, that is, in the direction of elements, categories, and numbers, or contract downwards, in the direction of proper names. [38]

Plato's system is much more abstract, and the rationale considerably more sophisticated than classfication by totemic organization, but it nonetheless exhibits a similar reification of abstractions, in this case genera and species.

Some scholars may be offended, if not incensed, at the suggestion that a thinker as brilliant as Plato could exhibit primitive modes of thought, but this might be less objectionable if one remembers that individuals, like societies, manifest many different levels of cognition, so that while Plato's reification of abstractions may be somewhat primitive, his thinking within that framework was nonetheless remarkably original and penetrating. More generally, although there is an understandable tendency to exalt Greek philosophical achievements because of their obvious superiority over the earlier mythical traditions, the level of development, measured by modern standards, is still relatively primitive. As Bochner states,

> the Greeks, for all their cleverness, were not able, or not yet able, to make abstractions that were more than idealizations from immediate actuality and "external" reality; their abstractions were rarely more than one-step abstractions, and as such they remained within the purview of what is called "intuitive" in an obvious and direct sense. They did not make second abstractions or, alternately, abstractions from intellectually conceived possibilities and potentialities, and such abstractions of higher order as were made remained rudimentary and operationally unproductive. [39]

Although this judgment is somewhat severe, it does counterbalance the inclination to inflate the intellectual capabilities and achievements of the ancient Greeks. But it should be remembered that intellectual attainments, like other accomplishments, progress by stages, and that secondary abstractions, and abstractions derived from conceived possibilities, could not be realized before the first stage of thinking in terms of primary abstractions had been reached. The difference between these stages of cognitive development can be illustrated concretely by contrasting Popper's notion of world 3, a reinterpretation of Plato's theory of Forms, with Plato's original theory.

Like Plato, Popper wants to account for theoretical discoveries in mathematics, logic, and science. In doing so, he also believes that mathematical and logical calculi, scientific problems and theories, as well as literary meanings and artistic import, have an objective, independent status. But while they were reified by Plato who ascribed to them a prior existence, that of subsisting beyond the empirical world, Popper conceives them to be man-made, usually embodied in physical objects, although possessing inherent, undisclosed meanings, implications, and significances that are the basis of later critiques and discoveries.

> Though Plato's world of intelligible objects corresponds in some ways to our World 3, it is in many respects very different. It consists of what he called "forms" or "ideas" or "essences" [. . .]. These ideas are conceived of as immutable, as timeless or eternal [. . .]. By contrast, our World 3 is man-made in its origin [. . .]. [40]

Popper distinguishes among three worlds: world 1 comprising the natural world of physical objects, world 2 consisting of human consciousness, thoughts, concepts, artistic aims, etc., and world 3 encompassing all the created products of human culture: e.g., manuscripts and books, machines and scientific instruments or apparatus, musical scores and performances, and the various creations of the visual arts. It is in world 3, actualized and implicit, that the rich heritage of human creation exists: epics, novels, poems printed in books; musical compositions inscribed on scores; paintings worked on panels, paper, and canvases; statues sculptured in clay, marble, and bronze; architecture constructed in brick, concrete, marble, steel and glass; scientific theories, mathematical and logical proofs presented in journals, etc. As products of human consciousness and craftsmanship these world 3 objects are man-made, but once created they attain an independent significance. It is their *embodiment* in or as world 1 objects that gives them their independent status, but it is the *medium* and the *form* in which they are created that endows them with their potential significance.

Once created, one can read and interpret novels and poems, critically examine and test scientific theories, reconstruct and check mathematical proofs, evaluate the artistic purpose and technique of a painting or a statue, follow a musical score, critique a work of architecture, etc. It is because of the rich implications of meanings inherent in the languages of literature, the theoretical frameworks of science, and the concepts and symbols of mathematics that make them inexhaustible sources of new insights, interpretations, and discoveries. While these contents are embedded in the product due to the creative abilities of the scholar or artist, working within a particular medium, such as a natural language, musical notes, mathematical symbols, paints, marble, etc., it is the subsequent generations of interpreters who discover, reveal, or bring out their significances. Art historians are still puzzling over the enigmatic smile of the Mona Lisa; Wagner's operas have been reinterpreted for their Freudian symbolism; Minkowski derived new implications from Einstein's relativity theory; the basis of calculus was revised in terms of Cantor's set theory; Mahler's symphonies are subject to numerous interpretations, and so forth.

Though in his last publication Popper referred to "The Reality of Unembodied World 3 Objects,"[41] which implies more of a convergence toward Platonic Realism than one would have expected from his earlier writings, this propensity to reify meanings can be avoided if one acknowledges the wealth of implications contained in certain concepts (e.g., substance, mind, God, time), geometrical forms, mathematical symbols, etc. All of the trigonometric functions, for example, are implicit in the six ratios (e.g., sine, cosine, tangent) derived from the relations of just three lines of variable lengths conjoined so as to form variously shaped triangles with their interior angles. The marvellous computations made possible by the development of calculus originate from the definitions of the derivative and the integral, definitions proved on the basis of previous mathematical concepts and operations. Rather than being "unembodied," all of these conceptual implications and discoveries are implicit in the symbolisms from which they are derived, although considerable ingenuity is required to uncover them.

Accordingly, solutions can now be given to the kinds of problems raised by Plato which avoid the reifying implications and metaphysical assumptions of his views. If it is asked why these latter consequences are less preferable, the reply is simply that from the contemporary perspective they raise more questions than they solve and are less explanatory than modern alternatives. But while Plato's solutions were necessarily more rudimentary, they exhibit an extraordinary element of foresight that accounts for their continual appeal to later thinkers. For example, consider his doctrine of anamnesis. In itself, the explanation of learning as the recalling of knowledge lost by the soul from a previous life will strike thinkers today as fantastic. Yet with a little emendation one can see Plato's explanation as foreshadowing the contemporary position that all learned knowledge is derived from inherited structures and capacities acquired as a result of the preexistent experiences of our species. Thus the general schema of Plato's view that all learning depends upon a prior endowment has reappeared in a revised form in such diverse theories of knowledge as those of St. Augustine, Descartes, Kant, and Piaget. Again, it is this remarkable capacity for readaptation that accounts for the seeming modernity of Plato's views and their continuing attraction to later scholars such as Whitehead.

PLATO'S CONCEPTION OF ASTRONOMICAL INQUIRY

After the theory of Forms, Plato's intellectual legacy stems mainly from his considerable influence on later astronomical developments, either indirectly because of astronomical research pursued by others in the Academy or directly owing to his own teaching and writings. While there is no question that the Academy had a strong impact on astronomical developments due especially to two of Plato's students and/or associates, Eudoxus and Heraclides of Pontus, the specific nature of Plato's own influence as a result of his conception of astronomical inquiry has been much debated. Though a definitive answer to this question is not possible, as indicated earlier, any assessment of Plato's contribution to the development of scientific rationalism must take into account this question.

There is no uncertainty, however, regarding the continued significance, indeed increasing importance, that Plato attributed to astronomy as an intellectual discipline. In the *Republic* it was considered, along with arithmetic, geometry (including steriometry or solid geometry), and harmonics, as one of the four sciences in the proper sense of that word, that of leading the mind away from the material world to a knowledge of the immutable and eternal Forms. Thus the role of astronomy in the education of the Guardians was primarily instrumental, that of guiding the intellect away from sensible appearances to the intelligible Realities lying above them. This is dramatized in the well-known passage where Socrates rebukes Glaucon for suggesting that the study of astronomy is especially suited to compel 'the mind to look upwards, away from this world of ours to higher things.'[42] This would depend, according to Socrates, on whether one means by 'higher things' the visible appearances of the heavens or the 'unseen reality' lying beyond, for it is only as regards the latter that one can gain true knowledge.

In the *Laws*, Plato's last dialogue except, perhaps, for the *Epinomis*, astronomy is assigned an important place in the general education of all of the citizenry, as well as an enhanced role in the education of the governing body, the 'nocturnal council.' All citizens are to be taught enough astronomy to understand the calendar and to observe the 'proper celebrations,' as well as to 'prevent blasphemy [. . .] and to ensure a reverent piety in the language of all our sacrifices and prayers.'[43] The 'noblest of all sciences' are now reduced to three: arithmetic, 'mensuration' (linear, superficial, and solid), and astronomy, the latter referred to as 'sublime.' Speaking as the Athenian Stranger, Plato says:

> What I am trying to say, I know, is startling, and might be thought unbecoming in a man of our years, but the plain truth is that a man who knows of a study which he believes sublime, true, beneficial to society, and perfectly acceptable to God, simply cannot refrain from calling attention to it. [44]

Furthermore, in the *Epinomis* (which either was the last work of Plato or at least was written in the immediate Platonic tradition), the study of astronomy was considered essential to attaining piety, now referred to as the highest virtue.

> There is no human virtue — and we must never let ourselves be argued out of this belief — greater than *piety*, and [. . .] I must tell you [. . .] how a man should learn piety, and in what it consists. It may seem odd to the ear, but the name *we* give to the study is one which will surprise a person unfamiliar with the subject — *astronomy*. Are you unaware that the true astronomer must be a man of great wisdom? [45]

Thus while Plato continued to advocate the study of astronomy not just for its own sake, but in terms of its beneficial effects upon the educational, moral, and religious development of the citizenry and rulers, he attached even greater importance to it in his later life. It was this emphasis on astronomy as an intellectual discipline, along with arithmetic and geometry (the inscription over the door of the Academy read, 'Let no one destitute of geometry enter my doors'), that had such a distinctive impact on the development of astronomy — the influence manifest in the famous challenge to later astronomers, attributed to Plato by Eudemus, to find 'what uniform and orderly motions can be postulated to save the appearances in the matter of the motion of the planets.' [46]

But what of Plato's conception of astronomical inquiry as such? Did his disparagement of beliefs derived from observations of the imperfect world of change and becoming and his conviction that knowledge as such must refer to the intelligible Realities lying beyond prevent his appreciating the importance of observation in astronomy, as many critics have claimed? For example, Neugebauer says of Plato that his "advice to the astronomers to replace observations by speculation would have destroyed one of the most important contributions of the Greeks to the exact sciences."[47] Or, as Dicks contends, had Plato arrived at the correct realization that astronomical observations had reached the limit of their effectiveness by the beginning of the 4th century and that what was needed was for astronomers to do "some hard thinking about astronomical theory, so as to make the best use of the observations they already had"?[48] — this "hard thinking" consisting of the attempt to derive a mathematically exact account of the irregularities in the motions of the planets (which are denied by Plato in the *Laws*) and of the inexact measurements of the days, months, and years. That is, rather than discounting entirely the need for observation was Plato envisioning a mathematical astronomy based on exact laws of motion of which the visible heaven is merely an imperfect representation?

In the *Republic* where the crucial passages occur, the inexactness of the expressions allows for both interpretations, perhaps indicating that the problem was not resolved in his own thinking. For example, after having rebuked Glaucon for suggesting that the mind can be enlightened merely by looking upward at the heavens, Socrates goes on to say:

> These intricate traceries in the sky are, no doubt, the loveliest and most perfect of material things, but still part of the visible world, and therefore, they fall far short of the true realities — the real relative velocities, in the world of pure number and all perfect geometrical figures, of the movements which carry round the bodies involved in them. These, you will agree, can be conceived by reason and thought, not seen by the eye. (529d)

Here Plato certainly appreciated one essential aspect of astronomy, the necessity of representing by 'numbers' and 'geometrical figures' the true motions of the celestial bodies not directly apparent in the visible sky. But did he also believe, as Kepler did, that the determination and proof of the true motions must be based on observational data (for Kepler, the data supplied by the exact observations of Tycho Brahe)? It is this question which is difficult to answer, for on the one hand he does refer to the visible or 'embroidered heaven as a model to illustrate our study' of the true mathematical realities, and yet on the other hand, he thinks it 'absurd to study them in all earnest with the expectation of finding in their proportions the exact ratio of any one number to another' (529e).

In other words, did Plato believe that the astronomical data available at the time were too imprecise to yield an exact mathematical description or did he believe that the astronomical bodies themselves were inherently too imperfect to allow for a precise mathematical representation? Unfortunately, however, what is never completely clarified is the relation or connection between the observations and the 'mathematical realities.' While he realized that the 'embroidered heaven' must be used as the model in astronomical inquiries, he still seemed to consider the domains of mathematics and observation *essentially independent*, the one having to do with the unchanging, exact, true Realities, the other with the changing, inexact world of imperfect material things.

> The genuine astronomer [. . .] will admit that the sky with all that it contains has been framed by its artificer with the highest perfection of which such works are capable. But when it comes to the proportions of day to night, of day and night to month, of month to year, and of the periods of other stars to Sun and Moon and to one another, he will think it absurd to believe that these visible material things go on forever without change or the slightest deviation, and to spend all his pains on trying to find exact truth in them. (530b)

As the passage indicates, Plato believed that it was the *inherent imperfection* of material bodies and their *consequent changes* that prevented finding an 'exact truth in them;' in contrast, Kepler rejected his earliest mathematical calculations even though they were accurate to 8' of arc because they did not conform precisely to Tycho Brahe's very exact observations. While modern scientists explain the deviations between observations and the mathematical formalism as due to imprecision in observations and experimental techniques (ignoring for the present the limitations of quantum mechanical investigations), Plato accounted for such deviations in terms of the inherent imperfection of physical reality itself. This would explain why he drew such an absolute distinction between the intelligible world of immutable Being and the perceptible world of imperfect Becoming.

To continue, Socrates then concludes the discussion with the highly contentious statement the exact meaning of which, if known, would probably resolve the whole question.

> If we mean, then, to turn the soul's native intelligence to its proper use by a genuine study of astronomy, we shall proceed, as we do in geometry, by means of problems, and leave the starry heavens alone. (530c)

The ambiguity pertains to what Plato meant by 'problems' and by 'leaving the starry heavens alone.' By the latter did he mean to deprecate astronomical observations in favor of a purely *a priori* mathematical approach, and thus retard the development of scientific astronomy? Or did he merely mean, as Dicks contends, that one should leave aside further astronomical observations and concentrate on an exact mathematical formulation of the laws of motion? Further, what did he mean by proceeding in astronomy as one does in geometry 'by means of problems'?

We shall probably never arrive at the exact interpretation, but the reference to "problems" is significant because, as Heath has pointed out, the rapid development of geometry during Plato's time depended largely upon investigating individual geometrical problems, such as the trisection of any angle and the duplication of the cube.[49] Plato himself took an interest in these problems, as shown in the *Meno*. So the phrase 'proceed by means of problems' seemed to mean that if one focused on the mathematical problems of astronomy, as mathematicians had concentrated on geometrical problems, one would achieve greater success than by continued observations. This, moreover, parallels the story of Plutarch, accepted by Heath, that Plato

blamed Eudoxus, Archytas and Menaechmus for trying to reduce the duplication of the cube to mechanical constructions by means of instruments, on the ground that 'the good of geometry is thereby lost and destroyed, as it is brought back to things of sense instead of being directed upward and grasping at eternal and incorporeal images'.[50]

Nonetheless, this advice still ignores the modern distinction between pure mathematical problems and the application of mathematics to astronomy. It is only if Plato believed, as contended earlier, that the mathematical aspect of astronomical inquiry was entirely independent of the observational that the analogy with mathematics would hold. Even so, it is the case that physicists have had either to look to the mathematician or to develop their own mathematical techniques (as Newton found it necessary to develop integral and Einstein tensor calculus) before they could make further headway in their physical inquiries. This may be what Plato intended, although it does not appear that he held the modern conception of the relation of mathematics and scientific inquiry expressed in the term "applied mathematics."

Moreover, the absolute distinction between the intelligible realm of Forms, and the visible empirical world, is reinforced in Plato's continued emphasis on the difference between a special 'faculty of the soul' capable of arriving at true knowledge, and our eyes which, however necessary and marvellous for observing the heavens, cannot discern truth. As Socrates says, 'it is quite hard to realize that every soul possesses an organ better worth saving than a thousand eyes, because it is our only means of seeing the truth [. . .]' (527e). This again implies a lack of an appreciation of the interrelation between observation and thought deemed so essential in modern science, although it does confirm the Platonic doctrine that our senses can only impede a special faculty of the soul in its purely intellectual quest for knowledge.

Even in the *Timaeus* where Plato attributed a greater importance to vision, asserting that 'sight [. . .] has been the cause of the greatest blessing to us,' since the Demiurge had created the sun to illuminate the movements of the heavens and had given us eyes to discern day and night, the month and the year, and thus enabled us to learn mathematics and philosophy, he still held that it was only by our intellects that we could come to know the true movements of the celestial bodies. Finally, if we consider the *Epinomis* an authentic work of Plato (or at least consistent with Platonism), there knowledge is clearly described as a 'revelation,' independent of sensory observation, of a unified mathematical order pervading all phenomena.

> To the man who pursues his studies in the proper way, all geometric constructions, all systems of numbers, all duly constituted melodic progressions, the single ordered scheme of all celestial revolutions, should disclose themselves, and disclose themselves they will, if, as I say, a man pursues his studies aright with his *mind's* eye fixed on their single end. As such a man reflects, he will receive the *revelation of a single bond of natural interconnection between all these problems.* If such matters are handled in any other spirit, a man, as I am saying, will need to invoke his luck.[51] (Italics added)

It is this sense of a direct 'revelation' of the structure of reality that seems to be uppermost in Plato's mind throughout all his writings. Moreover, the conviction that all natural phenomena are unified by 'a single bond' expresses a magnificent faith in the ultimate intelligibility of all reality, the kind of faith that sustained someone like Einstein (as well as Kepler and Newton) who, in spite of the well-intended criticism of his colleagues who felt that he had deserted them for the last forty years of his life in his isolated quest for a unified field theory, persisted in the search for the scientific grail. Thus Plato will always epitomize man's unending search for a rational understanding of reality, expressed in mathematical terms, however imperfect his own conception of this understanding was at the time.

1 This is a paraphrase of a "Greek epigram," the source of which is Scott Buchanan's "Introduction to Plato," *The Republic* (New York: World Pub. Co., 1946), p. 5.

2 A. N. Whitehead, *Process and Reality* (New York: The Social Science Book Store, 1929), p. 63.

3 Sir Thomas L. Heath, *A History of Greek Mathematics* (Oxford: At The Clarendon Press, 1921), Vol. 1, p. 316.

4 F. M. Cornford, *Plato's Cosmology* (New York: The Liberal Arts Press, [1937] 1957), p. 34. Cornford refers to both Whitehead and A. E. Taylor in this critical sense.

5 D. R. Dicks, *Early Greek Astronomy to Aristotle* (Ithaca: Cornell Univ. Press, 1970), p. 244, f.n. 212.

6 *Republic*, VII, 517. (Cornford trans.)

7 *Timaeus*, 29c-d. (Cornford trans.)

8 For Plato's "Autobiography," as well as an account of his life, cf. Paul Friedländer, *Plato*, Vol. I, trans. by Hans Meyerhoff, sec. ed. (Princeton: Princeton Univ. Press, [1958] 1969), Part I.

9 I. M. Crombie, *An Examination of Plato's Doctrines* (New York: The Humanities Press, 1963), Vol. II, p. 191.

10 For example, while I admire the comprehensive two-volume work of Crombie on Plato, it seems to me that he often interprets and analyzes Plato in terms of the "ordinary language analysis" approach of English philosophers, an approach which, in my opinion, is extremely limited and quite alien to Plato's main intent.

11 Cf. Cornford, *The Republic of Plato* (London: Oxford University Press, 1941), pp. 248-9; cf. Heath, *Aristarchus of Samos* (Oxford: At The Clarendon Press, 1913), pp. 136-7; cf. Dicks, *op. cit.*, pp. 104-5, cf. Paul Shorey, *Republic*, in *Plato*, ed. by Edith Hamilton and Huntington Cairns (New York: Bollinger Series LXXI, Pantheon Books, 1961), p. 762.

12 Cornford, *Plato's Theory of Knowledge* (New York: The Liberal Arts Press, [1934] 1957), p. 2.

13 *Phaedo*, 76d-e. (Tredennick trans.)

14 Cf. *Meno*, 81d. (Guthrie trans.)

15 *Met.*, 987b. (Ross trans.)

16 There is some disagreement as to the correct interpretation. Aristotle attributes an intermediate position to the objects of mathematics in Plato's system. Anders Wedberg, in *Plato's Philosophy of Mathematics* (Stockholm: Almqvist & Wiksell, 1955), pp. 61-68, agrees with Aristotle, distinguishing between mathematical and geometrical *ideas* that exist with the Forms and Ideal arithmetical and geometrical *objects* that have an intermediate existence between sensible objects and the supersensible Forms.

17 Aristotle, *Met.*, 987b. (Ross trans.)

18 *Parmenides*, 147d-e. (Cornford trans.)

19 *Phaedo*, 75c-d. (Tredennick trans.)

20 *Republic*, VI, 507. (Cornford trans.)

21 *Ibid.*, VI, 508.

22 *Ibid.*, V, 477-478.

23 *Ibid.*, VI, 508.

24 *Ibid.*, VII, 527c.

25 Archytas fragments, Diels 44B1. Quoted from S. Sambursky, *The Physical World of the Greeks* (*op. cit.*, ch. IV), p. 55.

26 *Met.*, 1078b-1079a. (Ross trans.)

27 *Republic*, VII, 532. (Cornford trans.)

28 Cornford, *Plato's Theory of Knowledge, op. cit.*, pp. 263-264.

29 *Sophist*, 249c. (Cornford trans.)

30 Cornford, *Plato's Theory of Knowledge, op. cit.*, p. 261.

31 For example, see the discussion by Crombie, *op. cit.*, pp. 247-473. Plato himself was quite aware of the various problems regarding the Theory of Forms, even submitting the theory to extensive examination and criticism, as in the *Parmenides*. Aristotle, however, was the most resolute critic of the independent existence (i.e., the Platonic status) of the Forms. For example, cf. *On Generation*, 335b and *Metaphysics*, Alpha, chs. 7, 8, 9; Beta, chs. 2, 4, 6; Kappa, chs. 1, 2; Mu, chs. 4, 5, 9.

32 Cf. Ludwig Wittgenstein, *Philosophical Investigations*, trans. by G. E. M. Anscombe (Oxford: Basil Blackwell, 1953), Part I, sec. 46.

33 Cf. Max Black, *The Nature of Mathematics* (Paterson: Littlefield, Adams & Co., 1959).

34 Cf. Willard Van Quine, *From a Logical Point of View* (*op. cit.*, ch. I), chs. I, VI.

35 Aristotle, *Met.*, 991a8-22. (Ross trans.)

36 Cf. *Phaedrus*, 265d-e and Cornford, *Plato's Theory of Knowledge, op. cit.*, pp. 252-273.

37 Wedberg, *op. cit.*, p. 42.

38 Previously quoted, ch. 2.

39 Salomon Bochner, *The Role of Mathematics in the Rise of Science* (Princeton: Princeton Univ. Press, 1966), p. 51.

40 Karl R. Popper and John C. Eccles, *The Self and Its Brain* (New York: Springer-Verlag New York, Inc., 1977), p. 43. Cf. also Karl R. Popper, *Objective Knowledge* (Oxford: At The Clarendon Press, 1972), chs. 3, 4.

41 Popper and Eccles, *op. cit.*, p. 46.

42 This and all following quotations from the *Republic* are from Bk. VII, Cornford translation.

43 *Laws*, 809d, 821d. (Taylor trans.)

44 *Ibid.*, 821b.

45 *Epinomis*, 989b-990a. (Taylor trans.)

46 An attribution owing to Eudemus, "commenting on Aristotle's *De Caelo*, as quoted by Simplicius." Cf. Crombie, *op. cit.*, p. 188, f.n. 1.

47 Otto Neugebauer, *The Exact Sciences In Antiquity*, 2nd ed. (New York: Harper Torchbooks, [1952] 1957), p. 152. For a list of other critics, cf. Dicks, *op. cit.*, pp. 234-235, f.n. 136, 143.

48 Dicks, *op. cit.*, p. 107.

49 Cf. Heath, *A History of Greek Mathematics*, Vol. I, *op. cit.*, ch. VII.

50 *Ibid.*, p. 287.

51 *Epinomis*, 991e-992a. (Taylor trans.)

CHAPTER XII

PLATO'S MYTHICAL COSMOLOGY

[. . .] *once in the world's history a supreme philosophical*
thinker should also have been a superb dramatic artist. [1]
Taylor

There is no doubt that Plato combined the role of a consummate philosophical thinker with that of a superb literary draftsman. This unique blend of talents is particularly evident in the *Timaeus* where he mixes on his philosophical palette metaphor and myth with ingenious insights and astute reasoning to paint an imaginative picture of reality. For unlike the Milesians, Atomists, and modern scientists, Plato did not believe that a strictly scientific explanation of the empirical world — the inchoate world of change and becoming — was possible, and hence a more faithful account, even if largely symbolic, would be one that described the world not as arising accidentally or from sheer necessity, but as fulfilling a rational purpose. The result is a "likely story," a conjectured myth of creation, including a narrative astronomical description and a fanciful cosmology. As Cornford asserts:

> The cosmology of the *Timaeus* is poetry, an image that may come nearer to conveying truth than some other cosmologies. But the truth to which it can approximate is not an exact and literal statement of 'physical laws', such as modern science dreams of; it is the truth, firmly believed by Plato, that the world is not solely the outcome of blind chance or necessity, but shows the working of a divine intelligence. Plato would have claimed that, considered as an explanation of sensible appearances, his own theory [. . .] was quite as plausible as the atomic theory of Democritus. [2]

Nonetheless, Plato's treatment of these scientific topics cannot come but as a disappointment after the serious attempts of the Milesians, Pluralists, and Atomists to give a more empirically satisfying and intellectually plausible account of the origin and structure of the universe. In place of Anaximander's attempt to explain celestial phenomena with models drawn from terrestrial experiences (e.g., the use of wheels and bellows to explain eclipses), or Anaxagoras' bold claim that the celestial bodies were molten stones with the moon containing plains and ravines similar to the earth, or the Atomists' explanation of the origin and structure of the universe along the lines of the nebular hypothesis and the steady-state theory — instead of these advances toward a more naturalistic science devoid of animism, we find Plato regressing to the mythical or anthropomorphic notions of a "maker" of the universe, of the soul as the "self-moving" origin of motion, of the conception of the stars and planets as "divine gods," and of explanations in terms of abstractions (the Forms) derived from philosophical and moral discourse. Yet mingled with these primitive notions are brilliant speculations and prescient anticipations of later developments in mathematical physics and astronomy so characteristic of Plato's remarkable imaginative powers. Once more one sees in his "many faceted intellect" the reflections of a mythologist, mathematician, metaphysician, physicist, astronomer, and mystic.

THE PRESUPPOSITIONS OF PLATO'S COSMOLOGY

Even prior to the *Timaeus*, Plato indicates in two earlier dialogues, the *Phaedo* and the *Sophist*, the general presuppositions on which his cosmological views would be based. In the *Phaedo* Plato recounts Socrates' early interest in natural science along with his subsequent disappointment in encountering so many conflicting explanations of phenomena that he became 'befogged by these speculations,' even to the extent of doubting what he previously thought he knew.[3] He took heart again, however, when he 'heard someone reading from a book by Anaxagoras' in which it was stated that mind (*Nous*) is the original cause of everything as well as the source of order in the universe.

> I assumed that he [Anaxagoras] would begin by informing us whether the earth is flat or round, and would then proceed to explain in detail the reason and logical necessity for this by stating how and *why it was better* that it should be so [. . .]. I was prepared also in the same way to receive instruction about the sun and moon and the other heavenly bodies, about their relative velocities and their orbits and all the other phenomena connected with them — in what way *it is better* for each one of them to act or be acted upon as it is. It never entered my head that a man who asserted that the ordering of things is due to mind would offer any other explanation for them than that *it is best* for them to be as they are.[4] (Brackets and italics added)

Upon reading Anaxagoras' book, however, he was shocked to find that instead of explaining phenomena in terms of 'what is best and the highest good,' as ordained by mind, Anaxagoras resorted to 'causes like air and aether and water and many other absurdities.' For Socrates believed in a different kind of causality, as he goes on to state:

> [. . .] see whether you share my opinion. It seems to me that whatever else is beautiful apart from absolute beauty is beautiful because it partakes of that absolute beauty, and for no other reason. Do you accept this kind of causality?
> Yes I do.
> Well, now, that is as far as my mind goes; I cannot understand these other ingenious theories of causation.[5]

Here is evidenced a primitive form of thought in which the causes of phenomena are explained magically by their 'partaking of' absolute Forms — reminiscent of Lévy-Bruhl's "law of Participation."[6] Furthermore, Socrates' rejection of Anaxagoras' explanations in terms of natural causes, because of his preferences for conscious intentions or purposes, recalls Lévy-Bruhl's contention that the primitive thinker is never satisfied with explanations in terms of secondary causes, but seeks more fundamental reasons as to the "why" of things.[7]

As evidence of the deep and lasting influence of Socrates on his thought, Plato incorporated each of these principles in his metaphysical cosmology: (1) that no cosmological explanation would be adequate that did not show that the order in the universe is good and therefore must have been created by an intelligent mind according to an ideal plan, and (2) that the ultimate cause of the nature of physical objects is their "participation in" the perfect, unchanging, eternal, hence ideally good, realm of Forms.

Later in the *Sophist* Plato indicates his dissatisfaction with a purely materialistic or atomistic account of the world in the allegory of the "Battle of Gods and Giants." In the battle, the Gods representing the Idealists, the believers in unseen intelligible entities as the true realities, are pitted against the Giants, the advocates of a view of reality conceived in terms of visual-tactual sensations, a reality of 'tangible bodies' — the model of reality that was reintroduced in the 17th century and that lasted for 300 years. As the Stranger recounts the story:

> What we shall see is something like a Battle of Gods and Giants [. . .]. One party is trying to drag everything down to earth out of heaven and the unseen [. . .] for they lay hold upon every stock and stone and strenuously affirm that real existence belongs only to that which can be handled and offers resistance to the touch. They define reality as the same thing as body [. . .]. [The other party] their adversaries are very wary in defending their position somewhere in the heights of the unseen, maintaining with all their force that true reality consists in certain intelligible and bodiless forms. [8] (Brackets added)

After repelling the atomic-materialism of the Giants, Plato forces the Idealists on the side of the Gods (i.e., Parmenides in particular as well as the Pythagoreans) to concede that reality must also include objects of change and becoming along with the eternal, unchanging Forms. Thus while rejecting a materialistic ontology, Plato indicates again that he will incorporate into his cosmology Heraclitus' world of becoming along with the One, unchanging Being of Parmenides. Accordingly, from the earliest to the latest dialogues Plato's thinking consistently exhibits the following tenets: (1) the rejection of a materialistic ontology, (2) the belief that the universe exhibits a type of order and purpose that attests to its having been created by mind, and (3) the assumption that soul antedates bodies and is the only proper source of motion. As he says in the *Laws*, written during the last thirteen years of his life:

> For on a shortsighted view, the whole moving contents of the heavens seemed to them [Anaxagoras and Democritus] a parcel of stones, earth, and other soulless bodies [. . .]. It was this that involved the thinkers in those days in so many charges of infidelity and so much unpopularity [. . .] but, as I told you, today the position has been reversed [. . .]. No son of man will ever come to a settled fear of God until he has grasped the two truths we are now affirming, the soul's dateless anteriority to all things generable, her immortality and sovereignty over the world of bodies, and moreover that presence among the heavenly bodies of a mind [. . .]. [9] (Brackets added)

COSMOLOGY AND ASTRONOMY

It is the *Timaeus*, of course, that contains the most comprehensive statement of Plato's cosmological, astronomical, and physiological views. The dialogue begins with an account of the creation and structure of the universe, including a detailed description of the motions of the stars and the planets, goes on to present a geometric reconstruction of the physical elements, fire, air, earth, and water, and ends with a description of the origin of man and his composition. Just as in the *Republic* Plato had based his conception of the organization of the state on the model of the tripartite structure of the soul, so in the *Timaeus* he seems intent on showing that the human soul reflects the divine order of the cosmos.[10] Plato is thus one source of that particularly Greek conception of the human being as an analogue of the cosmos, naturally in harmony with and at home in the universe, a conception incompatible with the supernaturalism of Christianity and negated by the gradual estrangement of man from the universe as a result of the development of modern science.

After an introductory conversation Timaeus states the general principles on which the cosmological account is to be based: (1) only that which is apprehended by thought is devoid of change and becoming and thus has eternal, real Being; (2) that which is the object of sensation and belief comes to be and passes away and thus presupposes a causal agency; (3) both heaven and earth can be seen and touched which indicates that they have come to be and therefore require a causal agent; (4) only that which has been created in the image or likeness (*eikon*) of an unchanging, eternal archetype will be good; (5) since it is obvious that 'the world is the best of things that have become,' the 'maker and father of this universe' clearly fashioned the world after an eternal model; (6) having been created according to an intelligible model, the world is comprehensible to rational understanding, although no more than a 'likely story' can be given of the account.[11] Thus Plato "introduced, for the first time in Greek philosophy," a common form of mythical explanation, that of the divine craftsman or artificer modeling the universe from preexistent materials according to a given ideal pattern. In Plato's mind the connection between 'cause' and 'maker' was so close that the meaning of one implied the other, as he states in the *Philebus*:

> Socrates: Well, is there anything more than a verbal difference between a cause and a maker? Wouldn't it be proper to call that which makes things and that which causes them one and the same?
> Protarchus: Quite proper. [12]

Accordingly, Plato's cosmology incorporates three independently existing components: (1) the eternal world of Forms as the model of creation symbolizing the perfect, unchanging, eternal order of Reality; (2) the Divine Craftsman or Demiurge symbolizing the rational causal agency inducing intelligent order, design, and purpose in a recalcitrant world; and (3) the Receptacle as the spatial container of the materials for the creative process symbolizing the sensory, imperfect, 'discordant' world of becoming. Since the Demiurge is good, and being good desired that all things be made as ideal as possible, he brought things from disorder into order, judging that to be better. Noting further that what conforms to intelligence is better than what does not, he 'fashioned' reason within the soul and soul within the body in order that nature be as perfect as possible.

> Desiring, then, that all things should be good and, so far as might be, nothing imperfect, the god took over all that is visible — not at rest, but in discordant and unordered motion — and brought it from disorder into order [. . .] he fashioned reason within soul and soul within body, to the end that the work he accomplished might be by nature as excellent and perfect as possible. This, then, is how we must say, according to the likely account, that this world came to be, by the god's providence, in very truth a living creature with soul and reason. [13]

Each of the components of Plato's cosmology will now be discussed separately.

The Demiurge: While the Demiurge is a creator god, little is actually known of his nature. Unlike the God of the Christian tradition, Plato's god is neither a religious figure nor an object of worship, nor does he create *ex nihilo*, but attempts to 'overcome Necessity with Reason' by 'persuading' the intractable aspects of chaos and disorder in the preexistent Receptacle to 'participate in' or 'mirror' the intelligible order exemplified by the world of Forms. Insofar as the disorder or 'Necessity' inherent in the Receptacle is never completely overcome, Plato's god is not omnipotent as is the Christian God. Neither the Receptacle nor the Forms are created by or exist in the Demiurge. It is not clear, however, as to how literal or figurative Plato intended the meaning of the Demiurge to be. According to Cornford:

> He is mythical in that he is not really a creator god, distinct from the universe he is represented as making [. . .]. On the other hand, there is no doubt that he stands for a divine Reason working for ends that are good. [14]

Since Timaeus says that 'it is a hard task to find the maker and father of this universe, and having found him it would be impossible to declare him to all mankind,' any further definition of Plato's god is largely futile.

The Creator's Model: The visible world having been portrayed as a living creature made after the likeness of an eternal intelligible model, this model is now described as a generic Form possessing as 'parts' the ideal types of all the various species of living creatures inhabiting our world. It is called the 'intelligible Living Creature' to distinguish it from the 'Living Creature' which is the visible world — the term 'intelligible' signifying that it is an eternal object of either divine or human rational thought. These generic forms comprise the structure of intelligible Reality which it is 'the method of Composition and Division' or the 'Science of Dialectic' to disclose and include the 'four main families' of living creatures: the heavenly gods (stars, planets, and earth), birds, fish, and animals. They therefore encompass all the species of living creatures among which Plato includes the celestial bodies as created gods. As there could be only one unique copy of the 'intelligible Living World,' the Demiurge did not make more than one world and that which he did make is everlasting in the sense that it is not destroyed and superseded by another, even though the elements in it are constantly undergoing change.

The World's Body: Preliminary to a more complete description of the composition of the visible world in the second part of the dialogue, Plato gives a brief sketch of the nature of 'the world's body' in the first part. Like Empedocles, he believes that the world consists of four primary entities, fire, air, water, and earth, but he goes beyond Empedocles in deducing reasons as to why these are the primary entities; fire is necessary that the universe be visible, earth that it be tangible, and air and water 'conjoined' between them as a 'bond' so that the world 'be solid in form' (31b-32c). Air is necessary as a bond between fire and earth and water adds three-dimensional solidity. Plato depicts the 'bond' holding the elements in the most perfect unity and harmony as a *geometrical proportion*, thereby substituting a mathematical harmony for mechanical forces as the power binding the primary entities together.

Along with this Pythagorean notion, he adopts the Eleatic conception that the world or living creature is 'a sphere, without organs or limbs, rotating on its axis.' It possesses no eyes or ears, feet or legs, or organs for breathing or eating:

> For nothing went out or came into it from anywhere, since there was nothing: it was designed to feed itself on its own waste and to act and be acted upon entirely by itself and within itself; because its framer thought that it would be better self-sufficient, rather than dependent upon anything else. (33c-d)

Like Xenophanes, Plato rejects a gross anthropomorphic conception of the visible world, but nonetheless conceives of it as a living creature, analogous to that of Parmenides and Melissus. This certainly is the most primitive aspect of Plato's cosmology, even when taken in a metaphorical or symbolic sense. The smooth spherical universe as a whole has one motion, rotation in one place, as this motion is most akin to the 'revolution of reason' — again, a conception taken from Parmenides.

The World-Soul: The purpose of the *Timaeus* being to show that the sensible world is the visible manifestation of a Divine Reason, and since reason 'cannot be present in anything apart from soul,' and, as argued also in the *Phaedrus* and later in the *Laws*,[15] the soul is the only conceivable self-moving source of motion, the visible universe which is both alive and in motion must possess soul. Plato therefore adopts the primitive animistic notion that the various motions of the universe, such as the rotation of the world as a whole and the revolutions of the heavenly gods must be due to a World-Soul which pervades the universe. This soul, extending from the center to the circumference of the spherical world, exists prior to body, although when combined with the World's Body the latter also becomes everlasting.

Nothing illustrates Plato's imaginative freedom (or fantasy) in his cosmological construction more than his choice of abstractions in describing the composition of the World-Soul. In the *Sophist* he had answered Parmenides' linguistic paradox as to how one can express negative statements without asserting meaningless judgments about Not-Being, by describing the way in which three kinds of Forms, 'Existence,' 'Sameness,' and 'Difference,' are connected in true affirmative statements and disjoined in negative statements.[16] Now he uses these linguistic or formal abstractions to describe the composition of the World-Soul.

> The things of which he [the Demiurge] composed soul and the manner of its composition were as follows: (1) Between the indivisible Existence that is ever in the same state and the divisible Existence that becomes in bodies, he compounded a third form of Existence composed of both. (2) Again, in the case of Sameness and in that of Difference, he also on the same principle made a compound intermediate between that kind of them which is indivisible and the kind that is divisible in bodies. (3) Then, taking the three, he blended them all into a unity, forcing the nature of Difference, hard as it was to mingle, into union with Sameness and mixing them together with Existence. (35a) (Brackets added)

The World-Soul is thereby compounded out of three abstract ingredients, intermediate 'Existence,' 'Sameness,' and 'Difference,' 'blended' into a unity. Their 'intermediate' nature distinguishes them from the 'indivisible existence' of the Forms and the 'divisible existence' of bodily entities, thus enabling the Soul, both Divine and human, to partake in and know both worlds, the unchanging intelligible Forms in the eternal world and the perishable sensible copies in the world of becoming.

The Demiurge then divides the mixture of soul-stuff into strips of different lengths with which he will represent the various motions of the universe and the celestial bodies. According to Cornford, in this mythical rendering of the construction of the motions of the heavens Plato had before him an actual model of an armillary sphere,[17] and imagined the Demiurge creating the motions of the celestial bodies as one would construct such a sphere; Dicks, on the contrary, denies that Plato could have had access to such a device, stating that "it is highly improbable that Plato had any other physical model at the back of his mind except a simple celestial globe."[18] In any case, having formed the soul-stuff into a single long band, the Demiurge then marks off the entire length into 'portions' measured by the numbers forming two series of four terms each: 1, 2, 4, 8 and 1, 3, 9, 27. (35b-c) These intervals in turn are further sub-divided according to arithmetical and harmonic means (the latter corresponding to the intervals of the diatonic musical scale), such that every portion has a 'rational' arithmetical and/or harmonic relation to every other.[19]

This long band with its marked off divisions is then slit lengthwise into two halves, each of which is connected in a circle with one placed inside the other, the outer circle positioned horizontally to represent the sidereal equator, the inner circle inclined to correspond to the Zodiac. The strip forming the sidereal equator is given a circular motion around its axis from East (Left) to West (Right), a motion corresponding to the diurnal rotation of the whole spherical solar system and called the 'movement of the Same.' The fixed stars will have this as their primary motion, although each star will also revolve on its axis to keep the same face always turned towards the earth: 'each always thinks the same thoughts about the same things' (40a). The band forming the Zodiac has an opposite motion from West to East, the direction of this motion corresponding to the common motion of the planets (which, though carried round daily with the whole solar system by the 'movement of the Same,' also gradually revolve in the opposite direction in orbits forming the month, year, etc.) and is called the 'movement of the Different.' As Cornford states, these "two original motions are motions of the World-Soul, associated with its cognitive faculty of making judgments involving sameness and difference."[20]

The composite motion of the Same and Different result in what Plato calls a 'spiral twist': 'For the movement of the Same [. . .] gives all their circles a spiral twist because they have two distinct forward motions in opposite senses [. . .]' (39a). According to Dicks, "recognition of these spiral courses of the planets argues no small astronomical insight on Plato's part, as there is no evidence that any such notion was entertained before him."[21] Thus Plato provides for the two predominant astronomical motions, the diurnal rotation of the entire solar system and the additional common motion of the planets (when introduced) in the reverse direction along the ecliptic within the band of the Zodiac (36b). These motions of the World-Soul are being described before the celestial bodies or gods have been 'enshrined' in it.

In order to include within the 'movement of the Different' the various individual or 'proper' motions of the seven planets (the term 'planet' meaning "wanderer" in reference to the particular orbits and velocities of the planets in contrast to the "fixed" position of the stars relative to one another), the Zodiacal strip is further divided 'into seven unequal circles,' the distances among them apparently corresponding to the intervals of the combined progression of integers composing the two series mentioned earlier, 1, 2, 3, 4, 8, 9, 27. These Zodiacal strips represent the individual orbits of the planets.

The diameters of these seven circular strips varying, each is nested within the other with a common center (coinciding with the center of the spherical universe) according to their diminishing size, though inclined in different angles. Cornford points out that the individual motions of the planets can be divided into two groups, one consisting of the Sun, Venus, and Mercury which have similar speeds (i.e., they complete their orbits in the same period), and the other consisting of the Moon, Mars, Jupiter, and Saturn which have speeds 'contrary' to one another and to the other three. Thus Cornford deduces that in addition to the double motion of the Same and the Different, imparted by the motion of the World-Soul, each planet will have a 'proper' motion owing to its being a living creature with a soul and thus having the power of self-motion. These individual motions account for the differences in time it takes each planet to complete its orbit and would explain also such apparent phenomena as retrograde motion (i.e., that at times all the planets except the sun and moon appear to come to a stop, move in the reverse direction, and then continue forward again).[22] This shows the degree to which the problem of the motions of the planets had been differentiated at the time of the Academy and Plato's own interest in the problem.

Cornford maintains further that each planet, like the stars, revolves on its axis to keep the same face toward the earth with the remarkable addition that at least Venus and Mercury have a *variable* velocity caused by the volition of their individual souls.[23] If true, this notion is extraordinary because one of the two inviolable assumptions of astronomical theorizing prior to Kepler was that of uniform motion (the other being circular orbits). In contrast to Ptolemy, Copernicus, and Galileo, Kepler was the first to challenge both of these assumptions in his description of the velocities of the planets as varying in speed along elliptical orbits. Plato also corrected the erroneous notion of Democritus that the nearest planets only appear to move faster than the more remote ones because of their smaller orbits: 'those in the lesser circles moving faster, those in the greater more slowly' (39a).

In summary, following Cornford's reconstruction, four motions will govern the movement of the planets, two "derived" from the motion of the World-Soul and two "proper" to the self-motion of each planet: (1) the 'movement of the Same' which carries round diurnally the entire spherical universe with all of its contents; (2) the 'movement of the Different,' the motion common to the seven planets along the Zodiac opposite in direction to the diurnal rotation of the entire spherical universe; (3) the proper motion of each of the planets in its individual orbit as it continues its common motion along the Zodiac; and (4) the rotation of each planet on its axis. Dicks, however, disagrees with Cornford's attribution of proper motions to all of the planets in addition to their common velocities caused by the movement of the different. According to Dicks, the evidence indicates that only Venus and Mercury have their own proper velocity to account for the fact that they are never seen far from the sun and crisscross in their orbits around it.[24] Dicks also denies that the planets have an axial rotation similar to the fixed stars. The evidence supports Dicks insofar as Plato states that 'in respect of the other five motions [i.e., the motion of the five planets, excluding the sun and moon, in addition to the motion of the Same and Different] he made each motionless and still, in order that each might be as perfect as possible' (40b).

Before filling in the scheme with the stars and planets themselves, the last stage in the process, the World-Body must be fitted to the World-Soul: "When the whole fabric of the soul had been finished to its maker's mind, he next began to fashion within the soul all that is bodily, and brought the two together, fitting them centre to centre' (36d-e). Now the gods or celestial bodies are 'enshrined' in their respective orbits, with the intent of creating time, the discovery of number, and the pursuit of philosophy.

> In virtue, then, of this plan and intent of the god for the birth of Time, in order that Time might be brought into Being, Sun and Moon and five other stars — 'wanderers' as they were called — were made to define and preserve the numbers of Time. Having made a body for each of them, the god set them in the circuits in which the revolution of the Different was moving [. . .]. (38c-e)

In completing the description Plato clearly indicates a knowledge of the motion of the inferior planets, Venus and Mercury, along with that of the moon and the sun, whereas he seems reluctant to give an account of the motions of the superior planets, Mars, Jupiter, and Saturn. As he states, Venus and Mercury are always found close to the sun, shuttling back and forth across the sun's disk, sometimes moving eastward with the sun, then regressing behind it, and finally reversing to overtake the sun again:

> [. . .] in seven circuits seven bodies: the Moon in the circle nearest the Earth; the Sun in the second above the Earth; the Morning Star (Venus) and the one called sacred to Hermes (Mercury) in circles revolving so as, in point of speed, to run their race with the Sun, but possessing the power contrary to his: whereby the Sun and the star of Hermes and the Morning Star alike overtake and are overtaken by one another. As for the remainder, where he enshrined them and for what reasons — if one should explain all these, the account, though only by the way, would be a heavier task than that for the sake of which it was given. (38c-e)

The 'bodies' of the heavenly gods were given a 'well-rounded shape' (establishing one of the important conclusions of later astronomical views), those of the stars and planets composed primarily of fire with a slight admixture of the other primary elements, air, water, and earth. In such a manner the Demiurge created the heavenly gods or celestial bodies, the stars and the planets, described as 'living creatures everlasting and divine.' By 'everlasting' Plato did not mean that they had the unchanging, eternal existence of the intelligible living Creature or Forms, but that they endured throughout time. Time, in fact, came into being only with the creation of the World-Soul and the World-Body, 'produced' by the celestial revolutions and therefore a feature of the heavenly framework — though also an image of the archetypical Form of eternal duration. Plato's conception of time thus incorporates the two aspects that will appear later in Newton's theory: (1) the "eternal duration" (called by Newton the "eternal flow" of time) that exists independently of any celestial or planetary movements, and (2) the empirical determination of time dependent upon the motions of the heavenly bodies: the universe as a whole and the moon and the sun.

> For there were no days and nights, months and years, before the Heaven came into being [. . .] but these have come into being as forms of time, which images eternity and revolves according to number [. . .]. The month comes to be when the Moon completes her own circle and overtakes the Sun; the year, when the Sun has gone round his own circle. The periods of the rest have not been observed by men, save for a few; and men have no names for them, nor do they measure one against another by numerical reckoning. (37e-39c)

And so the purpose of creation was fulfilled 'in order that this world may be as like as possible to the perfect and intelligible Living Creature; in respect of imitating its ever-enduring nature.' Moreover, the clock-like revolutions of the celestial bodies have served to arouse in man a study of nature and mathematics, along with the pursuit of philosophy. As Plato says, the sun was created especially so 'that he might fill the whole heaven with his shining,' and man was endowed with vision so that

> the sight of day and night, of months and the revolving years, of equinox and solstice, has caused the invention of number and bestowed on us the notion of time and the study of the nature of the world; whence we have derived all philosophy, than which no greater boon has ever come or shall come to mortal man as a gift from heaven. (47a-b)

An account of the universe hardly being complete that omitted our own terrestrial domicile, Plato now includes the earth with its soul and body among the heavenly gods, even depicting it as the 'first and most venerable' of them all: "And Earth he designed to be at once our nurse and, as she winds round the axis that stretches right through, the guardian and maker of night and day, first and most venerable of all the gods that are within the heaven" (40b-c). This quotation has generated considerable discussion among ancient as well as modern commentators, because if the earth rotated on its axis synchronously with the movement of the Same it could not be "the guardian of night and day," their common rotation canceling the observable daily movement of the heavens. So by 'winds round the axis' did Plato mean the earth *rotates* on its axis in the direction opposite to that of the movement of the Same, or did he mean that it merely remains *stationary* relative to the diurnal rotation of the world? Cornford holds the first view and Dicks the second;[25] however, there is insufficient evidence for deciding between the alternative interpretations.

This concludes Plato's fanciful account of the creation of the World-Soul and the World-Body, the 'handiwork' of the Demiurge himself. The creation of lesser souls he assigns to the heavenly gods, except for the immortal part of incarnated human souls which is 'mixed and blended' from the same ingredients used to make the World-Soul, 'only no longer so pure as before.' There are souls 'equal in number with the stars.' Plato discussed the necessity of believing in the traditional gods, in spite of the fact that their supporters 'speak without probable or necessary truths' (40e). But this aspect of the account will not be pursued further since, as Cornford states, "in all this section of the dialogue the veil of myth grows thicker again, and it is useless to discuss problems that would arise only if the statements were meant literally."[26]

THE RECEPTACLE

A description of the genesis of the nature and motion of the heavens and the earth having been given in the first part of the dialogue, Plato turns in the second to a more detailed account of the Receptacle and its contents; i.e., to the nature and organization of the empirical world of change and becoming. Here the Demiurge is literally portrayed as like a human craftsman who, given disorganized materials which he did not create but 'takes over,' attempts to construct them into an intelligible order according to an ideal pattern as represented by the Forms. So the account is primarily concerned with the Receptacle or 'nurse of Becoming' and the chaotic state ('even before the heavens came into being') of the discordant 'primary bodies' contained in it. It is in this part of the narrative, aside from the theory of Forms, that Plato displays his greatest imaginative powers.

In particular, Plato rejects the general schema of interpretation begun by the Milesians, continued by the Pluralists and Atomists, later formalized by Aristotle, and maintained throughout most of modern science; namely, that of explaining perceptible phenomena in terms of a restricted number of irreducible, *observable substances* (e.g., water, air, fire, etc.) or by underlying *insensible particles* such as seeds and atoms. As different, for example, as were the theories of Anaximander, Empedocles, Anaxagoras, Democritus, and Aristotle, each attempted to explain the nature and changes of the empirical world with respect to transformations or motions of some permanent *substance* or *elements*, either selected from the visible world or inferred as the irreducible substratum underlying all change. The latter distinction especially, the dualism between substance and attribute, appearance and reality, the secondary as compared to the primary qualities of things, referred to by Whitehead as "the bifurcation of nature," has permeated most of modern philosophical and scientific thought.

The reasons for the attraction of this mode of explanation are undoubtedly deep-seated and complex. One probably is linguistic, the grammatical distinctions between noun and adjective and subject and verb implying the categorical distinctions between substance and attribute or agent and action. Another is the assumption that no final knowledge of the world would be possible unless the causes of phenomena could be traced to a restricted number of unchanging, irreducible substances as the ultimate elements of explanation (the quest for such "basic elements" continues today). Still another reason is the fact that if one denies ultimate status to ordinary sensory qualities, such as hardness, colors, smells, etc., then it is natural to conceive of the elements of the universe in terms of those abstracted characteristics of macroscopic objects that still remain; i.e., solidity, extension, shape, spatial location, etc., which presuppose, as both Aristotle and Locke held, a substance in which they inhere.

Whatever the reasons for its appeal, Plato rejected this mode of interpretation. Reflecting on the transformation of the elements into one another, water evaporating into air, 'air becoming fire by being inflamed,' the process reversing itself when air 'condenses' to mist, mist to water, and water becoming 'compacted' into earth and stones, Plato concluded that no one of these elements as 'they transmit in a cycle the process of passing into one another [. . .] ever makes its appearance as the *same* thing [. . .]' (49c-d). In contrast to the Presocratics, Plato denied that any one of the visible substances could be considered the basic element subsisting through and reappearing in various transformations; instead, the qualities characteristic of these elements exist only as long as they are manifest, arising and ceasing with their appearance. This, at least, seems to be the meaning of the following obscure passage.

> Whenever we observe a thing perpetually changing — fire, for example — in every case we should speak of fire, not as 'this' but as 'what is of such and such a quality', nor of water as 'this' but always as 'what is of such and such a quality'; nor must we speak of anything else as having some permanence, among all the things we indicate by the expressions 'this' or 'that', imagining we are pointing out some definite thing. For they slip away and do not wait to be described as 'that' or 'this' or by any phrase that exhibits them as having permanent being. (49d-e)

What Plato appears to be asserting is that names like 'fire' and 'water' and demonstrative adjectives such as 'this' and 'that' do not refer to permanent substances, but designate transitory clusters of qualities within the receptacle. He thereby rejects an ontology of irreducible elements or atoms in favor of a conception of the world as a continuous, unsubstantial flux of events or shifting panorama of sensory qualities. According to Cornford:

> Plato's position was nearer to that of Heraclitus, who alone had rejected the notion of substance underlying change and had taught the complete transformation of every form of body into every other. We are now to think of qualities which are not also 'things' or substances, but transient appearances in the Receptacle. The Receptacle itself alone has some sort of permanent being. [27]

In contrast to the visible qualities, the Receptacle *can be* referred to as a 'this' or a 'that:' 'Only in speaking of that *in* which all of them [the qualities] are always coming to be, making their appearance and again vanishing out of it, may we use the words "this" or "that" [. . .]' (49e). Since 'everything that is must be in some place,' the Receptacle is necessary as a spatial locus for the occurence of the qualities, but not as a material stuff or underlying substratum. The simile often used is that of a mirror: the sensory world reflected in the Receptacle being an imperfect copy or image of the world of Forms. Today one would more readily think of a screen on which the panorama of the world passes as a sequence of sensory images. The Receptacle has no qualities or characteristics of its own to interfere with the qualities appearing in it and thus can be grasped only by a kind of 'bastard reasoning':

> [. . .] it is always receiving all things, and never in any way whatsoever takes on any character that is like any of the things that enter it: by nature it is there as a matrix for everything [. . .] while the things that pass in and out are to be called copies of the eternal things, impressions taken from them in a strange manner that is hard to express [. . .] . (50b-c)

The Receptacle has a permanence lacking to the visible appearances, while the visible appearances possess qualities lacking to the Receptacle which is also depicted as a kind of malleable material that can take on and pass off various forms or shapes. In the most precise description, Plato says of 'the mother and Receptacle of what has come to be' that we should 'not be deceived if we call it a nature invisible and characterless, all-receiving, partaking in some very puzzling way of the intelligible and very hard to apprehend' (51a-b). He resorts to the family image again when he describes all three aspects of the system: 'we may fittingly compare the Recipient to a mother, the model to a father, and the nature that arises between to their offspring' (50d).

Although the Receptacle may be 'invisible and characterless,' it is nonetheless spatial. In a summary of his schema, Plato distinguishes (1) the 'ungenerated and indestructible Forms,' from (2) the generated world in perpetual motion, from (3) the permanent Receptacle or Spatial frame:

> [. . .] there is, first, the unchanging Form ungenerated and indestructible, which neither receives anything else into itself from elsewhere nor itself enters into anything else anywhere, invisible and otherwise imperceptible [. . .].
>
> Second is that which bears the same name and is like that Form; is sensible; is brought into existence; is perpetually in motion, coming to be in a certain place and again vanishing out of it [. . .].
>
> Third is Space, which is everlasting, not admitting destruction; providing a situation for all things that come into being, but itself apprehended [. . .] by a sort of bastard reasoning [. . .]. (51e-52b)

Unlike Time which is 'produced' by the celestial revolutions and therefore is created by the Demiurge, the Receptacle is "finally identified with space, is treated as a given frame, independent of the Demiurge and a necessary condition antecedent to all his operations." [28] Like Parmenides' Being from which all sensory qualities were abstracted, Plato's Receptacle was left with the one property of dimension or spatiality. In this conception he foreshadows the later attempt by Descartes to reduce matter to mere extension.

The 'four kinds' or combinations of empirical qualities, air, earth, fire, and water located in the Receptacle are, of course, copies of the unchanging Forms of Air, Earth, Fire, and Water in the intelligible Realm. These latter Forms, in language antedating Kant, are what the four kinds 'are in themselves.'(51c). As to their existence in the Receptacle, the four kinds are further differentiated into opposite pairs of sensory qualities, hot, cold, moist, and dry, recalling the theories of the Milesians. According to Cornford's translation, these qualities in turn are described as 'powers' with the capability of acting on one another and of affecting our sense organs, causing sensations and perception.[29] If correct, this substitution of 'power' for 'substance' as the mark of the real being of anything is a further example of the striking originality of Plato. In the *Sophist* where this conception is first introduced and more fully developed, the Stranger says, "I suggest that anything has real being that is so constituted as to possess any sort of power either to affect anything else or to be affected [. . .]. I am proposing as a mark to distinguish real things that they are nothing but power." [30]

On this interpretation, the reality of things is not determined by their solidity or impenetrability, as materialism asserts, but by their *active power* to affect something else and the *passive capacity* to be affected by other things. The above quotation is so close to Locke's definition of "primary and secondary qualities" as "powers" within objects to affect the senses and cause alterations in other things (as well as objects having the passive capacity to be affected by them), that it would be difficult not to conclude that Locke derived this view from Plato.[31] But in this instance Plato was more sophisticated than Locke in that he rejected the latter's paradoxical notion of a "substance I know not what" underlying and supporting the sensory qualities.

More precisely, the qualities in the Receptacle are not really the independently existing sensory properties themselves, hot, cold, wet, and dry, but the 'powers' to affect the senses in such a way that they are perceived as qualities. Thus they exist only for a percipient when the 'powers' in the Receptacle affect the appropriate sense organ resulting in the sensation. As Cornford states, the Receptacle

> has 'every sort of diverse appearance to the sight', in the sense that, if there were a spectator with eyes to see it, it would cause in him sensations of various colours. But in the absence of any spectator, there are, strictly speaking, no colours — only changes capable of causing such sensa-

tions. Space is accordingly described as filled with 'powers' whose motions are in unordered and unbalanced agitation. [32]

The sensory qualities are not just subjective, however, because even though the Receptacle does not possess the qualities when no one is perceiving it, it does acquire them for as long as it is being perceived: i.e., during the interaction between the powers and the sense organ the Receptacle actually becomes hot, red, hard, etc. But while this interpretation is consistent with Plato's theory of perception presented in both the *Theaetetus* and the *Timaeus*,[33] it presents serious problems in that it is difficult to understand how the four qualities, which exist only for a percipient, nevertheless resemble the Forms from which they acquire their name. Furthermore, if as Cornford states, sensory qualities are "sensations" caused in the percipient, then how can they be qualities in or of the Receptacle? It appears that Plato's theory of perception is not entirely consistent with his conception of the Receptacle.

Also, it would be tempting at this point to attribute to Plato a prescient reduction of matter to pulsations of energy or vibratory activity, as in modern physics — to "force fields," "energy states," "electromagnetic vibrations," or even to Einstein's conception of matter as a "deformation of the space-time structure." But one must not forget that for Plato, as for primitive man, the soul is the ultimate source of motion. To repeat a constant reminder of Cornford: "The activity of soul [. . .] is the only possible source of the active powers of bodies — of their motion in space and of their power of altering one another qualitatively and affecting our sense organs.[34] The powers in the Receptacle corresponding to the qualities are therefore really animistic, derived ultimately from the motion caused by the World-Soul, but in this case an irrational motion.

Returning to the description of the state of the Receptacle confronting the Demiurge when he attempts to establish rational order in the universe, one must imagine it initially filled with a chaotic distribution of unruly motions resulting in disharmony and imbalance. This 'nurse of Becoming,' as Plato now refers to the Receptacle, because it was "filled with powers that were neither alike nor evenly balanced [. . .] [had] no equipoise in any region of it; but [. . .] was everywhere swayed unevenly and shaken by these things, and by its motion shook them in turn" (52d-e). The ancient model of the winnower's sieve is explicitly used to explain how the four kinds (yet to be described) will be separated in terms of their density and weight:

> [. . .] just as when things are shaken and winnowed by means of winnowing baskets and other instruments for cleaning corn, the dense and heavy things go one way, while the rare and light are carried to another place and settle there. In the same way at that time the four kinds were shaken by the Recipient, which itself was in motion like an instrument for shaking, and it separated the most unlike kinds farthest apart [. . .] and thrust the most alike closest together; whereby the different kinds came to have different regions, even before the ordered whole consisting of them came to be. (53a)

While Plato has not yet discussed how the four kinds or qualities of things were given a 'configuration by means of shapes and numbers,' and thus acquired the properties of volume, density, and weight, he here describes their distribution as a result of the winnowing-like motion of the Receptacle and the ancient principle of like being attracted to like. He adds that before this distribution 'all these kinds were without proportion or measure.'

This reference to the condition of the Receptacle before 'proportion or measure' was introduced illustrates one of the major theses of Plato, namely, that the divine craftsman attempts to instill order into the original 'chaos' by 'persuading' the tumultuous powers to accept a rational structure.

> For the generation of this universe was a mixed result of the combination of Necessity and Reason. Reason overruled Necessity by persuading her to guide the greatest part of the things that become towards what is best; in that way and on that principle this universe was fashioned in the beginning by the victory of reasonable persuasion over Necessity. (48a)

But Plato also acknowledges an inherent element of intransigent brute-fact in the universe which can never be completely eliminated or overcome by intelligent design. Although the Demiurge attempts to induce a rational order on the 'chaos' or 'Necessity' (*ananke*) inherent in the Receptacle, this is never fully successful, which accounts for the disharmony and imperfection in the world. As Plato says, it is only 'for the most part' that Reason can persuade Necessity. This view that reality includes a "rational and irrational given," an "irreducible surd" never completely rationalized, was developed by Edgar S. Brightman into the Christian concept of a finite-infinite God. [35]

The Receptacle having been filled with the disharmonious powers and qualities of the four kinds, Plato is not satisfied with accepting their existence, but attempts to explain how they acquired their 'distinct configuration by means of shapes and numbers': 'For to this day no one has explained their generation, but [. . .] speak as if men knew what fire and each of the others is, positing them as first principles [. . .]' (48b). Analogous to the Pythagoreans who derived material bodies from arithmogeometric figures, Plato describes how the Demiurge endowed the four kinds with the four regular solids of Greek antiquity (the solids used by the Pythagoreans and later by Kepler in describing the harmonic order of the celestial bodies), the theoretical construction of which had been completed by Theaetetus in the Academy: the tetrahedron (pyramid), octahedron, icosahedron, and the cube shown in Figure 1. [36]

Fig. 1. The five figures represent each of the five regular solids in the order discussed in the text: the fifth is the dodecahedron, the form of the universe as a whole.

Plato then demonstrates how each of the four solids can be reconstructed from two types of triangles: (1) the half-equilateral triangle obtained by drawing a line from any angle of the equilateral (which is the face of the first three of the regular solids) perpendicular to the opposite side, and (2) a half-square or right-angled isosceles triangle formed by dividing a square in half. Although Plato adds that these triangles could be analyzed further, presumably into points, lines, and angles, he accepts them as the two irreducible configurations necessary for the composition of the four solids believing, as the Pythagoreans, that solids can be constructed from planes, while the planes can be formed from triangles:

> [. . .] it is of course obvious to anyone that fire, earth, water and air are bodies; and all body has

depth. Depth, moreover, must be bounded by surface; and every surface that is rectilinear is composed of triangles. Now all triangles are derived from two, each having one right angle and the other angles acute. (53c)

Since equilateral triangles would have been sufficient to construct the surfaces of the first three solids and squares the fourth (the cube), Plato explains why he divided each into two further triangles; namely, to account for the *transformation* of the solids after their construction, this transformation depending upon the breakdown of the four solids into their constituent triangles with the latter regrouping to form new solids. In this way the (molecular) solids are decomposed into their (atomic) triangles which rejoin to form other (molecular) solids, each being of various sizes to account for some of the differences in things. Plato's account is even more complex and detailed than described here, but this presents the essential features of his scheme.

The manner of the construction of the four solids from the two types of triangles being determined, Plato assigns them to the four kinds according to their mobility and size.

> Now, taking all these figures, the one with the fewest faces (pyramid) must be the most mobile, since it has the sharpest cutting edges and the sharpest points in every direction, and moreover the lightest, as being composed of the smallest number of similar parts: the second (octahedron) must stand second in these respects, the third (icosahedron), third. Hence, in accordance with genuine reasoning as well as probability, among the solid figures we have constructed, we may take the pyramid as the element or seed of fire; the second in order of generation (octahedron) as that of air, the third (icosahedron) as that of water. (56a-b)

To earth he assigned the cubical figure, for of the four kinds 'earth is the most immobile and the most plastic of bodies' (55e). A fifth solid is added, the dodecahedron, which approaches the sphere in volume and therefore was used by the 'god [. . .] for the whole.' Since this figure coincides with the complete spherical universe it does not enter into the discussion of the transformation of the other solids.

Thus three-dimensional figures were added to the original powers and qualities of the four kinds to give them the 'bodies' they exhibit in ordinary perception. Although the first four geometric solids are themselves too minute to be perceived, when massed together they become observable.

> Now we must think of all these bodies as so small that a single body of any one of these kinds is invisible to us because of its smallness; though when a number are aggregated the masses of them can be seen. And with regard to their numbers, their motions, and their powers in general, we must suppose that the god adjusted them in due proportion, when he had brought them in every detail to the most exact perfection permitted by Necessity willingly complying with persuasion. (56c)

Plato then displays the theoretical possibilities of this schema in the account of the transformation of the four bodies. Apparently conceiving of this transformation along the lines of the upward and downward processes of Heraclitus, he describes how in the 'upward' phase earth is decomposed into earth figures and water is transformed into air and air into fire. When completed this process reverses itself in a 'downward' direction, the two directions corresponding to the winnowing-like agitation of the Receptacle wherein like is attracted to like and separated from the unlike. While water, air, and fire can be transformed into one another, earth can only be broken down into smaller earth figures because the half-equilateral triangles making up the other kinds cannot combine with the right-angled isosceles triangles composing the earth. However, the decomposition of the latter is brought about by the impact of the 'sharpness' of the edges of the pyramids of fire.

> Earth, when it meets with fire and is dissolved by its sharpness, would drift about [. . .] until its own parts somewhere encounter[ing] one another, are fitted together, and again become earth; for they can never pass into any other kind. (56d) (Brackets added)

The other elements, when 'overcome' by the principle agent fire, are reduced to their triangular surfaces which then 'drift about' in the space of the Receptacle until they meet other triangles and recombine into the same or different figures. Utilizing the number of the triangular surfaces making up the various solids, Plato describes in numerical terms how the transformation occurs: air as the eight-surfaced octahedra can be decomposed such that its eight surfaces regroup to form two four-surfaced fire pyramids; the twenty surfaces of the icosahedron as water can be broken down and rejoined as two eight-surfaced air octahedra and one four-surfaced pyramid, accounting for warm air; five four-surfaced fire pyramids can eventually recombine to form one twenty-surfaced water icosahedron, etc. Many combinations are possible except that water and air cannot be directly transformed into one another but must pass through intermediate stages. Owing to the different sizes of the figures, the combinations form 'an endless diversity.'

In this ingenious manner, Plato replaced a substantialistic interpretation of the four elements with a stereometric lattice structure according to which the physical properties and transformations of the four kinds are explained in terms of triangular surfaces and three-dimensional geometrical figures. Although the original idea was derived from the Pythagoreans, its development by Plato attests to the incredible imaginative power of his thought. Still, it is difficult to assess the merits of the conception in relation to Atomism, the theory Plato was primarily opposing. For while Plato's account appears in some respects closer to the modern atomic theory and its explanation of chemical transformations, the reasoning was purely *a priori*, lacking the slightest shred of empirical evidence.

In contrast, Leucippus and Democritus, although largely dependent upon *a priori* reasoning in their construction of the atomic framework to solve the theoretical paradoxes raised by Parmenides and Zeno, could adduce some empirical evidence in its support and thus allow for the *possibility* of predictive verification. For example, the analogy between insensible atoms and imperceptible motes (except when caught in a beam of sunlight) or invisible dust particles (caused by the crumbling of soft stones) does provide indirect empirical evidence for the atomic theory. The same type of examples and "transdictive" reasoning would be used by Boyle, Locke, Newton, and Gassendi in the 18th century to justify the corpuscular or atomic theory. [37] Also, presuming we had "microscopical vision" (Locke's term), one could imagine perceiving the atoms on the analogy of Brownian Motion.

Thus while the Atomists could claim indirect empirical evidence in support of their theory, Plato's geometric lattice structure could only be described as a 'likely story' or 'probable' account. But "probable" in relation to what? Plato seems to have meant by a probable account one that conceivably could explain the phenomena and, more importantly, fulfilled his criterion that the universe could not have arisen by chance or necessity, since it exhibits the kind of order and perfection that could result only from its having been created by an intelligent deity. But this criterion is arbitrary in that it imposes subjective, *a priori* conditions on the explanatory framework rather than deriving the principles of explanation from the phenomena themselves. Moreover, what evidence could Plato adduce for his theory? Could he provide any analogue in experience to account for the solidity and transformation of things in terms of triangular surfaces and three-dimensional empty figures? Could two-dimensional planes float about in space and adhere to one another? Could qualities and powers attach themselves to spatial configurations devoid of solidity or materiality?

The answer to these questions up to the 20th century generally would be "no" — whatever cannot be imagined as possessing both tactual and spatial qualities cannot be real. Today, however, the equivalence of mass and energy, the interpretation of matter as a manifestation of vibratory, electronic forces, and the attempted reduction of the physical world to a "four-dimensional continuum of events" refutes the irreducibility of tactual-visual properties, such as mass, solidity, hardness, etc. Still, empirical evidence, however indirect, is essential for the formulation and confirmation of such theories, while it is entirely absent in Plato's account. This lack of empirical evidence and any possibility of verification probably accounts for the fact that while Greek atomism has had a long history of further development and elaboration, continuing today in atomic physics, chemistry, and molecular-biology, there is no evidence that Plato's purely *a priori* stereometric conception had any effect on later scientific developments.

CHAPTER XII

[1] A. E. Taylor, *Plato, The Man And His Works*, 6th ed. (London: Methuen & Co., Ltd., [1926] 1949), p. 23.

[2] F. M. Cornford, *Plato's Cosmology* (*op. cit.*, ch. XI), p. 30.

[3] Cf. *Phaedo*, 96a-e.

[4] *Ibid.*, 97e-98a. (Tredennick trans.)

[5] *Ibid.*, 100c-d.

[6] Cf. ch. II.

[7] *Ibid.* For an explicit example in Plato, see his supplementation of the 'secondary causes' of the function of our eyes in vision with an explanation in terms of 'their highest function for our benefit, for the sake of which the god has given them to us.' *Timaeus*, 47a. (Cornford trans.)

[8] *Sophist*, 246a-c. (Cornford trans.)

[9] *Laws*, 967c-e. (Taylor trans.)

[10] Cf. Cornford, *op. cit.*, p. 6. It will be apparent throughout how much I owe to Cornford's excellent exposition of the *Timaeus*.

[11] Cf. *Timaeus*, 27d-29d.

[12] *Philebus*, 26e-27. (Hackforth trans.)

[13] *Timaeus*, 30a-c. (Cornford trans.) Unless otherwise indicated, all further translations from the *Timaeus* are from Cornford.

[14] Cornford, *op. cit.*, p. 38.

[15] 'You mean that the selfsame reality which has the name *soul* in the vocabulary of all of us has *self-movement* as its definition?' 'I do.' *Laws*, 896a. (Taylor trans.)

[16] Cf. *Sophist*, 258c-259d. Also Cornford, *op. cit.*, p. 61.

[17] According to Cornford, there were such spheres in use in the Academy at that time. Cf. Cornford, *op. cit.*, pp. 74-75.

[18] D. R. Dicks, *Early Greek Astronomy to Aristotle* (*op. cit.*, ch. XI), p. 119ff.

[19] For a detailed analysis of this conception see Cornford, *op. cit.*, pp. 66-74.

[20] *Ibid.*, p. 78.

[21] Dicks, *op. cit.*, p. 129.

[22] Cf. *Timaeus*, 38d and Cornford's discussion, *op. cit.*, pp. 80, 89.

[23] Cf. Cornford, *op. cit.*, pp. 106-108 and Dicks, *op. cit.*, p. 127.

[24] Cf. Dicks, *op. cit.*, p. 127.

[25] Cf. Cornford, *op. cit.*, pp. 120-134; cf. Dicks, *op. cit.*, pp. 132-137.

[26] Cornford, *op. cit.*, p. 143.

[27] *Ibid.*, p. 178.

[28] *Ibid.*, pp. 102-103.

[29] Crombie denies the existence of such powers. Cf. I. M. Crombie, *An Examination of Plato's Doctrine*, Vol. II (*op. cit.*, ch. XI), pp. 222-224.

[30] *Sophist*, 247e. (Cornford trans.)

[31] Cf. John Locke, *An Essay Concerning Human Understanding*, Bk. II, ch. 23, sec. 9.

[32] Cornford, *op. cit.*, p. 205.

[33] Cf. *Theaetetus*, 156e and *Timaeus*, 61d-62c.

[34] Cornford, *op. cit.*, p. 206.

[35] Cf. Edgar S. Brightman, *A Philosophy of Religion* (New York: Prentice-Hall, Inc., 1940), pp. 336-341.

[36] These figures were adapted from a diagram in Thomas Kuhn, *The Copernican Revolution* (*op. cit.*, Introduction), p. 218.

37 Cf. Maurice Mandelbaum, *Philosophy, Science and Sense Perception: Historical and Critical Studies* (Baltimore: The Johns Hopkins Press, 1964).

CHAPTER XIII

ARISTOTLE'S EMPIRICAL METHODOLOGY

[. . .] the master of them that know.[1]
Dante

Aristotle has been best known to large segments of the Western world as "the philosopher," the title conferred upon him by St. Thomas Aquinas and the Scholastics. For from the 13th century when Aristotle's writings began to make their way back to the Latin West by means of the Arabic conquest of Spain, until the scientific revolution of the 17th century, Aristotle's philosophy and scientific inquiries and theories, as interpreted and taught by the Scholastics, represented the dominant world-view of Western man. To few individuals has it been granted to mold men's minds for nearly five centuries. The ascendence of Aristotle's cosmology received its most elevated expression in Dante's *Divine Comedy*, while its impending dissolution was lamented by John Donne in *The Anatomy of the World*, for the rise of the "new philosophy" of mechanics depended upon the fall of the old cosmology of Aristotelian-Scholasticism. Thus Aristotle's philosophy is important not only as the culmination in many respects of the genius of Hellenic thought and as the cosmological basis of late medieval and early Renaissance thought, but also as the foundation on which the structure of modern science was built. Just as most cathedrals have been constructed on the foundation of previous *églises*, however different their architectures, so modern science was built on the foundations of the razed ediface of Aristotelianism.

Born in 384 B.C. in the city of Stagira in Thrace on the northern coast of the Aegean Sea, Aristotle is often referred to as "the Stagirite." Since Stagira was an Ionian colony speaking a variety of the Ionic dialect, he was appropriately linked with the scientific tradition of the Ionians. His father, a member of the Asclepiad clan or guild named after the legendary physician Asclepios, served as court physician to King Amyntas II of Macedonia. Although his parents died when he was still a boy, one may reasonably suppose that the atmosphere of medical science in which he was brought up had a lasting effect on his interests and thought, as evinced by his exacting biological studies and the overall organismic character of his cosmological system.

Apparently little more is known of Aristotle's life until he entered Plato's Academy at the age of 17 or 18. He was attracted to the Academy because it offered the best education in Greece and remained in its intellectual fraternity for 20 years until Plato's death in 348-7. An outstanding student, he was called by Plato 'the reader' and 'the mind of the school.' As Jaeger demonstrated in his classic book on Aristotle, the impact of Plato was such that Aristotle began his philosophic career as a Platonist and retained much of this influence throughout his life, even though he subsequently became the foremost critic of Plato's most original contribution, the theory of the transcendent or 'separate' existence of the Forms, and pursued more empirical research than was generally characteristic of the members of the Academy. Also, in contrast to the previously accepted view that Aristotle had formed his philosophical system very early in opposition to Plato's and retained it unmodified throughout his life, Jaeger has shown that like most thinkers, Aristotle changed his views as he developed and matured. In fact, although at the time that Aristotle entered the Academy Plato had already written his Socratic dialogues and had begun writing the series of more expository or didactic essays of his later period, Jaeger has pointed out that Aristotle's earliest writings imitated the dialogue form and even the subject matter and titles of Plato's early dialogues.[2] As incredible as it now seems with mainly the repetitious, ponderous treatises available to us, these early dialogues earned from Cicero and Quintilian the plaudits of the "golden stream" and "sweetness of eloquence."

Shortly after Plato's death Aristotle left Athens accompanied by Xenocrates, a fellow Academic who later became director of the Academy during part of the time that Aristotle was head of the rival Lyceum. Various reasons have been offered to explain Aristotle's leaving, especially his opposition to the tendency of Plato's nephew Speusippus, who succeeded Plato as head of the Academy, 'to turn philosophy into mathematics' (as recently there has been a tendency by Quine and Kripke to turn philosophy into logic), an orientation uncongenial to the more empirical interests of Aristotle. Whatever the reasons, Aristotle left Athens and settled in Assos among a small circle of Platonists. There he married Pythias who bore him a daughter of the same name. Apparently Pythias died during his later stay in Athens whereupon he formed a lasting though "unlegalized" relationship with Herphyllis, a native of Stagira, with whom he had a son Nicomachus. It is after this son that he named his later, more famous work on ethics.

His legal wife, Pythias, was the niece and adopted daughter of Hermias, the ennuch and former slave who had risen to become the powerful ruler of Atarneus and Assos in Mysia. Earlier Plato had written two former members of the Academy, Erastus and Coriscus, who had settled in their native town of Scepsis near Atarneus, urging them to establish a friendship with Hermias so that they might achieve the kind of political reform taught in the Academy by persuading him to adopt the philosophical life. These expectations were fulfilled with Hermias studying geometry and dialectic and becoming something of a philosopher-king, changing his tyranny 'into a milder form of constitution.' He subsequently presented Erastus and Coriscus with the town of Assos where Aristotle settled in the company of these Platonists and where he is reputed to have become the intellectual leader of this group of Academics. Later in 341 when Hermias was deceived into capture by the Persian General, Mentor, refusing under torture to divulge his secret political alliance with Philip of Macedon, he was crucified. It is said that under torture "the king caused him to be asked what last grace he requested. He answered: 'Tell my friends and companions that I have done nothing weak or unworthy of philosophy.' "[3] Aristotle wrote a very beautiful, moving poem to commemorate the tragic death of this heroic man who had risen from a lowly origin to become one of the most powerful but just rulers of Asia Minor.

After remaining in Assos for three years Aristotle moved to Mitylene, in Lesbos, perhaps at the urging of Theophrastus, his distinguished student, to whom he later entrusted the leadership of the Lyceum. There, in addition to teaching, he seems to have spent considerable time in biological research.

> To his stay at Assos, and even more to his stay at Mitylene, belong many of his enquiries in the region of biology; his works refer with remarkable frequency to facts of natural history observed in the vicinity, and more particularly in the island lagoon of Pyrrha.[4]

Then in 343-2 Aristotle accepted an invitation from King Philip of Macedon to come to the court in Pella as tutor to the Prince. This invitation may have been due to his acquaintance with Philip when Aristotle's father was physician to the court or, more likely, to the political alliance between Hermias and Philip which placed the philosopher in a position of confidence between them. In any case, Aristotle became tutor to the Prince, then thirteen years of age, who was soon to become the legendary Alexander the Great. There is hardly any episode in the entire romance of human history better suited to excite the imagination than that of Aristotle, on the verge of becoming one of the greatest philosophers of all time, tutoring Alexander, who was destined to become one of the greatest military conquerors and political visionaries in the history of man. What influence did Aristotle have on the character and thought of the subjugator of much of the known civilized world? What effect, in turn, did Aristotle's acquaintance with Alexander (first as tutor, then as friend and counselor) have on his own philosophy, especially on the *Politics*? Unfortunately, however, very little is known regarding the period of tutelage. As Ross states, "little or nothing is known of the education imparted by him to his distinguished pupil." It is nearly certain, however, that Aristotle composed the works "Monarchy" and "Colonies" for Alexander who later reciprocated by providing funds to help establish a museum of natural history at the Lyceum, instructing his subordinates to supply it with specimens of scientific interest.

As further evidence of Alexander's regard for Aristotle, after Philip's death "Alexander fulfilled his teacher's dearest wish by rebuilding his birthplace Stagira, which had been devastated by Philip's troops during the Chalcidic war."[5] Also, probably as a concession to Aristotle and his disciple Theophrastus, the latter's birthplace Eresus was spared when Macedonian troops took the island. Later, however, Aristotle's regard for Alexander must have been severely affected when Aristotle's nephew Callisthenes, who had helped him complete the list of Delphic victors and then had joined the King in his campaign in Asia to write the history of the saga, fell out of favor as a result of court intrigue and was executed. Moreover, as a firm advocate of the superiority of the Greeks over the 'barbarians,' Aristotle wrote in opposition to Alexander's ambitious attempt to fuse the races by combining the Hellenic and the Oriental cultures.

In the year 335-4, following Philip's death and Alexander's accession to the throne, Aristotle returned to Athens. This must have been something of a triumphant return after an absence of about a dozen years, and the beginning of the most productive period of his life. As Jaeger states,

> Aristotle came to Athens as the flower of Greek intellect, the outstanding philosopher, writer, and teacher, the friend of the most powerful ruler of the time, whose rapidly rising fame raised him with it even in the eyes of persons who stood too far from him to understand his own importance. [6]

Once in Athens Aristotle rented some buildings (Athenian law prevented an alien from owning land) in a grove often frequented by Socrates outside the city, a grove sacred to Apollo Lyceius and a sanctuary of the Muses, and established his own school. With its expansion later there is reference to a garden, houses adjoining the garden, lecture halls, and a large building which probably served as the museum of natural history and perhaps housed Aristotle's large collection of manuscripts and maps — the model for the later libraries at Alexandria and Pergamon.

The school, called the Lyceum after the sacred grove, was in competition with the Academy but under the protection of Aristotle's powerful Macedonian friend Antipater, whom Alexander during his campaigns in Asia had left as regent of Macedonia and Greece. We are told that Aristotle gave his lectures on more difficult philosophical subjects such as logic, physics, and metaphysics in the morning in the *Peripatos*, a covered walk or *loggia*, with students following as he walked and lectured, hence the name "peripatetic" or "follower of Aristotle." In the afternoon he lectured to a larger audience on less difficult subjects such as rhetoric, ethics, and politics. The contrast in subject matter between the morning and afternoon lectures is the basis for the later misleading distinction between the 'esoteric' and 'exoteric' doctrines; i.e., between the later treatises containing the results of Aristotle's more mature research and the earlier dialogues written in imitation of Plato.

During the twelve-year period of his leadership of the Lyceum Aristotle directed scientific investigations (e.g., into the history and classification of plants and animals), organized historical research (e.g., encyclopaedic histories of philosophy and of physical research), directed and aided in the compilation of documents (e.g., the collection of 158 constitutions), encouraged the dissection of animals, resolved the question of the cause of the flooding of the Nile in the *Meteorology*, and lectured or wrote on practically every area of recognized knowledge. It is these lectures and notes that compose the body of his writings which we possess today. In at least one case, that of logic, his was the first systematic treatment of the subject which was so thoroughly worked out that it remained essentially unchanged until the present century. Small wonder, then, that Aristotle came to be called "the master of them that know" and "the philosopher."

Although he was not as prescient in his physical and cosmological theories as Democritus and did not possess the literary imagination or brilliant intuitive power of Plato, his range and penetration of knowledge was truly astounding. As vague as the term is, his could be called the most powerful intellect of all time. He established in the Lyceum a program of organized research characteristic of the best modern universities. While he was not alone in emphasizing the importance of *sustained* empirical research, this being characteristic also of the schools of medicine, such as the Hippocratic school at Cos, he was the first to organize and institutionalize research on such a broad scale. As Jaeger says:

> We do not hear of many disciples of Aristotle by name, but what Greek is there who wrote during the next hundred years on natural science, on rhetoric, on literature, or on the history of civilization, and was not called a Peripatetic? Lavish as the grammarians are with this title, it is easy to see that the intellectual influence of the school soon extended over the whole Greek-speaking world [...]. In the Lyceum Plato's communal life [...] became a university in the modern sense, an organization of sciences and of courses of study. [7]

In 323 news reached Athens of the death of Alexander. Unlike previous false accounts, this report proved to be true. Since Antipater, the friend of Aristotle and protector of Macedonians was in Asia Minor on his way to join the King, there was no one to contain the anti-Macedonian feeling that erupted in Athens. An "absurd charge of impiety," based on the "hymn and [...] epitaph which he had written to Hermeias," [8] was brought against Aristotle similar to that brought against Socrates, so he withdrew from Athens 'lest the Athenians sin twice against philosophy.' He left the school in the hands of Theophrastus and fled to Chalcis where he settled in the paternal property of his deceased mother. There he remained the following months until his death in 322 at the age of sixty-three, dying apparently from an intestinal disorder from which he had long suffered.

Before his death he drew up a will which is indicative of his character, showing a concern for the welfare of others, an abiding trust in his friends, and his respect for the wishes of his deceased wife — in short, the traits of a just, kind, generous, and considerate person. For example, he made provisions for the security of his daughter Pythias, still a minor; he left property to Herpyllis (along with the instruction that 'if she desires to be married [. . .] she be given to one not unworthy'), the woman with whom he lived after the death of his wife; he provided for his son Nicomachus born of his relation with Herpyllis; he arranged for the freedom of some of his servants and left money to others; and he requested that he be buried with his first wife according to her wishes. [9] The document is a touching memorial to a man whom tradition has usually divested of any human qualities.

ARISTOTLE'S EMPIRICISM

Aristotle might well be considered the greatest empiricist of all time, if one means by empiricism that all knowledge must be grounded in sensory observation. Although it is true that scientific knowledge for Aristotle consists of demonstrations from first principles or axioms, and therefore must be necessary and certain, the original principles are themselves derived by induction from perception of particular objects. Over and over he repeats that 'we *see*' that such and such is the case, that we must accept 'the evidence of our senses,' that 'our eyes tell us' something is so, [10] especially when he is criticizing the rationalistic speculations of the Pythagoreans, Eleatics, and Platonists (such references to observation are entirely missing from his metaphysical writings, however). In one instance he uses unusually harsh language in condemning his predecessors' willingness to accept 'dialectical' arguments over the testimony of their senses.

> Reasoning in this way [. . .] they were led to transcend sense-perception, and to disregard it on the ground that 'one ought to follow the argument': and so they assert that the universe is 'one' and immovable [. . .] that it is infinite [. . .]. [A]lthough these opinions appear to follow logically in a dialectical discussion, yet to believe them seems next door to madness when one considers the facts. For indeed no lunatic seems to be so far out of his senses as to suppose that fire and ice are 'one' [. . .]. [11]

This empiricism is also evident in his acute interest in the entire range of human experience and natural phenomena, as the following expressions, 'all knowledge begins with curiosity' and 'all men by nature desire to know' indicate. It also lies behind his exhortations to his colleagues in the Lyceum not to disdain empirical research in favor of more exalted studies just because the former deals with ordinary perishable objects while the latter investigates the eternally perfect, celestial bodies.

> Of things constituted by nature some are ungenerated, imperishable, and eternal, while others are subject to generation and decay. The former are excellent beyond compare and divine, but less accessible to knowledge [. . .]. Both departments, however, have their special charm. The scanty conceptions to which we can attain of celestial things give us, from their excellence, more pleasure than all our knowledge of the world in which we live [. . .]. On the other hand, in certitude and in completeness our knowledge of terrestrial things has the advantage. Moreover, their greater nearness and affinity to us balances somewhat the loftier interests of the heavenly things that are the objects of the higher philosophy [. . .]. We therefore must not recoil with childish aversion from the examination of the humbler animals. Every realm of nature is marvellous [. . .] for each and all will reveal to us something natural and something beautiful. [12]

Unfortunately, however, "Aristotle was too much of an empiricist," [13] as Randall asserts, relying too uncritically on the evidence of ordinary observations and common sense convictions. Because of this, even though his approach to problems and analyses initially seem quite obscure, owing to his strange terminology and to the fact that the paradigms of scientific research have radically changed, his overall cosmology is not as difficult to understand as that of some of the Presocratics and of Plato, or even of Newton, since it represents a more natural, unsophisticated, common sense view of the world. As Kuhn aptly states:

> Part of the authority of Aristotle's writings derives from the brilliance of his own original ideas, and part derives from their immense range and logical coherence, which are as impressive today as ever. But the primary source of Aristotle's authority lies, I believe, in a third aspect of his thought, one which it is more difficult for the modern mind to recapture. Aristotle was able to express in an abstract and consistent manner many spontaneous perceptions of the universe that had existed for centuries before he gave them a logical verbal rationale. In many cases these are just the perceptions that, since the seventeenth century, elementary scientific education has increasingly banished from the adult Western mind. Today the view of nature held by most sophisticated adults shows few important parallels to Aristotle's, but the opinions of children, of the members of primitive tribes, and of many non-Western peoples do parallel his with surprising frequency. [14]

This is not intended to denigrate Aristotle's tremendous philosophical and scientific achievements, but to emphasize that the base of reference in his philosophizing, as important as was the theoretical influence of Plato, was the world of common experience as disclosed by immediate perception. In contrast to the Atomists and Plato (as well as modern scientists) who held that the ordinary physical world is merely a world of appearances, Aristotle believed that the macroscopic world, consisting of sensory objects composed of substance and form *is the real world*. While Plato maintained that knowledge (*episteme*) must have as its referent an Ideal Reality transcending the empirical world, Aristotle claimed that knowledge consists of knowing the real essences inherent in perceptible things. Thus his common sense realism represents an advance over Plato's quasi-mythical, narrational interpretation of nature, although from the standpoint of modern science he was naive in his uncritical acceptance of the evidence of ordinary experience.

But the fact that Aristotle was able to attain such an *objective attitude* towards the physical world, *even though this objective standpoint proved to be only an initial phase in the development of science*, was itself a necessary stage in the growth of thought. Before one could begin to correct the deceptive perceptions and concepts of our immediate or egocentric frame of reference, which significantly began with the Copernican Revolution, one had to penetrate the veil of myth to attain this direct perspective. Not that Aristotle achieved this in every instance, but he came considerably closer than many of his predecessors, explaining why Aristotelianism provided the necessary background for the development of modern science. Even Atomism, a more fruitful theoretical framework in the long run for interpreting the physical world, could not take hold until a more empirical and experimental attitude toward nature had developed. This was even more true of the fanciful, *a priori* number cosmology of the Pythagoreans and of Plato's *Timaeus*.

Thus while Atomism stood for the essential theoretical component of science and Pythagoreanism and Platonism represented the crucial mathematical factor, neither aspect could attain fulfillment until accompanied by the empirical. It was precisely the latter element of empirical research that Aristotle initiated and cultivated during his leadership of the Lyceum. As Jaeger indicates:

> To us moderns the scientific study of minutiae is no longer unfamiliar. We think of it as the fruitful depth of experience from which alone genuine knowledge of reality flows. It needs a lively historical sense, such as is not often found, to realize vividly at this time of day how strange and repellent this mode of procedure was to the average Greek of the fourth century, and what a revo-

lutionary innovation Aristotle was making [. . .]. He needed unspeakable labour and patience to lead his hearers into new paths. It cost him many efforts of persuasion and many biting reprimands to teach the young men, who were accustomed to the abstract play of ideas in Attic verbal dueling, and understood by a liberal education the formal capacity to handle political questions with the aid of rhetoric and logic, or at best perhaps the knowledge of 'higher things' [. . .] to teach them to devote themselves to the inspection of insects and earth-worms, or to examine the entrails of dissected animals without aesthetic repugnance. [15]

SCIENTIFIC KNOWLEDGE

Unlike Plato who continually stressed the importance of freeing rational thought from contact with and dependence upon the senses, Aristotle constantly emphasizes the importance of sense perception for knowledge. While he states that 'wisdom is knowledge about certain principles and causes' and thus superior to sense perception which is common even to animals, he also asserts:

All men by nature desire to know. An indication of this is the delight we take in our senses; for even apart from their usefulness they are loved for themselves; and above all others the sense of sight [. . .]. The reason is that this, most of all the senses, makes us know and brings to light many differences between things. [16]

In the *De Anima* he claims that 'the primary form of sense is touch,' since it is common to all animals and the most essential for survival, but sight reveals more of the quali tative differences among things, and thus is more important for knowledge. For though 'scientific knowledge' depends upon demonstration rather than observation, such knowledge would be impossible without sense perception.

It is also clear that the loss of any one of the senses entails the loss of a corresponding portion of knowledge, and that, since we learn either by induction or by demonstration, this knowledge cannot be acquired [. . .] [for] demonstration develops from universals, induction from particulars [. . .]. [17]

Scientific knowledge consists of demonstration from universals, but these universals are known by induction from perception of particular objects. Induction is defined very generally in the *Topics* as 'a passage from individuals to universals' and more explicitly in the *Posterior Analytics* as 'exhibiting the universal as implicit in the clearly known particular.' [18]

As to how the universal is exhibited in the particular, at the very end of the *Posterior Analytics* Aristotle offers his well-known explanation. First, sense perception itself is 'a congenital discriminative capacity' possessed by all animals; that is, since universals are general qualities or attributes, such as whiteness, triangularity, animality, etc., initially one must have the capacity to discriminate among the phenomena presented to the senses those qualities or characteristics which resemble each other more than they differ. Thus one white object is more like another white object in its color than either is to any black object, but colors resemble each other more than the other sensory qualities, such as smell, hearing, or touch. Aristotle illustrates how the initial identification of one entity allows one to discriminate a group of such entities with an analogy: 'It is like a rout in battle stopped by first one man making a stand and then another, until the original formation has been restored. The soul is so constituted as to be capable of this process.' He then completes the analogy as follows:

When one of a number of [. . .] particulars has made a stand, the earliest universal is present in the soul: for though the act of sense-perception is of the particular, its content is universal — is man, for example, not the man Callias. A fresh stand is made among these rudimentary universals, and the process does not cease until the indivisible concepts, the true universals, are established: e.g. such and such a species of animal is a step towards the genus animal, which by the same process is a step towards a further generalization. [19]

Initially, that is, an animal or an infant is confronted by a flux of sensations, but once the senses begin to discriminate among various objects and their qualities, these discriminated universals become a rallying point or standard for collecting or grouping other objects. Then, from a repetition of such discriminations, there occurs the capacity 'to retain the sense-impression in the soul,' that is, the beginning of memory and of recognition. The conjunction of repeated perceptions and persistent memories gives rise to experience, the more explicit recognition and organization of universals into a recognizable, stable world. Thus the discrimination, recognition, classification, and organization of objects and qualities depends upon universals becoming first 'explicit' and then 'stabilized' in the soul. Furthermore, men have the additional capacity of making more abstract 'conceptual' distinctions and groupings, thus forming successively broader classifications. It is these latter discriminations of essential properties into genera and species that are fundamental to Aristotle's theory of knowledge.

This account is the basis of Aristotle's empiricism. He explicitly denies 'innate knowledge' for the same reasons that Locke will, that it would be strange if we possessed innate knowledge from birth without recognizing it except through perception,[20] thereby rejecting the "nativist thesis" advanced by such thinkers as Plato, Descartes, Kant, and Chomsky. He also denies the Platonic view that the recognition of universals presupposes their separate existence or subsistence beyond the particulars of which they are general qualities. As he says,

demonstration does not necessarily imply the being of Forms nor a One beside a Many, but it does necessarily imply the possibility of truly predicating one of many; since without this possibility we cannot save the universal, and if the universal goes [. . .] demonstration becomes impossible. We conclude, then, that there must be a single identical term unequivocally predicable of a number of individuals. [21]

That is, demonstration presupposes the predication of universal qualities, but since these qualities are known from (but not by) perception, it does not presuppose or require their transcendent existence.

While Plato in the *Sophist* posited an abstract matrix of classifications into Ideal Species and Genera as a separate Reality, Aristotle believed that objects themselves possess 'essential attributes' or 'commensurate universals' belonging to an appropriate genus and species: 'the apprehension that animal is an element in the essential nature of man is knowledge; the apprehension of animal as predicable of man but not as an element in man's essential nature is opinion: man is the subject in both judgments, but the mode of inherence differs.'[22] It is this natural existence of real kinds defined in terms of essential attributes necessarily inhering in particular objects that constitutes the basic intelligibility of the world for Aristotle. An example in modern science would be the taxonomic system of biology based on the classifications of Linnaeus; for Aristotle, however, this taxonomic structure is not limited to biology but pervades all of nature.

Each particular object or event is intelligible because it exhibits, in addition to certain individual or 'accidental' characteristics, necessary properties which 'cause' it to be the kind of thing it is. Having scientific knowledge means being able to demonstrate that an essential quality necessarily inheres in an object because the object belongs to a certain genus which is defined in terms of that essential property. What is crucial to being human is not that one is of a certain color, sex or height, but that one has the general attributes of an animal with the more specific quality of being rational. If one asks why man is mortal, one answers that man is an animal and all animals are mortal. One must know that animals possess the essential characteristics of mortality, and that man exhibits the attributes that belong to the genus animality, and then one can demonstrate why men are mortal. As Randall has pointed out, Aristotle's conception of scientific inquiry is not that of a method of investigation, but a method of proof.

The proof consists of demonstrating by syllogistic reasoning that something necessarily is the case. Having invented the syllogism in the *Prior Analytics* as a method of proof, Aristotle utilizes this procedure in the *Posterior Analytics* as the proper method of arriving at scientific knowledge. The aim of demonstration is to show that certain conclusions follow from certain premises by deductive reasoning: that the premises 'produce' or are the 'cause of' the conclusion. Today we would not claim that premises "cause" or "produce" conclusions, asserting instead that the conclusion "follows from" or "is deduced from" the premises. For Aristotle, however, the premises produce the conclusion (a statement of fact) in the sense that they contain or exhibit the cause of the fact. Just as we now assume that an event is explained if we can specify its cause, so Aristotle believed that a demonstration is the cause of the conclusion because it exhibits the reason for the conclusion being what it is. One begins with 'a fact' and ends with 'a reasoned fact.'

In addition, for the demonstration to be productive of *scientific knowledge* the premises of the syllogism must fulfill certain conditions: 'Assuming then that my thesis as to the nature of scientific knowing is correct, the premises of demonstrated knowledge must be true, primary, immediate, better known than and prior to the conclusion [. . .]. [23] The premises must be true or it would not necessarily follow that the conclusion was true; the premises must be primary and indemonstrable or they would require further proof, possibly leading to an infinite regress; and they must be prior and better known than the conclusion, not because they are the first to be learned or are more familiar or more easily understood, but in the sense that though they are more difficult to learn and hence are learned later, they are of such a fundamental nature that they make other facts intelligible. What is 'prior and better known to man' is often different from 'what is prior and better known in the order of being,' just as objects of sense are prior and better known than scientific laws or theories, although the latter explain the behavior of the former.

Now that we know what conditions the premises or 'basic truths' must fulfill in order to produce scientific knowledge, it is necessary to indicate more precisely 'what is their character' and how they produce the desired result. The basic premises of demonstrations are definitions, the difference between a definition and a demonstration being that a definition defines the essential nature of something, while a demonstration deductively proves that an attribute attaches or does not attach necessarily to a subject. Since 'demonstration proves the inherence of essential attributes in things,' the premises must assert that certain properties necessarily belong to their subjects. Demonstration proceeds in two ways, either (1) by revealing that the attributes are essential elements in the nature of their subjects, or (2) by showing that their subjects are essential conditions of the attributes. An example of the former would be possessing sensation and being warm blooded, which are essential attributes of man; an example of the latter would be animal since being an animal is essential for being warm blooded and having the capacity for sensation.

Such attributes must be 'necessary as well as consequentially connected with their subjects' and are called 'commensurately universal' because they belong to every instance of their subject. Moreover, '[s]ince it is just those attributes within every genus which are essential and possessed by their respective subjects as such that are necessary,' the subjects of the premises must be genera.

> For there are three elements in demonstration: (1) what is proved, the conclusion — an attribute inhering essentially in a genus; (2) the axioms, i.e., the axioms which are premises of demonstration; (3) the subject-genus whose attributes, i.e., essential properties, are revealed by the demonstration. [24]

If we know 'the fact' that men are mortal, we can arrive at the 'reasoned fact' by demonstrating that men are mortal because they are animals, the genus animal (in contrast to the genus godly) possessing the essential attribute of mortality. Thus what is proved, the conclusion, consists of an attribute (mortality) belonging to the class of men, the proof dependent upon two axioms, each of which contains the subject-genus (animals) whose essential property, mortality, is revealed in the demonstration. The truth of the conclusion is not based on the meaning of mere words, however, but is recognized by the soul: 'all syllogism, and therefore a fortiori demonstration, is addressed not to the spoken word, but to the discourse within the soul [...].'

This analysis of syllogistic demonstration would be incomplete without an explanation of the important role of the 'middle' term. The 'middle,' in fact, is called by Aristotle 'the cause' of the deduction because of its crucial importance. To take the above example, which is in the first figure of the syllogism, and hence 'the most scientific:'

$$A \qquad B$$
All men are animals.

$$B \qquad C$$
All animals are mortal.

$$A \qquad C$$
Hence: All men are mortal.

The general form of the syllogism is: all A's are B's, all B's are C's, hence all A's are C's. The deduction is possible through the middle term 'animal,' the term that is common to the two premises, since one proves that men are mortal by demonstrating that all men are animals and all animals possess the essential attribute of mortality. It is the 'subject-genus' animality (with its essential property of mortality) that is both the cause of the conclusion and the cause of the fact that men are mortal.

> Where demonstration is possible, one who can give no account which includes the cause has no scientific knowledge. If, then, we suppose the syllogism in which, though A necessarily inheres in C, yet B, the middle term of the demonstration, is not necessarily connected with A and C, then the man who argues thus has no reasoned knowledge of the conclusion, since this conclusion does not owe its necessity to the middle term; for though the conclusion is necessary, the mediating link is a contingent fact. [25]

Aristotle does not confine the use of syllogistic demonstration merely to object-attribute relationships, as the above example might suggest, but applies it also to demonstrations of why events do or do not occur; however, to make this application he must interpret the events within the same subject-attribute schema (in his later explanations of physical and astronomical phenomena he usually bypasses this restrictive schema of syllogistic deduction altogether, even though the explanation still depends upon essential attributes and generic distinctions). For example, if one wishes to explain why the planets do not twinkle, Aristotle argues as follows:

$$A \qquad B$$
The planets are proximate bodies.

$$B \qquad\qquad C$$
Proximate bodies do not twinkle.

$$A \qquad\qquad C$$
Hence: The planets do not twinkle.

Or, If one wants to prove why the moon 'always has its bright side turned toward the sun.'[26]

$$A \qquad B$$
The moon is lighted by the sun.

$$B \qquad\qquad C$$
Whatever is lighted by the sun has its bright side turned

toward the sun.

$$A \qquad\qquad C$$
Hence: The moon has its bright side turned toward the sun.[27]

In each of these cases, the 'syllogism is effected by means of three terms,' proving 'that A inheres in C by showing that A inheres in B and B in C.' This figure, the first, is preferred because it results in an affirmative universal conclusion, whereas the second figure results in a *negative* universal conclusion, and the third in an affirmative *particular* conclusion. In both the *Prior Analytics* and the *Posterior Analytics*, Aristotle recognizes only three figures of the syllogism while we recognize four (the fourth resulting in a *negative particular* conclusion).

Aristotle is always concerned to emphasize that scientific knowledge is never possible through perception alone, since the latter is always of the particular and the former of a universal. For example, even if we were in a position to directly observe an instance of a lunar eclipse, this experience would not constitute knowledge of why the eclipse occurred. To possess scientific knowledge of such an occurrence one would have to know that the eclipse *always* and *necessarily* occurs 'because of the failure of light through the earth's shutting it out.'

Seeing, therefore, that demonstrations are commensurately universal and universals imperceptible, we clearly cannot obtain scientific knowledge by the act of perception [. . .]. So if we were on the moon, and saw the earth shutting out the sun's light, we should not know the cause of the eclipse; we should perceive the present fact of the eclipse, but not the reasoned fact at all, since the act of perception is not of the commensurate universal.[28]

However, he does not mean to 'deny that by watching the frequent recurrence of this event we might, after tracking the commensurate universal, possess a demonstration, for the commensurate universal is elicited from the several groups of singulars.'

Still, there are instances where sense perception and the awareness of the commensurate universal or cause are so connected that the one occurs almost instantaneously with the other:

> [...] there are cases where an act of vision would terminate our inquiry, not because in seeing we should be knowing, but because we should have elicited the universal from seeing; if, for example, we saw the pores in the glass and the light passing through, the reason of the illumination would be clear to us because we should at the same time see it in each instance and intuit that it must be so in all instances. [29]

Occasionally, then, the perceived occurrence reveals the cause so clearly that the soul intuits the universality and necessity of the cause immediately, the grasp of this necessity being essential for any scientific knowledge.

> To sum up, then: demonstrative knowledge must be knowledge of a necessary nexus, and therefore must clearly be obtained through a necessary middle term; otherwise its possessor will know neither the cause nor the fact that his conclusion is a necessary connection. [30]

One could hardly read this without thinking of Hume. Hume maintained that we neither could perceive nor demonstrate any "necessary connection" among observable phenomena because the sense impressions themselves do not disclose any necessity for their occurrence or necessary relation to antecedent sensory impressions. Aristotle, on the contrary, maintains that while *perception itself* does not directly disclose a necessary connection among phenomena, the *structured relation* among objects and events is such that, on repetition, the mind 'intuits' the reason for their necessary connection. The difference between the two views marks the contrast between Aristotelian and modern science (as well as that of Ancient Atomism). While Aristotle held that the soul can intuit the essential natures of, and necessary causal connections among, observable objects from ordinary perception, modern science (like Ancient Atomism) claims that the real essences of things lie hidden in their microscopic structures, and since we cannot perceive or know the real nature of these structures, we cannot ascertain the ultimate causes of phenomena, though we can infer some approximate causes and effects. Living at a time when the modern scientific conception was emerging, Locke clearly evinces this skepticism.

> Had we senses acute enough to discern the minute particles of bodies, and the real constitution on which their sensible qualities depend, I doubt not but they would produce quite different ideas in us, and that which is now the yellow colour of gold would then disappear, and instead of it we should see an admirable texture of parts of a certain size and figure [...]. Nay, if that most instructive of our senses, seeing, were in any man a thousand or a hundred thousand times more acute than it is now by the best microscope, things several millions of times less than the smallest object of his sight now would then be visible to his naked eyes, and so *he would come nearer the discovery of the texture and motion of the minute parts of corporeal things*, and in many of them probably get ideas *of their internal constitutions*: but then he would be in a quite different world from other people: nothing would appear the same to him and others: the visible ideas of everything would be different. [31] (Italics added)

How divergent this view is from Aristotle's confidence in the attainment of complete scientific knowledge: 'If [...] one is aiming at truth, one must be guided by the real connections of subjects and attributes.'[32] Not that Aristotle was unaware of the alternative, as well as the difficulties posed by Locke. Quite the contrary. He explicitly antici-pates Locke in describing how such knowledge based on a perception of the mechanical combination of imperceptible particles would vary with the visual acuity of the observer. As he states, if composition merely means 'composition of small particles,' 'the constitu-ents will only be "combined" relatively to perception: and the same thing will be "com-bined" to one percipient if his sight is not sharp, [but not to another,] while to the eye of Lynkeus nothing will be "combined".' [33]

Before concluding this discussion it is necessary to say a little more about the origin of the premises of scientific deduction. We have seen how one comes to have inductive knowledge of commensurate universals essential for deduction, but how in the various sciences does one come to know the original premises in which these commensurate universals occur? Just as Aristotle had said that if we could see in each instance the light passing through the pores of glass we 'would intuit that it must be so in all instances,' he declares that scientific knowledge depends upon 'rational intuition' defined as 'an originative source of scientific knowledge [. . .]' [34] Particular to each science are its own basic truths as the premise of scientific deduction which themselves cannot be demon-strated or proved, but that can become manifest to 'rational intuition.'

Aristotle is not unaware of the difficulty of determining, in any particular case, whether we truly know the basic truths of the science.

> It is hard to be sure whether one knows or not; for it is hard to be sure whether one's knowl-edge is based on the basic truths appropriate to each attribute — the differential of true knowledge. We think we have scientific knowledge if we have reasoned from true and primary premises. But that is not [necessarily] so: the conclusion must be homogeneous with the basic facts of that science. [35] (Brackets added)

In other words, we could have valid deductive knowledge which still would not constitute scientific knowledge because it would not apply to the true attributes of that science.

Aristotle denies that a *unified* deductive science is possible, such as that envisioned by mathematical physicists like Eddington and Einstein or by the "unity of science move-ment" of the Positivists. For Aristotle, the basic truths of each science remain separate and irreducible.

> It is no less evident that the peculiar basic truths of each inhering attribute are indemonstrable; for basic truths from which they might be deduced would be basic truths of all that is, and the science to which they belonged would possess universal sovereignty [. . .]. But, as things are, demonstration is not transferable to another genus with such exceptions as we have mentioned of the application of geometrical demonstrations to theorems in mechanics or optics, or of arith-metical demonstrations to those of harmonics. [36]

Again, it is his strong empiricistic bent that leads him to deny the rationalistic vision of a unified deductive science and affirm the irreducible separateness of the various disci-plines.

COMPARISON OF ARISTOTLE' S METHOD WITH THAT OF MODERN SCIENCE

As regards the origin and foundation of scientific knowledge, it would seem that Aristotle's account is quite sound. Whether it be the various senses as Aristotle maintains, or the senses plus the cerebral cortex, at any rate, tracing the origin of knowledge to the 'congenital capacity' to discriminate among (or as we would say today, to "decode") sensory stimuli would seem to be correct. As Aristotle maintains, no experience or knowledge would be possible without the ability to differentiate colors, sounds, tactual sensations, shapes, etc. Furthermore, Aristotle's general theory as to how experience and knowledge arise from this preliminary discrimination would also seem to be sound, except that now we realize that our higher cognitive functions depend upon complex *neurological* structures and processes. Along with the capacity for sensory discrimination, one must acknowledge that certain objects and qualities can be grouped according to their similarities, which presupposes at least an implicit recognition of universal qualities. Then, given a repetition of similar perceptions, the soul (or in today's terms, the mind or the psycho-physical organism) must be able to 'retain' these sense impressions in such a way that they can be used to recognize, remember, and anticipate similar occurrences. This, in turn, would seem to be the basis of concept formation and of learning a language. All of this is described in such a way as to obviate positing innate concepts, although Aristotle attributes to the soul, not the senses, the ability to form the commensurate universals, necessary for all knowledge, on the basis of repetitive experiences.

Regarding the adequacy of Aristotle's conception of scientific knowledge, however, a different assessment is required. In spite of Randall's favorable evaluation of Aristotle's account of scientific knowledge, except for the latter's overconfident reliance on direct observation, it is not difficult to point out the inadequacies of this conception from the modern point of view. Even Randall admits that Aristotle's account of scientific knowledge "is not, of course, a statement of scientific method, of the procedures to be followed in inquiry and discovery" because "the syllogism is in no sense a method of investigation, but a method of proof."[37] However, Randall then goes on to say that Aristotle's conception of the task of scientific demonstration "to fit [. . .] observed facts into a system of knowledge [. . .] to formalize our observations" is still the ideal of modern science and "states precisely what Newton did, in discovering the mathematical *archai*, the *principia mathematica*, from which the facts observed by Copernicus, Kepler, and Galileo could be demonstrated."[38] In a very general sense what Randall says is true, but in his appreciation of Aristotle (understandable in the light of the usual hostility) he seems to overlook the difference between what "formalizing our observations" meant for Aristotle and what it meant for Newton, though Randall's own statement points to the difference. Whereas Aristotle's conception of "formalized knowledge" was that of fitting phenomena into a taxonomic system such as one finds in biology, Newton's formalism consisted of deducing mathematical laws and discovering quantifiable forces with which one could explain the phenomena. As Newton says in the "Preface" to the first edition of *Principia Mathematica*, "I offer this work as the mathematical principles of philosophy, for the whole burden of philosophy seems to consist in this: from the phenomena of motions to investigate the forces of nature, and then from these forces to demonstrate the other phenomena [. . .]."[39] There is a world of difference between this conception and that of Aristotle.

It was precisely because Aristotle's conception of formalized knowledge depended upon discovering the essential attributes differentiating various genera that he adopted the syllogism as his method of scientific demonstration, generally (though not entirely as his frequent references to 'ratios' and to the application of mathematics to mechanics, optics, and harmonics indicate) ignoring the importance for scientific knowledge of discovering quantifiable laws and correlations among occurrences. The typical designation of Aristotelian science as qualitative rather than quantitative is not without basis. In contrast to Aristotle, both the Pythagoreans and Plato realized the necessity of discerning mathematical relationships among processes and intuiting an ideal mathematical pattern in terms of which observable phenomena, such as the musical scale and planetary motions, could be explained. It is no accident that the scientists mentioned by Randall were strongly influenced by the mathematical ideal of Plato (although Galileo was also influenced by the empiricism of Aristotle and the experimental approach of Archimedes). In words similar to those of Plato, Copernicus sought a more harmonious mathematical account of the celestial motions "wrought for us by a supremely good and orderly Creator," [40] while Kepler used the Platonic solids in his search for the fundamental laws describing planetary motion, both endeavors expressed by Galileo in his (Platonic) statement that "the language of nature is mathematics."

One could also argue that insofar as Aristotle explicitly emphasized induction and deduction he incorporated in his conception of scientific knowledge the two fundamental aspects of scientific method. Here again, however, what is significant is not the similarity of the terms used, but the difference in their meanings. Even as regards induction, a necessary aspect of any scientific method, modern science means by this not so much 'discerning the universal among particulars,' as discovering a common pattern of behavior among a group of objects that can be expressed as a general law. More important than essential attributes or qualities is the fact that certain phenomena manifest a law-like behavior under certain conditions: e.g., the elliptical orbits of the planets, the universal gravitational attraction of masses, the uniform acceleration of free falling objects, etc. The modern scientist tries to quantify the various factors involved and express their functional correlations in the form of a mathematical equation and then explains the correlations by some theory. In this way the laws of Kepler, Galileo, and Newton were formulated, which Newton then tried to explain in terms of such theoretical concepts as inertial, gravitational, and centrifugal forces.

In fairness to Aristotle, he did recognize the first step in any explanation, that of observing some uniformity in nature, such as the occurrence of eclipses. He also recognized the importance of discovering the cause: in the case of eclipses, that of the obstruction of light by the imposition of another celestial body. His reference to essential attributes also could be seen as a preliminary stage before quantification, since one has to isolate and define the controlling factors in any investigation before he can submit these to measurement. Finally, his syllogistic demonstration could be seen as a formalization of deductions from general laws. Why did the moon suffer eclipse? Because its light from the sun was shut out by the earth, and all eclipses are caused by the obstruction of light owing to the imposition of another celestial body. Thus the present eclipse could be seen as an instance of a more general law of nature.

Nevertheless, there still are crucial differences between Aristotle's conception of scientific explanation and that of modern science. Generally speaking, there are three ways in which a modern scientist tries to explain phenomena: (1) by discovering that something occurs as an effect of observable, antecedent causes, (2) by showing that an occurrence is an instance of a natural law expressed as a mathematical equation, and (3) by deducing the phenomena from a general theory which posits unobservable entities (such as atoms, light waves, gravitational forces, etc.) as the causes of the phenomena. Taking the examples in inverse order, as pointed out earlier, Aristotle did not recognize the need for theoretical explanations by means of inferred scientific constructs (as the Greek Atomists did), because he thought nature was intelligible in terms of observable qualities and generic distinctions. Unlike modern scientists, he did not conceive of nature as a manifestation of underlying, unobservable elements and structures which may be more real than their qualitative manifestations. In fact, as the Periodic Table of Mendeleev illustrates, the modern classification of elements is not based on observable qualities but on experimentally discoverable and theoretically defined properties such as atomic weights and numbers. Secondly, Aristotle's aversion to the abstract mathematical speculations of the Platonists led him to minimize the significance of mathematics for science in favor of qualitative explanations and syllogistic deductions. Thirdly, although he did refer to causes, he thought of these causes in terms of essential attributes or subject-genera which serve as the middle term in syllogistic deductions. In contrast, modern science thinks of causes not in terms of unchanging essential attributes, but in terms of antecedent physical conditions that bring about the phenomena. It is the search for these antecedent conditions that requires experimentation and leads to the discovery of new knowledge.

Furthermore, though in a previously quoted passage form his biological work on *De Partibus Animalium*, Aristotle exhorted younger members of the Lyceum to engage in the examination and dissection of animals, he nowhere refers to experimentation as a method of acquiring new physical knowledge (as opposed to biological) or of testing theories. Rather, his was a method of demonstrating as *conclusively true* what one already *believed*, but did not *know*, to be a fact. As such, this method does not have the self-corrective aspect of experimental testing that is such a crucial element of modern science. This explains why Galileo and Fabricius, in the areas of physics and biology respectively, often criticize Aristotle for not having taken the trouble to test some of his assertions and theories.[41] So while Aristotle's conception of scientific knowledge could be applied in biology, it has very little application in the other sciences, and even in biology classification is less important today than explanations in terms of biomolecular concepts and causes.

None of this is said to minimize the magnificent achievements of Aristotle. It would be unreasonable to expect him at that time to have advanced a conception of science which was not formulated until about twenty centuries later. But neither should one attribute to Aristotle a modern conception of scientific knowledge based on very general similarities. What we particularly owe to Aristotle is his insistence on giving due regard to the 'evidence of the senses' in contrast to the *a priori* theorizing of the Presocratics (especially Parmenides) and the quasi-mythical speculations of Plato. At that period, especially, without such respect for empirical evidence one's theorizing tended to result in fanciful or fruitless conjectures. Unfortunately, because of the use made of Aristotle by the Scholastics who were not as empirically oriented, this aspect of Aristotle is usually overlooked.

But it should not be forgotten that during the Renaissance when there was so much criticism of Aristotle it was still the latter's empirical observations and theories which were the starting point for further critical research — not the theories of Parmenides or even of Democritus or Plato. Galileo pointedly distinguished between the Scholastics and Aristotle, often declaring his admiration for the latter, "considering him as I do a man of brilliant intellect [. . .]."[42] And as great an empirical scientist as Darwin, writing as late as the middle of the 19th century, could still speak with awe about Aristotle: "From quotations which I had seen, I had a high notion of Aristotle's merits, but I had not the most remote notion what a wonderful man he was. Linnaeus and Cuvier have been my two gods [. . .] but they were mere school-boys compared to old Aristotle."[43] Anyone approaching Aristotle with an open mind could not help but admire the power and range of his intellect — perhaps never equalled by another thinker.

NOTES

CHAPTER XIII

1   Quoted from John Herman Randall, Jr., *Aristotle* (New York: Columbia Univ. Press, 1960), pp. 1-2.

2   Cf. Werner Jaeger, *Aristotle*, second ed., trans. by Richard Robinson (Oxford: At The Clarendon Press, 1948), p. 30.

3   *Ibid.*, p. 117. The original Greek has been omitted from the quotation.

4   W. D. Ross, *Aristotle*, fifth ed., rev. (London: Methuen and Co., Ltd., 1949), pp. 3-4. Also, see Randall, *op. cit.*, pp. 219-224.

5   Jaeger, *op. cit.*, p. 123.

6   *Ibid.*, p. 312.

7   *Ibid.*, p. 316. For a detailed description of "The Organization of Research" in the Lyceum, see Jaeger, *op. cit.*, ch. XIII.

8   Ross, *op. cit.*, p. 6.

9   For a translation of the will, see Jaeger, *op. cit.*, pp. 322-323.

10   For examples note the following references to 'seeing' in Aristotle: *Phys.*, Bk. VIII, ch. 3, 254a-36b1; *De Caelo*, Bk. I, ch. 3, 270b11-12: ch. 5, 272a5-7, Bk. III, ch. 7, 306a17, Bk. IV, ch. 4, 311b21; *De Gen. et Corr.*, Bk. I, ch. 1, 314b15, Bk. II, ch. 9, 335b21: ch. 10, 336b18; *De Anima*, Bk. III, ch. 7, 432a7.

11   Aristotle, *De Gen. et Corr.*, Bk. I, ch. 8, 325a14-23. (Joachim trans.)

12   Aristotle, *On the Parts of Animals*, Bk. I, ch. 5, 644b23-645a23. (D'Arcy Thompson trans.) Unless otherwise indicated, all the quotations are from the Richard McKeon edition of *The Basic Works of Aristotle* (New York: Random House, 1941).

13   Randall, *op. cit.*, p. x.

14   Thomas S. Kuhn, *The Copernican Revolution* (*op. cit.*, Introduction), p. 95.

15   Jaeger, *op. cit.*, pp. 336-337.

16   *Met.*, I, 980a22-28. (Ross trans.)

17   *An. Post.*, I, ch. 18. (Mure trans.)

18   *Topics*, I, 105a13. *An. Post.*, I, 71a7.

19   *An. Post.*, II, ch. 19.

20   Cf. *Ibid.*, II, 99b25-29.

21   *Ibid.*, 77a5-9.

22   *Ibid.*, 89a35-39.

23   *An. Post.*, I, 71b19-22.

24   *Ibid.*, I, 75a39-75b2.

25   *Ibid.*, I, 74b28-31.

26   Cf. *An. Post.*, I, ch. 13.

27   *Ibid.*, ch. 34.

28   *Ibid.*, I, 87b34-88a.

29   *Ibid.* 88a13-17. The word 'illumination' is substituted for 'kindling' in the original translation.

30   *Ibid.*, I, 75a12-15.

31   John Locke, *An Essay Concerning Human Understanding*, Bk. II, ch. 23, Sec. 11 & 12.

32   *An. Post.*, I, 81b22-23.

33   *De Gen. et Corr.*, Bk. I, 328a12-16. (Joachim trans.) Brackets in original.

34   *An. Post.*, I, 88b36.

35   *Ibid.*, 76a27-32.

36   *Ibid.*, 76a17-27.

37   Randall, *op. cit.*, p. 41.

38   *Ibid.*, p. 42.

[39] Sir Isaac Newton, Preface to the First Edition of *Principia Mathematica*, trans. by Andrew Mott, 1729.

[40] Kuhn, *op. cit.*, p. 141.

[41] Cf. Galileo Galilei, *Dialogues Concerning Two New Sciences*, trans. by Henry Crew and Alfonso de Salvio (New York: McGraw-Hill Book Co., Inc., [1914] 1963), p. 60ff.

[42] Galileo Galilei, *Dialogue Concerning the Two Chief World Systems* (*op. cit.*, Introduction), p. 321.

[43] Charles Darwin to William Ogle, on the occasion of the publication of the translation of Aristotle's *The Parts of Animals*, 1882. *Life and Letters of Charles Darwin*, ed. by Francis Darwin (New York: 1896), II, p. 427. Quoted from Randall, *op. cit.*, p. 219.

ARISTOTLE'S PRINCIPLES OF EXPLANATION AND ORGANISMIC COSMOLOGY

> *The name of Aristotle suggests* [. . .] *intellectual sovereignty over the whole world of abstract thought throughout long stretches of history* [. . .][1]
>
> Jaeger

An analysis of Aristotle's concept of scientific method and the difference between his method and that used by modern scientists having been presented in the last chapter, we shall turn now to his actual manner of conceptualizing physical problems and to a description of his overall cosmological view. Again we shall see how different his "principles of explanation" are from those of modern scientists and how they lead to different results. Although Aristotle disclaims being an astronomer, he does consider himself a physicist, so the discussion will focus on his physical principles of explanation.

He begins the *Physics* with the assertion that the science of nature, like any other area of inquiry, depends upon discovering the underlying principles, conditions, or elements of explanation. As he stated in the *Posterior Analytics*, one begins with those 'confused masses' which are more familiar to us at first, proceeding to those principles that are 'clearer and more knowable by nature;' namely, those theoretical principles in terms of which the phenomena in question can be explained. This preliminary examination of the basic *archai* of explanation is undertaken also in various other works, such as Book II, *On Generation and Corruption*, and particularly Books Alpha (I), Beta (II), and Lambda (XII) of the *Metaphysics*.

What first strikes the reader is how different this discussion is from that found in modern textbooks on physics. While Aristotle, like the modern physicist, is concerned primarily with an analysis of motion, one is surprised to find him discussing under the category of motion not only displacement in space, or locomotion as he calls it, the primary form of motion, but also various kinds of change ('metabole') such as alteration, generation or destruction, and the addition or reduction of quantity. Moreover, one is puzzled by his use of examples, such as the musical coming from the unmusical or the not-white becoming white, as in the following passage.

> Our first presupposition must be that in nature nothing acts on, or is acted on by, any other thing at random, nor may anything come from anything else, unless we mean that it does so in virtue of a concomitant attribute. For how could 'white' come from 'musical', unless 'musical' happened to be an attribute of the not-white or of the black? No, 'white' comes from 'not-white' — and not from *any* 'not-white', but from black or some intermediate colour. Similarly, 'musical' comes to be from 'not-musical', but not from *any* thing other than musical, but from 'unmusical' or any intermediate state there may be. [2]

One is tempted to ask in exasperation, "what has all this to do with physics, with an explanation of various kinds of motions in terms of certain forces?"

The perplexity increases when Aristotle starts interpreting the 'principles' of natural science in such terms as 'subject,' 'form,' 'contrary,' 'privation,' 'potentiality,' 'actuality,' etc., terms which are never encountered in modern physics. What on earth could he have meant by saying "if 'man' is a substance, 'animal' and 'biped' must also be substances," or by referring to imagination and thought as motions, or by defining motion itself as 'the fulfillment of what exists potentially, insofar as it exists potentially'? The answer to these puzzles lies, of course, in the realization that Aristotle wrote in the 4th century B.C., about two thousand years before the 'principles' of modern classical physics were formulated, 'principles' such as force, mass, gravitational attraction, etc. Even the categorical distinction between *changes* such as growth and alteration in quality, *actions* like sitting, and the physical *motions* of projectiles and planets, though recognized at that time, were not conceptualized as they would be today; instead, they were grouped together as different species of motion.

To understand Aristotle one must realize that the scientific tradition that he encountered was that of Plato and the Presocratics, not that of the founders of modern classical mechanics, such as Galileo and Newton. Although he constantly called attention to natural phenomena that everyone can see, the intellectual traditions and theoretical frameworks Aristotle had to work with were radically different from those of the modern physicist. From this perspective, his selection and definition of problems, as well as his methodological and conceptual approach to them, begin to make sense. One must realize how diversified and wide open the possibilities of interpretation and explanation were at that time, as exemplified in the following statement: 'For some make hot and cold, or again moist and dry, the conditions of becoming; while others make odd and even, or again Love and Strife [. . .].'[3] The possible principles of explanation were as divergent as the four elements of Empedocles, the Numbers of the Pythagoreans, Parmenides' One, the atoms of Democritus, or Plato's Forms.

Like his predecessors, Aristotle was confronted with the task of explaining generation, change, and motion without violating Parmenides' canon that 'Being cannot come from not-Being nor can it cease to be.'

> The first of those who studied science were misled in their search for truth and the nature of things by their inexperience [. . .]. So they say that none of the things that are either comes to be or passes out of existence, because what comes to be must do so either from what is or from what is not, both of which are impossible. For what is cannot come to be (because it *is* already), and from what is not nothing could have come to be (because something must be present as a substratum). So too they exaggerated the consequence of this, and went so far as to deny even the *existence* of a plurality of things, maintaining that only Being itself is.[4]

Aristotle refuses to take this conclusion seriously because it is so contrary to experience: 'to investigate whether Being is one and motionless is not a contribution to the science of Nature.' There is no point in trying to refute positions which, merely for the sake of argument, deny the existence of common phenomena, such as change and motion: 'accept one ridiculous proposition and the rest follows — a simple enough proceeding.' 'We physicists,' he says, 'must take for granted that the things that exist by nature are, either all or some of them, in motion — which is indeed made plain by induction.' Who, for example, could seriously deny that the sun and the moon alter their positions, that objects vary in quality, changing from cold to hot, moist to dry, green to black, that things sometimes increase in magnitude, as when a child grows and a house is built, or decrease in quantity as when an older person shrinks in size and a statue is broken. Such irrefutable instances of change and motion must not be explained away, but should lead to a search for the appropriate 'principles' in terms of which they can be understood.

What happens in such instances? First, there is the appearance of something new and different: the sun is in another position, the olive is now black, the child has grown, the house is large, Socrates is shorter, Praxiteles' statue has lost an arm. Secondly, in acquiring its new feature the entity has lost its former characteristic: the sun's previous position, the olive's green color, the child's smaller size, etc. Is one justified, therefore, in concluding that something has come to be from nothing and that something else has passed into nothingness? Aristotle answers that such a conclusion is based on a confused or false interpretation of what the phrases 'coming to be' and 'ceasing to be' or 'being' and 'not being' mean.

In all change (and motion) there are three principles involved: (1) the 'substratum' or 'matter' which persists throughout the change and is the subject of the changing attributes (in the previous examples, the sun, olive, Socrates, etc.), (2) an initial quality (such as green or smallness) which is also the privation of (3) the new contrary quality (such as black or largeness) that replaces the former quality. Thus by virtue of the 'privation' inherent in it due to its nature, an object can acquire a 'contrary quality' if acted on by an appropriate cause. This is Aristotle's conception of the fundamental principles of explanation in the science of nature: 'The causes and the principles, then, are three, two being the pair of contraries of which one is definition and form and the other is privation, and the third the matter.'[5] Thus the ancient doctrine of opposites reappears in Aristotle in a much more abstract form.

In this conceptual analysis of change he has attempted to account for a number of factors: (1) that in all change there is something, the underlying substratum or matter, which persists throughout the change, (2) that during the change this matter both loses and acquires some definable quality, structure, or 'form,' (3) that changes are not haphazard or random but usually (unless accidental) occur among natural 'opposites,' and (4) that the new 'form' replaces a 'privation' inherent in the nature of the matter prior to the change. As he says:

> Sensible substance is changeable. Now if change proceeds from opposites or from intermediates, and not from all opposites (for the voice is not-white [but it does not therefore change to white]), but from the contrary, there must be something underlying which changes into the contrary state; for the *contraries* do not change. Further, something persists, but the contrary does not persist; there is, then, some third thing besides the contraries, viz. the matter.[6]

Aristotle's mode of explanation which we initially found so puzzling should now be intelligible, since it is apparent that he is trying to provide a schema for interpreting what he considers to be the essential aspects involved in all change. As such, the schema is not so much an explanation as it is a conceptual clarification. He accepts the fact that changes occur in the world, but these changes are not completely random or haphazard. There is repetition and order among phenomena: acorns develop into oak trees, not into elephants; objects placed in the sun's rays become hot not cold; olives turn from green to black while tomatoes turn from green to red; men become musical not from being black or white but from being unmusical, while objects become white from a contrasting color, not from being musical; patients become healthy from being unhealthy, not from being sculptors or philosophers.

The material of each thing because of its inner nature has a 'privative' or 'potential' for certain kinds of changes and not for others; given the appropriate cause, this privation is replaced or the potentiality realized by the actualization of a new form. The contraries are evidence, therefore, of a constraining or regulative effect on the possible kinds of changes any particular matter can undergo by virtue of its nature. In effect, they represent the fundamental principles of order in the universe which today are expressed by scientific laws: 'change proceeds from a contrary to a contrary or to something intermediate: it is never the change of any chance subject in any chance direction [. . .].' Although his interpretation is different, Aristotle is as concerned as Plato was to eliminate chance as a basic principle in explaining natural phenomena.

Unlike the ancient Atomists and modern physicists, Aristotle does not attempt to explain, in terms of inferred, imperceptible elements or atoms, how these contrary changes are brought about; in fact, he explicitly rejects such explanations. Thus from the modern point of view his 'principles of explanation' are apt to appear superficial or merely verbal, but he was convinced that these three principles allowed him to conceptualize change without violating Parmenides' canons. Nothing comes to be from nothing or ceases to be absolutely, since an underlying substratum persists throughout the change: 'For my definition of matter is just this — the primary substratum of each thing, from which it comes to be without qualification, and which persists in the result.'[7] Even in 'substantial change' involving 'coming-to-be and passing-away,' as when a leaf burns or bread is digested, an underlying or 'prime matter' still persists throughout the change, although its identity is not perceptible.

Before any change occurs the 'form' that will be actualized could be said to exist in a state of privation or not-being, because 'privation in its own nature is not-being.' Thus in *one sense* one could say that the actualization arises from a state of not-being — but only in this sense. In *another sense*, since the privation is an inherent state of the underlying matter, the actualization does not arise from not-being, but from the privation existing in the matter. Analogously, one could say in *one sense* that the form that ceases to be when the object changes enters a state of not-being. In *another sense*, however, it merely ceases to be an attribute of that particular matter, while the matter still 'persists in the result,' so that no state of not-being ensues. It was the confusion of these two senses that misled the Parmenideans, in Aristotle's opinion.

> It was through failure to make this distinction that those thinkers gave the matter up, and through this error that they went so much farther astray as to suppose that nothing else comes to be or exists apart from Being itself, thus doing away with all becoming.
> We ourselves are in agreement with them in holding that nothing can be said without qualification to come from what is not. But nevertheless we maintain that a thing may 'come to be from what is not' — that is, in a qualified sense. For a thing comes to be from the privation, which in its own nature is not-being — this not surviving as a constituent of the result [. . .].
> In the same way we maintain that nothing comes to be from being, and that being does not come to be except in a qualified sense. [8]

On a more fundamental level, nothing absolutely comes to be or ceases to be because the basic matter of the universe, the 'prime matter,' always has existed and will exist: it was not created nor can it be depleted. This is Aristotle's expression of the law of the conservation of matter. Moreover, as we shall find, since the ultimate source of motion is eternal and continuous, he also maintains the law of the conservation of motion (or energy in today's terminology); as he says, 'neither the matter nor the form comes to be — and I mean the last matter and form.'[9] Also, though the 'essential attributes' or 'commensurate universals' may cease to exist in a *particular* entity, because they are themselves unmodifiable and undoubtedly qualify *some object* at any one time, they constitute the 'unchanging forms' characterizing all changes. Thus as Aristotle repeats so often, 'man begets man.' If the different forms (qualities, genera, species, etc.) of things did not already exist either actually or potentially in preexisting things, no uniform changes could occur. As Randall claims, for Aristotle there was no problem as to which came first, the chicken or the egg: the individual chickens must exist first before the genus and specific differentia could be passed on through the fertilization of the egg. There is no evolutionary modification of forms; the different types of forms have always existed and are merely newly actualized in different particulars. Novelty occurs as a result of the particular and therefore unique actualization of forms at any one time, not because of a transformation of the forms themselves.

In the *Treatise On Generation and Corruption*, Aristotle presents a detailed explanation as to how 'coming-to-be,' 'alteration,' and 'transformation' occur. This discussion comprises more than a conceptual clarification of the 'principles' to be used in *understanding* change, attempting to *explain* how transformation itself occurs. Like Empedocles, he states that the primary elements are Fire, Earth, Air, and Water and that all physical changes are brought about by the interaction of these primary elements. However, while these are the irreducible elements underlying all changes, they can be *analyzed further* into prime matter and the primary opposites or 'contrarieties' (as he now calls them), though these analyzed aspects *never exist separately*. As he says: 'We must reckon as an "originative source" and as "primary" the matter which underlies, though it is inseparable from, the contrary qualities [. . .].'[10] The latter qualities are the hot and the cold, the wet and the dry.

Each of the four elements is composed of prime matter plus a pair of the contrary qualities: 'For Fire is hot and dry, whereas Air is hot and moist (Air being a sort of aqueous vapour), and Water is cold and moist, while Earth is cold and dry.' Transformation among the simple elements is explained as due to an alteration among the 'couplings' of the contrary qualities.[11] Those elements which have one quality in common, for example Air and Water, the first being moist and hot and the second being moist and cold, will convert most easily, since only one quality has to be 'overcome' for the conversion to occur: if the hot in Air is 'overcome' by the cold then Water will result, but if the cold in Water is 'overcome' by the hot then Air will be produced. Those elements are the most difficult to convert, such as Fire and Water and Earth and Air, which have no qualities in common because both pairs of qualities must be overcome. For example, for Fire to become Water, the hot and the dry in Fire must be overcome by the cold and the wet; similarly with Earth and Air. Though more difficult, such changes do occur. The other qualities in things, like 'heavy-light, hard-soft, viscous-brittle, rough-smooth, coarse-fine,' are derived from these more basic qualities: e.g., the heavy is an additional property of Earth and Water while the light is a further property of Fire and Air in virtue of the 'tendency' of the former 'to move toward the center' and that of the latter 'to move to the limit' or outer circumference of the cosmos.

THE FOUR CAUSES

We have described Aristotle's selection of the basic 'principles' to be used in *understanding* change and motion, along with his *explanation* of empirical transformations, but still have not discussed his 'doctrine of the four causes,' one of his most celebrated theories. It is a doctrine to which he frequently returned, often reintroducing it at the beginning of each scientific treatise.[12] His conception of causality, like that of children and primitive peoples discussed earlier, presupposes that explanations of all phenomena must include the reasons *why* they occur: 'Knowledge is the object of our inquiry, and men do not [. . .] know a thing till they have grasped the "why" of it (which is to grasp its primary cause).'[13] The following is the most concise statement of his position:

> There are four causes underlying everything: first, the final cause, that for the sake of which a thing exists; secondly, the formal cause, the definition of its essence (and these two we may regard pretty much as one and the same); thirdly, the material; and fourthly, the moving principle or efficient cause. [14]

Since in the above work he is discussing the generation of animals, in which purpose plays such an important role, he lists the final cause first, though usually the material and formal causes in that order precede the discussion of the efficient and the final causes.

The *material cause*, as the name implies, consists of the underlying substance, stuff, or material of which anything is made. As the supporting material, its nature includes both the 'privation' of, and the 'potentiality' for actualizing, certain attributes or forms. Bronze, for example, has the potentiality for becoming a statue, wheat for becoming bread, and acorns for becoming oak trees. Living things particularly have the potentiality for actualizing their species and genus. 'In one sense, then (1) that out of which a thing comes to be and which persists, is called "cause", e.g. the bronze of the statue, the silver of the bowl, and the genera of which the bronze and the silver are the species.'[15] All but the last clause in this definition is clear. The material is that stuff out of which the thing is made, the bronze of the statue and the silver of the bowl, but because bronze and silver are species of the genus metal, the genus is also a material cause since the object can be said to 'come to be' from the genus metal as well as from the species bronze and silver. As he says, 'the classes that include the [. . .] cause are also causes,' or, alternately, the cause may be seen either 'as the individual, or as the genus.'

The designation of something as a material cause is relative: while the bronze is the material of which the statue is made, bronze itself is made from the base metals copper and tin or, according to Aristotle, water, since he believed that all metals ultimately come from water. Similarly, protoplasm could be said to be the materials of cells, cells of tissues, tissues of organs, organs of systems (e.g., muscular, nervous, etc.), and systems the material of the body. But even protoplasm consists of a mixture of the four basic elements which themselves, as we have found, can be analyzed into 'prime matter' and the four qualities. In any case, however, any change presupposes a supporting material that persists throughout the change of opposite qualities or contraries:

> [. . .] the *substratum* is the material cause of the continuous occurrence of coming-to-be, because it is such as to change from contrary to contrary and because, in substances, the coming-to-be of one thing is always a passing-away of another, and the passing-away of one thing is always another's coming-to-be. [16]

The *formal cause* consists of the 'essence,' 'definition,' 'formula,' or 'form' which determines the kind of thing the object is. While matter has a 'potentiality' for form, it is the 'form' that 'actualizes' the matter: 'For the form exists actually [. . .] but the matter exists potentially [. . .].' The shape of the statue and the bowl is the actualization of the bronze statue and the silver bowl as the form of oak tree is the actualization of the substance of the acorn, etc. 'In another sense (2) the form or the archetype, i.e. the statement of the essence, and its genera, are called "causes" (e.g. of the octave the relation 2:1, and generally number), and the parts in the definition.'[17] The formal cause is thus a formula, sometimes a mathematical ratio but usually a verbal definition, defining the nature of the thing or the thing's essence by including the 'parts' or the essential attributes. Again the genera are included in the formal cause since they constitute the kind of thing anything is. For example, Aristotle as an individual was the biological cause of Nicomachus his son, but the fact that Aristotle belongs to the genus animal and the species man also make these classes formal causes of Nicomachus, as transmitted by Aristotle: 'no one makes or begets the form, but it is the individual that is made, i.e., the complex of form and matter that is generated.'

The significance of genera and species for determining the essential nature of things is brought out in a different way in one of Aristotle's earliest works, the *Categories*. There while indicating that individuals constitute substances in 'the primary and most definite sense,' he also says that the genera and species could be called 'substances' in a 'secondary' sense. Although denying the Platonic thesis that genera and species have a separate existence, his classification of them as 'substances,' even if 'secondary,' shows the continued influence of Plato on his thought.

> Substance, in the truest and primary and most definite sense of the word, is that which is neither predicable of a subject nor present in a subject; for instance, the individual man or horse. But in a secondary sense those things are called substances within which, as species, the primary substances are included; also those which, as genera, include the species. For instance, the individual man is included in the species 'man', and the genus to which the species belongs is 'animal'; these, therefore — that is to say, the species 'man' and the genus 'animal' — are termed secondary substances. [18]

Just as (theoretically) there is a 'prime' or unformed matter which would be pure potentiality, so there is (actually) pure form or pure actuality, devoid of all matter and potentiality, as the culmination of the hierarchical organization of generic forms. Unlike prime matter, however, pure form is not just a theoretical abstraction, but exists as the highest form of being and perfection.

The *efficient cause*, that which brings about change or motion, is the one most familiar to us because it is the closest to our modern scientific meaning of cause. Since neither the form nor the matter itself can actualize the objects consisting of both, an external or efficient cause is required: e.g., the sculptor, the silversmith, the baker, and for the acorn, the necessary natural conditions such as the sun and moisture. The efficient cause is defined as '(3) the primary source of the change or coming to rest; e.g., the man who gave advice is a cause, the father is cause of the child, and generally what makes of what is made and what causes change of what is changed.' [19] Aristotle's examples are inadequate since none refers to motion as such, yet motion is the original or 'primary' cause, for it is the eternal motion of the heavens which causes the circular motion of the stars and the planets, while the revolution of the sun (or 'generator' as he often calls it) is the source of change in the terrestrial world; i.e., the source of the changing seasons and of the transformations, previously described, of the four elements.

> The result we have reached is logically concordant with the eternity of circular motion, i.e., the eternity of the revolution of the heavens [. . .]. Thus, from the being of the 'upper revolution' it follows that the sun revolves in this determinate manner; and since the sun revolves *thus*, the seasons in consequence come-to-be in a cycle [. . .] and since they come-to-be cyclically, so in their turn do the things whose coming-to-be seasons initiate. [20]

The sun's successive approach and recession from the earth as it makes its way annually along the ecliptic not only causes the changing seasons, it also is the source of the alternate cyclical ascent and descent of Water, Air, and Fire to their natural places as occurs in storms. This turbulence in turn brings the simple elements in contact causing the 'overcoming' of a contrary quality resulting in the transformation of the elements.

> The cause of this perpetuity of coming-to-be [. . .] is circular motion: for that is the only

motion which is continuous. That, too, is why all the other things — the things, I mean, which are reciprocally transformed [. . .] e.g. the 'simple' bodies — imitate circular motion. For when Water is transformed into Air, Air into Fire, and the Fire back into Water, we say the coming-to-be 'has completed the circle', because it reverts again to the beginning. [21]

This continuous cyclical transformation in the terrestrial world, imitating the eternal circular motion of the celestial, prevents the elements from becoming ' dissevered from one another in the infinite lapse of time;' i.e., prevents their separating and remaining static in their own natural places.

The last cause is the *final cause,* the 'end result' or 'fulfillment' or 'actualization' attained by the change: the cause '(4) in the sense of end or "that for the sake of which" a thing is done, e.g. health is the cause of walking about.'[22] For Aristotle, no explanation would be adequate that did not refer to 'the end,' 'the essence' or 'the form,' 'that for the sake of which' the change is taking place. In fact, the two previous causes, the efficient and the formal, often coincide with the final cause in bringing about the change or in actualizing the nature of something. This is especially true of Aristotle's cosmology where an ultimate Being, the Prime Mover, combines within itself each of these causes: (1) the efficient cause as the ultimate, eternal source of motion, (2) the formal cause as pure actuality, and (3) the final cause as 'that for the sake of which' all motion or change occurs.

As has often been pointed out, while the originators of modern classical dynamics retained in a revised form Aristotle's first three causes, there was no place in their mechanistic science for a final cause. The material cause was retained in the form of the reduction of matter to mass or atoms, the formal cause retained as mechanical definitions and mathematical laws, and the efficient cause as mechanistic forces and, fundamentally, the kinetic motion of the basic particles (caused ultimately by God's will). But in a universe in which all changes and motions were thought to take place by mechanical necessity, any purpose or design had to be built into the cosmic machine at the very beginning and manifested mechanically. Yet, contrary to the claims of Randall and contemporary physicists such as Heisenberg,[23] who see a resemblance between Aristotle's concept of the actualization of potentialities and the quantum mechanical notion of indeterminate states attaining actuality under various experimental conditions, there was no more possibility of real novelty in Aristotle's system than in Newton's. To be sure, Aristotle emphasized the importance of 'chance' and 'hypothetical necessity' in the world, which meant that antecedent causes were not the only causes in nature. But since in his interpretation of natural processes there were no changes among the generic forms and since all alterations in the world were limited to these eternal, preexistent forms, any emergence of novelty or evolutionary modifications were as definitely ruled out by Aristotle as by Newton. As Jaeger clearly states: "In spite of its uninterrupted change nature has no history according to Aristotle, for organic becoming is held fast by the constancy of its forms in a rhythm that remains eternally the same . . ."[24]

The difference between the two systems consists in the different metaphors as to how nature operates: for Aristotle nature operates as an organism while for Newton it operates as a machine. There is an element of irony in attributing less possibility of chance or novelty in a mechanistic system than in Aristotle's organismic one, since Aristotle explicitly rejected *mechanical necessity* as an adequate interpretation of natural changes precisely because it would allow for *too much* chance and spontaneity and *not enough* regularity and order in nature. As a biologist, Aristotle was thoroughly imbued with the regularity of changes and the constancy of forms exhibited 'for the most part' by nature. As he expresses the dilemma:

[...] why should not nature work, not for the sake of something, nor because it is better so, but just as the sky rains, not in order to make the corn grow, but of necessity? What is drawn up must cool, and what has been cooled must become water and descend, the result of this being that the corn grows [...]. Why then should it not be the same with the parts in nature, e.g. that our teeth should come up *of necessity* — the front teeth sharp, fitted for tearing, the molars broad and useful for grinding down the food — since they did not arise for this end, but it was merely a coincident result; and so with all other parts in which we suppose that there is purpose? Wherever then all the parts came about just what they would have been if they had come to be for an end, such things survived, being organized spontaneously in a fitting way; whereas those which grew otherwise perished and continue to perish [..].[25]

In this quotation Aristotle states the essential features of Darwin's theory of evolution; i.e., that certain (genetic) changes occur 'coincidently' or 'spontaneously' (that is, not for a purpose), those resulting in a beneficial function tending to survive, those occurring otherwise perishing. But Aristotle rejects this explanation because he does not believe that it can account for the purposeful sequences or recurrent orderly developments in nature.

Yet it is impossible that this should be the true view. For teeth and all other natural things either invariably or normally come about in a given way; but of not one of the results of chance or spontaneity is this true. We do not ascribe to chance or mere coincidence the frequency of rain in winter, but frequent rain in summer we do [...]. If then, it is agreed that things are either the result of coincidence or for an end, and these cannot be the result of coincidence or spontaneity, it follows that they must be for an end [...]. Therefore action for an end is present in things which come to be and are by nature [26]

By the terms 'coincidence' or 'spontaneity' Aristotle is not referring to *uncaused* occurrences, but to accidental or fortuitous *necessary occurrences* in contrast to those which occur purposefully for a specific end. Just as in intelligently planned actions each step is taken to bring about a result consistent with the desired end, Aristotle believes that the coordinated, unconscious striving of nature to realize certain ends indicates 'that nature is a cause, a cause that operates for a purpose.' As he says in a famous passage:

If then it is both by nature and for an end that the swallow makes its nest and the spider its web, and plants grow leaves for the sake of the fruit and send their roots down (not up) for the sake of nourishment, it is plain that this kind of cause is operative in things which come to be and are by nature. And since 'nature' means two things, the matter and the form, of which the latter is the end, and since all the rest is for the sake of the end, the form must be the cause in the sense of 'that for the sake of which.'[27]

There is a difference between a mechanical and a teleological production of ends, but insofar as the possible kinds of ends were as fixed for Aristotle as they are in a mechanical world, though for quite different reasons, his universe was just as devoid of novelty. Aristotle was still too much influenced by Parmenides to admit that anything radically or basically new could emerge in the universe (and in fact we still find it difficult to conceptualize this today). Thus he says explicitly that our 'first presupposition must be that in nature nothing acts on, or is acted on by, any other thing at random [...].'

There is the continuous coming to be and passing-away of forms as the world changes, but since the *forms themselves* never change, only the particular manifestations of them, nothing really new is ever brought into existence. But this is true also of a mechanistic system: there can be an endless rearrangement of particles, but insofar as every state develops necessarily out of an antecedent state, nothing entirely new can arise. It is only as a result of quantum mechanics that contemporary scientists, generally with much trepidation and reluctance, are even willing to *consider* the possibility of chance, spontaneity, novelty, and creativity as inherent aspects of the universe itself. Just the admission of such a possibility is usually regarded as equivalent to a denial of intelligibility, as Einstein himself stated, while it could be looked upon as a new category of interpretation.

This, then, concludes the discussion as to the various kinds of causes that must be taken into account in explaining 'the why of things.'

> It is clear then that there are causes, and that the number of them is what we have stated. The number is the same as that of the things comprehended under the question 'why' [. . .]. Now, the causes being four, it is the business of the physicist to know about them all, and if he refers his problems back to all of them, he will assign the 'why' in the way proper to his science — the matter, the form, the mover, 'that for the sake of which.' The last three often coincide [. . .]. [28]

The principles discussed in the last section, substance, contraries, privation, potentiality and actuality, represent the general conceptual schema for interpreting change and motion, while the four causes provide a more specific explanation of *why* something changes. The doctrine of the four causes, therefore, supplements the use of the principles in attaining scientific knowledge. Thus it is necessary to speak of causes in several different senses:

> e.g. the cause of man is (1) the elements in man (viz. fire and earth as matter, and the particular form), and further (2) something else outside, i.e. the father, and (3) besides these the sun and its oblique course, which are neither matter nor form nor privation of man nor of the same species with him, but moving causes. [29]

It should be kept in mind that Aristotle's own account is usually somewhat more involved than the description of it presented here, but this description serves the purpose of selecting the major features of his position and presenting them in a concise fashion.

ARISTOTLE'S ORGANISMIC COSMOLOGY

Aristotle's cosmology represents the culminating integration of his scientific principles, particularly as they refer to various kinds of natural motions, as well as the best illustration of his unique blend of empiricism with scientific rationalism and cosmological speculation. While the empirical basis of his theorizing is evident throughout ('in the knowledge of nature is the unimpeachable evidence of the senses as to each fact'), as any scientist he had to fit these empirical observations into a conceptual framework. The result was a theoretical paradigm or cosmological model that dominated the development of Western thought for nearly five hundred years. As Randall states:

> Aristotle's cosmology proved to be of momentous historical significance [. . .]. From the thirteenth until well into the seventeenth century, it provided the main outlines of the universe educated Europeans conceived themselves to be living in. Hence it colored the way men felt as well as thought about the cosmic scene of human life; it was reflected in their common way of speaking, and embodied in their highest imaginative achievements. For anyone who wants to understand the long history of thought and feeling in the Western tradition, a knowledge of Aristotle's cosmology is indispensable. [30]

This cosmology is based on (1) Aristotle's description of the various kinds of natural motions, both terrestrial and celestial, (2) his analysis of the different kinds of simple substances of which the above motions are essential manifestations, and (3) his attempt to give a complete explanation of the necessary causes or reasons for these motions. With the possible exception of Descartes, never in the history of science before or after has one been as confident as Aristotle that one could give a complete explanation of practically all phenomena, demonstrating why things must be as they are. His primary concern, like that of all physicists, was to explain motion: 'the science of the natural philosopher deals with the things that have *in themselves* a principle of motion.'[31] This 'theoretical science' is distinguished from the practical and the productive which are concerned with objects that have the source of their motion outside themselves, e.g., healing and building.

The primary works in which he presents his cosmology are Book VIII of the *Physics*, *De Caelo* (*On the Heavens*), and Book *Lambda* of the *Metaphysics*. In Book VIII of the *Physics* he begins by discussing the theories of motion of his predecessors, especially the view of Heraclitus that everything is in flux and of Parmenides that all things are always at rest: is either of these views true or is it the case that different things at different times are sometimes in motion and sometimes at rest? Moreover, is motion such that it had a beginning and could perish, or is it eternal? His answers, based on observation, are that within the *terrestrial* world (the domain extending from the center of the earth to the inner concave orbital sphere of the moon) 'we *see* some things that are sometimes in motion and sometimes at rest,' while the heavens or the celestial world is always in continuous motion. As for motion itself, it is definitely everlasting: 'motion is eternal and cannot have existed at one time and not at another: in fact, such a view can hardly be described as anything else than fantastic.'

Within the terrestrial world there are two contrary forms of natural motion, upward and downward, corresponding to the four simple elements making up this domain: earth and water naturally gravitate downward to the center of the earth (which coincides with the center of the universe), while air and fire naturally rise upward toward the inner circumference of the celestial realm. These directional motions define absolute light and heavy: 'By absolutely light, then, we mean that which moves upward or to the extremity, and by absolutely heavy that which moves downward or to the centre.'[32] Things are called 'relatively light or heavy' when, 'endowed with weight and equal in bulk' to another body, the speed of their natural downward motion is either less or greater than that of the other body: e.g., wood is relatively light compared to bronze.

Not content with describing the absolute motions, Aristotle also attempts to explain them:

— how can we account for the motion of light things and heavy things to their proper situations? The reason for it is that they have a *natural tendency* respectively towards a certain position: and this constitutes the essence of lightness and heaviness, the former being determined by an upward, the latter by a downward, *tendency*. [33] (Italics added)

These 'natural tendencies' are correlated with the 'influence' exerted by the various proper places on the simple elements or the 'potency' inherent in a natural place to receive its appropriate element; i.e., the stationary earth in the center of the universe surrounded successively by water, air, and fire, the latter element extending to the innermost orbital sphere carrying the moon, the division between the terrestrial and the celestial worlds.

The natural state of things would be for each of these elements to reach its natural place and remain there motionless: 'it is a law of nature that earth and all other [terrestrial] bodies should remain in their proper places and be moved from them only by violence [...].' This violent or unnatural motion comes about because of the mixture of the elements and/or the displacement of an element from its proper place. When fire moves downward as in a flash of lightning or physical objects are carried upward by a gust of wind, this is violent or unnatural motion. Another form of unnatural motion is projectile motion. Having explicitly rejected the concept of inertial motion because it is not supported by observation (i.e., ordinary objects put in motion do not continue in motion indefinitely but naturally come to rest), Aristotle had difficulty explaining the continued motion of an object when the source of the motion is no longer in contact with the object; either the cause of motion must be simultaneous with all its effects or there must be a succession of movers. But the latter alternative, the only plausible one, does not solve the problem: how can something put in motion continue to be a source of motion when *its* source is no longer active? Aristotle purports to answer the problem by maintaining that 'it ceases to be in motion at the moment when its movent ceases to move it, but it still remains a movent, and so it causes something else consecutive with it to be in motion [. . .].'[34] Exactly how this is possible, he does not explain. The concepts of "impetus" and "inertia" will be formulated later in an attempt to provide a better explanation, yet the precise *modus operandi* will be no clearer than in Aristotle's account.

The two contrary forms (upward and downward) of natural terrestrial motions are rectilinear. Rectilinear or straight-line motion itself has a contrary, circular, rotary motion: 'if there is a contrary to circular motion, motion in a straight line must be recognized as having the best claim to that name.' (There are also combinations of both forms.) Although both rectilinear and circular motions are species of locomotion, circular motion is primary because it is continuous and represents a more perfect geometrical figure. That is, while all rectilinear motion has two contraries, a beginning and an end, along with a reversible direction, circular or rotary motion has no beginning or end point and therefore no contrary form; thus it can be continuous and eternal.

> Now it is plain that if the locomotion of a thing is rectilinear and finite it is not continuous loco-motion: for the thing must turn back, and that which turns back in a straight line undergoes two contrary locomotions [. . .].
>
> On the other hand, in motion on a circular line we shall find singleness and continuity [. . .]. Again, a motion that admits of being eternal is prior to one that does not. Now rotary motion can be eternal: but no other motion [. . .] since in all of them rest must occur, and with the occurrence of rest the motion has perished. [35]

The argument from perfection occurs in the *De Caelo*: 'circular motion is necessarily primary [. . .] [because] the perfect is naturally prior to the imperfect, and the circle is a perfect thing.'[36] As Aristotle continually maintains, 'we always assume the presence in nature of the better, if it be possible [. . .].'

Rotary motion having been described in the *Physics* as the most perfect form of locomotion, hence continuous, eternal, and prior, in the *De Caelo* he identifies it with the motion of the heavens: 'our eyes tell us that the heavens revolve in a circle, and by argument also we have determined that there is something to which circular motion belongs.'[37] Moreover, the uniqueness of circular motion presupposes an additional element beyond those already mentioned, namely, the 'aither,' a weightless substance even finer than fire. The heavens being immutable, the fifth element aither, unlike the four terrestrial elements, is 'ungenerated and indestructible,' 'exempt from increase and alteration,' 'unaging,' and hence 'eternal' and 'divine.'

> For in the whole range of time past, so far as our inherited records reach, no change appears to have taken place either in the whole scheme of the outermost heaven or in any of its proper parts

[i.e., the various spheres]. And so [. . .] the primary body [. . .] beyond earth, fire, air, and water [. . .] [has been given] a name of its own, *aither*, derived from the fact that it 'runs always' for an eternity of time. [38] (Brackets added)

It was this Aristotelian view of the perfect, immutable heavens that Galileo argued against in his controversial essays, "The Starry Messenger" and "The Assayer" (*Il Saggiatori*),[39] attempting to demonstrate that recently observed comets occurred in the translunar realm and thus controverted the immutability of the celestial world. For one of the main obstacles against acceptance of the Copernican heliocentric view was the persistent belief, derived primarily from Aristotle, in the qualitative distinction between the celestial and the terrestrial worlds. Copernicus' revolutionary hypothesis involved an interchange of the earth and the sun, but how could a weightless celestial body composed of aither be at rest in the center of the terrestrial realm, while a heavy terrestrial body revolved in the celestial orbs? It can readily be seen that such a view immediately upset Aristotle's hierarchical conception of the universe in which each simple element had its own natural motion and proper place. Yet in fairness to Aristotle, such phrases as 'at least with human certainty' and 'so far as our inherited records reach,' as well as his constant references to what 'we can see,' indicate that were he shown the evidence of the comets he undoubtedly would have changed his views. As Galileo says, "I declare that we do have in our age new events and observations such that if Aristotle were now alive, I have no doubt he would change his opinion." [40]

Following his description of the nature and motion of the heavens in Book I of the *De Caelo*, Aristotle turns to consider the earth, its position, its shape, and whether it be at rest or in motion, in Book II. This book especially shows his detailed knowledge of astronomical phenomena, as well as the manner in which his theoretical assumptions or 'physical principles' influence his explanations. The fact that his arguments or demonstrations often depend upon what seem today to be arbitrary assumptions or principles has led some commentators to refer to his arguments as *a priori*.[41] In fact, however, his reasoning is no more *a priori* than that of Galileo or other scientists — it is just that what he accepts as explanatory presuppositions and principles (necessary in any scientific demonstration) is different from those of later scientists — not that his is *a priori* and theirs *a posteriori*.

He begins, typically, by reviewing and refuting the views of his predecessors (as 2000 years later Galileo will begin his scientific inquiries by considering and rejecting the views of Aristotle). Regarding the position of the earth, he denies the Pythagorean view that the earth revolves around a central fire thereby creating night and day — criticizing the Pythagoreans for not giving due consideration to 'observed facts' and 'observations,' as Galileo will chide him for not having tested some of his empirical conclusions ("I greatly doubt that Aristotle ever tested by experiment [. . .]."[42]): 'in all this they are not seeking for theories and causes to account for the observed facts, but rather forcing their observations and trying to accommodate them to certain theories and opinions of their own.' [43]

The facts to which Aristotle appeals in denying that the earth revolves with separate motions, an orbital revolution and an axial rotation or 'rolling motion,' are based on ordinary observations as well as more subtle arguments, not all of which can be reproduced here. As an example of the former:

The observed facts about earth are not only that it remains at the centre, but also that it moves to the centre. The place to which any fragment of earth moves must necessarily be the place to which the whole moves; and in the place to which a thing naturally moves, it will naturally rest. [44]

The earth rests at the center of the universe and, in contrast to Newton's theory of gravitation, detached fragments of the earth are not attracted to the center of the earth as such, but to the center of the universe with which the center of the earth coincides. As he states: 'If one were to remove the earth to where the moon now is, the various fragments of earth would each move not towards it but to the place in which it now is.' [45]

The more subtle arguments are as follows. If the earth moved, either rotating in orbit or revolving on its axis, such circular motion would be contrary to the observed natural rectilinear motion of pieces of the earth to the center, and thus would be 'constrained,' i.e., a forced, unnatural motion. 'Being, then, constrained and unnatural, the movement could not be eternal. But the order of the universe is eternal.' This sounds like begging the question, but better arguments follow. If the earth moved with either of the assigned motions, one should observe a displacement of the fixed stars: 'there would have to be passings and turnings of the fixed stars. Yet no such thing is observed.' In modern terminology, there is no observed parallax as regards the fixed stars. This was an important argument also in the 17th century against the Copernican view — one which convinced Tycho Brahe — an argument that could be countered only by assuming a universe of vastly greater dimensions than was generally conceived at the time of Aristotle (for the angle of parallax diminishes with the distance of the observed object).

Aristotle then introduces the important argument that if the earth rotated objects thrown straight upward would not return to their starting point, because during the time of their ascent and descent the earth would have moved, especially if the distance were great: 'the earth must be at the centre and immovable [. . .] because heavy bodies forcibly thrown quite straight upward return to the point from which they started, even if they are thrown to an infinite distance.' [46] This was one of the crucial arguments that Galileo was compelled to refute in defending Copernicus' thesis that the earth revolves around the sun.

> Aristotle says, then, that a most certain proof of the earth's being motionless is that things projected perpendicularly upward are seen to return by the same line to the same place from which they were thrown, even though the movement is extremely high. This, he argues, could not happen if the earth moved, since in the time during which the projectile is moving [. . .] the place from which the projectile began its motion would go a long way toward the east, thanks to the revolving of the earth, and the falling projectile would strike the earth that distance away from the place in question. [47]

Galileo offers several counter arguments, the most decisive of which is based on the experiment of dropping objects from the masthead of a sailing ship. As he affirms, the objects fall parallel to the mast whether the ship is at rest or in motion — providing the motion is uniform!

> For anyone [. . .] will find that the experiment shows exactly the opposite of what has been written; that is, it will show that the stone always falls in the same place on the ship, whether the ship is standing still or moving with any speed you please. Therefore, the same cause holding good on the earth as on the ship, nothing can be inferred about the earth's motion or rest from the stone falling always perpendicularly [. . .]. [48]

This illustrates Galileo's discovery of what Einstein called the "restricted principle of relativity," namely, that the laws of nature are the same in all inertial systems (i.e., in all systems at rest or moving uniformly).

Aristotle next takes up the question of the shape of the earth. Again, he counters the evidence of his predecessors who claimed that the earth was drum-shaped or flat.

Those who maintained that it was flat did so because the horizon line of the earth across the sun as the latter rises and sets looks straight, not curved. But Aristotle adduces correctly that such arguments 'leave out of account the great distance of the sun from the earth and the great size of the [latter's] circumference, which, seen from a distance on these apparently small circles appears straight.'[49] The sphericity of the earth *is* evident, however, from the curved reflection of the earth on the moon's surface during lunar eclipses, since 'the form of this line will be caused by the form of the earth's surface, which is therefore spherical.'

He also argues correctly that the earth's dimension is insignificant because as one travels north or south different stars are observable, some seen to rise and set that previously were constantly visible. 'All of which goes to show not only that the earth is circular in shape, but also that it is a sphere of no great size: for otherwise the effect of so slight a change of place would not be so quickly apparent.' He records the mathematicians' calculation of the size of the earth's circumference as 400,000 stades, or 9,987 geographical miles, a computation much too slight since the actual figure is 24,901 miles at the equator. Because of the relative smallness of the earth's size he says we should not be too incredulous of the belief that 'there is continuity between the parts about the pillars of Hercules and the parts about India, and that in this way the ocean is one.' Randall remarks that this "passage was the basis of Columbus' confidence that he could reach the Indies by sailing westward"[50] and also the reason the indigenous people on the continent discovered by Columbus were called "Indians."

Following his description of the terrestrial world, Aristotle presents his theory of the heavens. Since he was not a mathematician, he based his astronomical view on the mathematical theory of "concentric spheres" developed by Eudoxus, a conception that Heath asserts "may be said to be the beginning of scientific astronomy."[51] Eudoxus studied geometry under the famous Pythagorean mathematician Archytas, becoming a brilliant mathematician on his own, inventing the theory of proportions and discovering the method of exhaustion, the pre-calculus method used by the Greeks for calculating various areas and volumes. Aristotle also worked with another astronomer, Callippus, correcting and completing the discoveries of Eudoxus, though he indicates that one could not expect a final or 'necessary' account of the movements of the heavenly bodies.

Eudoxus' system consisted of twenty-seven concentric spheres to account for the various motions of the planets: one for the diurnal rotation of the fixed stars, three each for the moon and the sun, and four each for the remaining planets, Mercury, Venus, Mars, Jupiter, and Saturn. The outermost two spheres of each of the seven planets have the same motion, the first duplicating the rotation of the fixed stars and thereby causing each planet's diurnal rotation from east to west, while the second sphere has its axis perpendicular to the zodiac so that as it revolves its equator bisects the Zodiac. The various rotational speeds of this sphere from west to east along the ecliptic account for the different orbital periods of the planets: e.g., one month for the moon, one year for the sun, twenty-nine years for Saturn, etc.

The inclination of the axes as well as the directions and the speeds of the third spheres of the sun and moon and the third and the fourth spheres of the other five planets were determined so as to account for other astronomical phenomena, such as variations in longitude and latitude and retrograde motion. Each of the planets is carried round on the equator of its innermost sphere, its annual motion being a composite of the motions of all its spheres, except that of the diurnal rotation of the outermost. The axes or poles of each successive sphere are attached to the inner concave surface of the nearest outer sphere and so are rotated by that sphere, the whole cosmos consisting of a successive number of nesting spheres rotating with various inclinations, directions, and speeds around a common center. To further "save the phenomena" or provide a more accurate description of the movements of the planets, Aristotle states that Callippus added two more spheres each to the sun and moon, and an additional sphere to Mercury, Venus, and Mars, assigning the same number to Jupiter and Saturn as Eudoxus had. Thus the total number of spheres for Callippus was thirty-four, including the sphere of the fixed stars.

While Eudoxus and Callippus were content to try to provide an accurate geometrical description of the motion of each of the planets separately, without trying to give a mechanical explanation of their interrelated motions as a totality, as a physicist Aristotle felt qualified to take on this latter task. He realized that the motion of each planet not only was determined by the motion of its own spheres, but also was affected by the spheres above it. Eudoxus and Callippus had described the motion of the planets individually, beginning with the motion of the fixed stars, without considering the effects of each planet's motion on its successor, from Saturn to Jupiter to Mars, etc. Aristotle realized that in order to compensate for these successive motions, one had to introduce 'counteracting spheres' so that the motion of the planets could be determined in relation to the original motion of the fixed stars, rather than that of the preceding planets. As Heath states,

> if n is the number of the deferent spheres of a planet, the addition of $n-1$ reacting spheres inside them neutralizes the operation of $n-1$ of the original n spheres and prevents the inner set of spheres from being disturbed by the outer set. The innermost of the $n-1$ reacting spheres moves [. . .] in the same way as the sphere of the fixed stars. [52]

So Aristotle adds one less number of counteracting spheres to each group assigned by Callippus (except for the lowest planet, the moon, which need not be counteracted since there is no planet below it), for a total of twenty-two. These twenty-two counteracting spheres added to the previous thirty-three posited by Callippus (omitting the sphere of fixed stars) comprise a total of fifty-five. [53] However, Aristotle would seem to have been guilty of an elementary arithmetical error when he concluded: 'if one were not to add to the moon and the sun the movements we mentioned [i.e., those of Callippus], the whole set of spheres will be forty-seven in number.'[54] The discrepancy is due to the fact that, if one were to subtract from the fifty-five spheres those added by Callippus, namely two each for the sun and the moon, plus the two additional counteracting spheres for the sun, making a total of six, this would leave forty-nine rather than forty-seven spheres. Whether this error is actually due to a simple mathematical slip as Heath states or could be attributed to a scribe's error as Dicks suggests, we shall never know.

Here is where the discussion might well have ended. The composition, structure, motion, and shape of the universe being accounted for, one would think that everything had been explained. The heavens are composed of an immortal, divine element, the aither, that has a continuous circular motion which is the source of all the other motions and changes in the cosmos ('for the eternal motion, by causing "the generator" to approach and retire, will produce coming-to-be uninterruptedly'), so no further explanation would seem to be required. That this is not the case is due to the fact that Aristotle accepts as a primary principle that 'all things that are in motion must be moved by something,' and as this principle applies not only to unnatural motions but to natural motions as well, he is compelled to explain the cause of the eternal motion of the heavens. Moreover, the relentless logic of the argument requires that the explanation terminate in an *unmoved mover* — hence the source of Aristotle's famous doctrine of the 'Prime Mover.'

The argument is presented in various forms in Book VIII of the *Physics* and in Book

Lambda of the *Metaphysics* (although, surprisingly, it is not present in the *De Caelo*), but the most concise statement is the following:

> There is, then, something which is always moved with an unceasing motion, which is motion in a circle; and this is plain not in theory only but in fact. Therefore the first heaven [i.e., the outermost sphere of the fixed stars] must be eternal. There is therefore also something which moves it. And since that which is moved and moves is intermediate, there is something which moves without being moved, being eternal, substance, and actuality. And the object of desire and the object of thought move in this way; they move without being moved. [55] (Brackets added)

In Greek (as in English) there are two senses of the phrase 'to be moved,' one designating a mechanical movement or displacement in space, the other referring to an affective state, as when one is 'moved' by an artistic production or by wanting or desiring a contemplated object. For Aristotle the object of desire or of contemplation, being a cause of motion in what can be so affected or moved, is an 'unmoved mover.' Accordingly, while the sphere of the fixed stars transmits its motion mechanically to the outermost sphere of Saturn, which then transmits its motion to the next innermost sphere, and so on, the original, prime, Unmoved Mover, by being an object of desire or by being loved, causes the sphere of the fixed stars itself to move: i.e., the Prime Mover moves the sphere of the fixed stars affectively or movingly, as a beloved object moves the lover! 'The final cause, then, produces motion as being loved, but all other things move by being moved.'

This concept of the 'Prime Mover' represents a culmination of all of the principles of explanation mentioned earlier. As 'the producer of motion by being loved,' it is the *final cause*. As pure form devoid of all matter and potentiality, it is *pure actuality*, the only pure actuality in the universe: 'one actuality always precedes another in time right back to the actuality of the eternal prime mover.' Finally, as the unmoved source of motion it is the necessary *efficient cause*:

> [...] since there is something which moves while itself unmoved, existing actually, this can in no way be otherwise than as it is. For motion in space is the first of the kinds of change, and motion in a circle the first kind of spatial motion; and this the first mover *produces*. The first mover, then, exists of necessity; and in so far as it exists by necessity, its mode of being is good, and it is in this sense a first principle. [56]

'On such a principle, then, depend the heavens and the world of nature.'

The Prime Mover is 'eternal and unmovable,' 'without parts and indivisible,' 'impassive and unalterable,' and 'separate from sensible things.' Although there is no empty space or void beyond the spherical universe, the Prime Mover 'occupies the circumference' of the outermost sphere of the fixed stars. Since in the *De Caelo* Aristotle had concluded that 'beyond the heaven there is no body at all,' the fact that the Prime Mover 'occupies the circumference' might seem a contradiction; however, as the Prime Mover is pure actuality devoid of matter and potentiality, it is not a 'body,' and hence there is no conflict on Aristotle's presuppositions. As he maintains in the *Metaphysics*, the Prime Mover is Divine Thought thinking on itself and therefore does not require a place in which to exist.

A final problem remains, however: is there only one Prime Mover or is there a plurality of Movers? Aristotle faced the question both in the *Physics* and the *Metaphysics*, presenting the famous arguments against an infinite series of movers that were to become, after the reformulation by St. Thomas Aquinas, a basis of the "rational proofs" for the existence of God. The first argument denies an infinite series of movers on the grounds that since an infinite series would have 'no first term,' there would be no beginning of the series culminating in the present movements: 'Now since of an infinite number of terms there is not a first, the first in this series will not exist, and therefore no following term will exist. Nothing, then, can either come to be or move or change.[57] Accordingly, 'it is impossible that there should be an infinite series of movents.'

That is, beginning with the present motion of the heavens, if one were to attribute to it a retrogressive series of causes *ad infinitum*, since an infinite series never comes to an *end*, then the series could never include a *beginning* cause, and hence would not explain how the series of causes could have culminated in the present motion. Moreover, for Aristotle an infinite series only exists potentially, in the *possibility* of never exhausting or of reaching a last term in the series. Thus an infinitely remote cause would only exist potentially, but a potential cause could not be the cause of the actual motion of the heavens. Therefore the first cause must be pure actuality and logically (though not temporally) prior to the motion of the heavens, for the potential always presupposes the actual: 'Obviously, then, actuality is prior both to potency and to every principle of change.'

The second argument is based on the assumption that motion has merely a contingent status. However, if all motion were contingent or accidental, it would be possible at some time for moving things to cease to be in motion, but this consequence is contrary to the eternity of both time and motion. Hence motion is not an accidental attribute but a necessary one.

> If then it [motion] is an accidental attribute, it is not necessary that that which is in motion should be in motion: and if this is so it is clear that there may be a time when nothing that exists is in motion, since the accidental is not necessary but contingent. Now if we assume the existence of a possibility, any conclusion that we thereby reach will not be an impossibility, though it may be contrary to fact. But the non-existence of motion is an impossibility: for we have shown that there must always be motion. [58]

Aristotle concludes the argument by saying that 'it is reasonable, therefore, not to say necessary, to suppose the existence of [. . .] that which causes motion but is itself unmoved.'

Having denied an infinite series of movers, does he accept a single mover or a limited number of movers? Continuing the argument in the *Physics*, he raises the possibility of a plurality of movers, but then decides in favor of one.

> Motion, then, being eternal, the first movent, if there is but one, will be eternal also: if there are more than one, there will be a plurality of eternal movents. We ought, however, to suppose that there is one rather than many, and a finite rather than an infinite number. When the consequences of either assumption are the same, we should always assume that things are finite rather than infinite in number, since in things constituted by nature that which is finite and that which is better ought, if possible, to be present rather than the reverse: and here it is sufficient to assume only one movent, the first of unmoved things, which being eternal will be the principle of motion to everything else. [59]

In Book *Lambda* of the *Metaphysics*, however, Aristotle decides in favor of a plurality of unmoved movers, the exact number of which is determined by the total number of spheres necessary to explain the motion of the planets, since he there claims that each sphere requires an unmoved mover as a source of its motion. Thus in addition to the Prime Mover which causes the motion of the fixed stars, there will be whatever number of unmoved movers are necessary to account for the motion of the planets:

> [. . .] since we see that besides the simple spatial movement of the universe, which we say the

first and unmovable substance produces, there are other spatial movements — those of the planets — which are eternal [. . .] each of *these* movements also must be caused by a substance both unmovable in itself and eternal [. . .]. Evidently, then, there must be substances which are of the same number as the movement of the stars [i.e., of *both* the fixed stars and the planets] [. . .].[60] (Brackets added)

Thus the planets really have three sources of motion: (1) the *primum movens* which moves the sphere of the fixed stars, which in turn carries with it the total heavens in a diurnal rotation; (2) the mechanical motion transmitted successively from the sphere of the fixed stars to each of the planetary spheres; and (3) the lesser unmoved movers which are an additional source of motion of the planets.

Regarding the last two motions, it is to be noted that while the *axis* of each sphere is carried round mechanically by the adjacent outer sphere, the *revolving motion* of each sphere on its axis would seem to be caused by an unmoved mover, thereby accounting for the different directions and speeds of each planet's motion. Aristotle does not explicitly say this, but such an interpretation would be natural and consistent.[61] The major flaw, recognized at the time, of the whole theory of concentric spheres was its inability to account for such visible astronomical phenomena as the variations in brightness of Venus and Mars and the changes in size of the sun and moon as they revolve in their orbits. Neither phenomenon could be explained on the assumption that the distance of the planets from the center of the universe was invariant or constant. Nonetheless, the attempt to provide a coherent theory of planetary motions based on the simplest hypothesis of uniform, spherical, and concentric orbital revolutions was an initial approximation to observable data, and therefore a necessary stage in the development of planetary astronomy. From this point on astronomers could introduce more refined variations in the theory such as epicycles, eccentrics, deferents, etc., to "save the phenomena" mentioned above.

As to the actual nature of the stars and the planets, Aristotle says very little. They are composed of the same elements as the spheres, the aither, and possess the same characteristics except that their shape is globular. No attempt is made to explain this nor is a very adequate account given of their visibility. He does offer an explanation of the source of the heat and light of the sun, both arising from the friction between the sun's revolving spheres and the air (whether this is the mixture of air in the upper translunar realm or the air in the lower sublunar realm is not made clear). In the *Meteorologica* he discusses in considerable detail the nature of shooting stars, comets, the Milky Way, etc., as well as more common meteorological phenomena such as atmospheric conditions and storms. However, his discussion of these matters does not exhibit the same degree of sophistication as his treatment of other natural phenomena, particularly the biological.

METAPHYSICS

Although the twentieth century has seen her deposed and exiled to a meaningless status, through the ages metaphysics had come to be known as the "queen of the sciences" owing to Aristotle's treatment of the subject. For above and beyond such sciences as mathematics, biology, physics, astronomy, etc., disciplines which investigated a particular segment of being, Aristotle wrote of an additional theoretical science, one that dealt with 'being' comprehensively in terms of the most fundamental questions; i.e., those pertaining to the *ultimate* first principles, the *highest* causes, and particularly to 'being qua being' or substance.

> There is a science which investigates being as being and the attributes which belong to this in virtue of its own nature. Now this is not the same as any of the so-called special sciences; for none of these others treats universally of being as being. They cut off a part of being and investigate the attribute of this part; this is what the mathematical sciences for instance do. Now since we are seeking the first principles and the highest causes, clearly there must be some thing to which these belong in virtue of its own nature [...]. Therefore it is of being as being that we also must grasp the first causes. [62]

This science he refers to as 'first philosophy' or 'theology' (in the *De Anima* the designations 'first philosopher or metaphysician' occur). Above all else man seeks Wisdom, and only an understanding of the ultimate, universal principles of knowledge and of the categories of being can constitute Wisdom.

Aristotle's own *Metaphysics* illustrates his search for such a discipline, exhaustively reexamining the various principles of explanation and ontological categories used by his predecessors and by him in his previous works, in order to arrive at those which are ultimate. While much of this discussion duplicates that of his earlier treatises, his primary intent is to show that just as the cosmos includes a unique kind of substance, the divine substance of the unmoved movers, so there must be a special science, first philosophy or theology, that investigates this unique substance or being:

> [...] if there is something which is eternal and immovable and separable, clearly the knowledge of it belongs to a theoretical science — not, however, to physics (for physics deals with certain movable things) nor to mathematics, but to a science prior to both [...]. There must, then, be three theoretical philosophies, mathematics, physics, and what we may call theology [...]. Thus [...] if there is an immovable substance, the science of this must be prior and must be first philosophy [...]. [63]

In the *Physics* he had justified the existence of an eternal unmoved mover, while in the *De Caelo* he described the unmoved mover as a divine substance; beyond that, however, he had had little to say about the exact nature of the divine substance. This deficiency is made up in Book *Lambda* of the *Metaphysics*, generally regarded as a late addition to the *Metaphysics* which itself is a prototype of Aristotle's conception of first philosophy or theology.

As described in the previous section, the logic of Aristotle's argument that 'whatever moves requires a mover' drove him to posit, as the source of the motions of the heavens, a plurality of unmoved movers. However, as these movers are themselves unmoved they could not move the heavens physically or mechanically, but produced motion affectively by being 'objects of desire' or by 'being loved.' But the fact that the heavens are moved out of a desire for the divine unmoved movers, and that these prime movers are the efficient, formal (ultimate actuality), and final causes of the cosmos, still does not define their essential natures. Now, however, he states that 'as the primary objects of desire and of thought are the same,' the 'real good' being 'the primary object of rational wish,' the essence of the divine substances is contemplative thought. Moreover, since the object of divine thought cannot arise from any source outside itself because then it would not be a formal actualization of its own nature, and since it must represent what is most perfect and good, divine thought contemplates itself: 'Therefore it must be of itself that the divine thought thinks (since it is the most excellent of things), and its thinking is a thinking on thinking.' [64]

Although this conception was to have a tremendous influence on Christian theology and Western philosophy from St. Thomas to Hegel, it still is quite obscure. But the rationale for the doctrine is to be found, in part at least, in the *De Anima* where Aristotle discusses the nature and function of the soul and of the mind or intellect (the active intellect being the rational part of the soul, the only part that is eternal and immortal). There he states that the soul is the source of motion in the living body as well as being its essence and final cause,[65] just as the divine substances or unmoved movers are the sources of motion in the heavens as well as being the highest actualities and ultimate final causes. (There is an essential difference between the two, however, in that the divine unmoved movers are *substances*, though not bodies, whereas the soul and mind are not substances but *powers*.) Moreover, the crucial doctrine that the ultimate source of motion is not produced mechanically, but induced by desire or love, pervades the *De Anima*. For Aristotle the action of all living organisms has its origin in appetition or desire which is a kind of motion operating within the organism. In addition, as thinking is an activity or *motion* of the soul, it is perhaps easier to understand how he could conclude that a *psychic* motion could be a cause of *physical* motion; i.e., even though they are different *kinds* of motion, they still are both *motions*. 'It follows that there is a justification for regarding these two as the sources of movement, i.e. appetite [desire] and [. . .] thought [. . .].'[66]

The *De Anima* also provides additional insight as to how the mind can be an object to itself: 'in the case of objects which involve no matter, what thinks and what is thought are identical; for speculative knowledge and its object are identical.'[67] In the *Metaphysics* he similarly describes the nature of *divine* thought and its object: 'Since, then, thought and the object of thought are not different in the case of things that have not matter, the divine thought and its object will be the same, i.e. the thinking will be one with the object of its thought.'[68]

Then, in a further characterization of the divine thought, wherein Aristotle commends his 'forefathers in the most remote ages' for holding the 'inspired' view (if one separates the essential content from the mythical form) that 'these bodies are gods and that the divine encloses the whole of nature,' he also equates the divine thought with God.

> Therefore the possession rather than the receptivity is the divine element which thought seems to contain, and the act of contemplation is what is most pleasant and best. If, then, God is always in that good state in which we sometimes are, this compels our wonder; and if in a better this compels it yet more. And God *is* in a better state. And life also belongs to God; for the actuality of thought is life, and God is that actuality; and God's self-dependent actuality is life most good and eternal. We say therefore that God is a living being, eternal, most good, so that life and duration continuous and eternal belong to God; for this *is* God. [69]

In this statement one sees, once again, the consistent interrelation and culmination of so many Aristotelian principles and concepts. After reading such a statement it is not difficult to understand how St. Thomas was able to synthesize Aristotle's "pagan philosophy" with the religious doctrines of Christianity. Also, one can see in this statement the highest rationalization of the primitive conception that only spirit, soul or mind can be a source of life and motion.

CONCLUSION

Because Aristotle's significance is usually viewed through the distorted perspective of the 17th century controversy engulfing his cosmology, his actual contribution has often been misjudged. Thus the magisterial authority of Aristotle in contrast to the discredited status of the latter-day Peripatetics poses an intriguing question of interpretation. Even today there is confusion between the unparalleled achievements of Aristotle and the often dogmatic, contentious, and peevish opposition of the Scholastic Aristotelians to the new scientific evidence and ideas reaching fruition in the 17th century. But it is significant that this attitude was not shared either by Galileo himself, who continuously distinguished between Aristotle and the Aristotelians, or by the great 19th century naturalist and innovator, Darwin, as we have noted.

What is it that accounts for the predominance of Aristotle for so many centuries, despite the obvious animistic, egocentric, and uncritically commonsensical features — especially in contrast to the Atomists — of his cosmology? The answer lies in three interdependent aspects: (1) his unequivocal adherence to observable, empirical evidence as the final arbiter of conflicting theories, (2) his brilliant analytical powers that enabled him to reduce complex problems to their essential features (to a certain degree of approximation and level of abstraction), and (3) the extraordinary comprehensiveness of his thought that resulted in the most inclusive, coherent system of knowledge of all times. There are other historical and sociological factors that contributed to his unparalleled stature as a thinker throughout an extended period of intellectual history, particularly the amenability of his cosmology to a theistic interpretation as achieved in the grand synthesis of Thomas Aquinas, but these factors, notwithstanding their importance, are derivate and hence dependent upon the intellectual.

Of the aspects accounting for his acclaim, the overriding respect for empirical evidence is fundamental to explain both the special nature of his contribution and the predominance of his views during the scientific revolution. For in spite of the mathematical and astronomical influence of the Pythagoreans (Galileo usually referred to the supporters of Copernicanism as Pythagoreans), Plato, and Ptolemy, along with the experimentalism of Archimedes during the 17th century, it was Aristotle's observations and scientific concepts that mainly provided the empirical facts or background for the current investigations and controversies. For example, in Galileo's *Dialogue Concerning the Two Chief World Systems* there are 153 references to Aristotle, 37 to Ptolemy, 9 to Plato, and 7 to Archimedes.[70] Even in his later book dealing primarily with physical theories, he still cites Aristotle twice as often as Archimedes and nearly five times more than Plato, although the references are less numerous.

It was precisely because he advocated accepting the testimony of the senses over the 'likely stories' of Plato and the 'dialectical arguments' of the Parmenideans and the Pythagoreans, who 'forced their observations' into preconceived theories, that *his* doctrines rather than theirs were usually the starting point of the new scientific inquiries, particularly in physics and biology. Just as the conviction of investigators such as Fabricius, Harvey, Kepler, Galileo, and Newton that the final test of truth must be *experimental* verification was a turning point in intellectual history, so Aristotle's acceptance of *empirical* evidence over the dialectical arguments of his predecessors marks a radical shift in cognitive outlook. The former turn is evident in Galileo's frequent criticism of Aristotle for not having tested his hypotheses and in the following typical assertion of Newton: "in natural philosophy, the investigation [. . .] consists in making experiments and observations and in drawing general conclusions from them by induction, and admitting of no objections against the conclusions but such as are taken from experiment, or other certain truths."[71] Except for the reference to "experiments," this statement could have been made by Aristotle.

Though the recognition of the crucial function of experimentation in scientific inquiry does mark an essential difference between the conception of scientific method as defined by Aristotle and as practiced by the founders of modern science, it is imperative to understand that *the awakening to the decisive role of experimentation in science could not have occurred prior to the realization of the importance of empirical observations in scientific investigations.* It was precisely this latter awareness, with all its implications for developing verifiable scientific theories, that Aristotle contributed to the growth of scientific rationalism. As his persistent criticism of the doctrines of his teacher Plato and other predecessors testifies, unlike the Scholastics, Aristotle did not accept "authority" as the arbiter of truth, but observable evidence. Thus if Galileo can be called the "father of modern science" (in spite of the important contributions of Copernicus, Bacon, and Kepler), Aristotle should be called the "grandfather of modern science" (in spite of the considerable contributions of the Pythagoreans, Democritus, and Plato).

Secondly, Aristotle's genius for analyzing complex phenomena and problems into clearer distinctions and more natural categories also explains why his conceptual interpretations usually were the basis of later investigations; e.g., his distinction of motions into natural and unnatural, with his law that heavy bodies accelerate proportional to their weights, provided the stimulus for Galileo's investigations of free falling objects, while his attempt to explain unnatural or violent projectile motion motivated further inquiry into the problem by such thinkers as Buridan and Oresme in the 14th century, eventually leading to the concepts of 'impetus' and 'force;' his definition of speed as a ratio of distance traveled per time and his distinction between uniform and accelerated motions were the initial basis of Galileo's analysis of these concepts; his definition of time 'as the measure of motion' and place as 'the innermost boundary of what contains' are repeated almost verbatim by Newton in the discussion of these concepts in the *Principia*; his explicit definition and discussion of inertial motion, though he rejected the concept, served to identify the possibility of this kind of motion; his concept of substance as that of which attributes can be predicated but 'which is neither predicable of a subject nor present in a subject' had a considerable impact on later science as well as on such philosophers as Descartes, Locke, and Kant; his concepts of induction, rational intuition, and scientific deduction, along with the doctrine of the four causes, were indispensable contributions to the modern conception of scientific method, etc. — the examples could be continued indefinitely. These analyses, like his empirical investigations, were only a first approximation to the later resolution of the problems, but such initial progress is a necessary condition for further corrections, refinements, or replacements of concepts.

Finally, the fact that Aristotle seemed to be able to correct the conclusions of his predecessors, incorporating the diverse results of his empirical and analytical inquiries into a comprehensive theoretical framework in which each aspect appeared to have its natural place consistent with his theoretical principles of explanation, was a colossal achievement that understandably left a profound impression on all others faced with similar problems. This explains why Aristotle was called "the philosopher" and why for such an extended period of time he was considered the final authority on almost all matters — the more so as with the passing centuries people were conditioned to think about the universe in Aristotelean terms. Thus while initially the major attraction of his cosmology was that it seemed to fit observable reality so closely, as anyone could "see," with the passing of time only those phenomena would be accepted as real which could be identified by and assimilated into his theoretical framework. Though certainly not his intention, this explains why his cosmology eventually became the intellectual fortress that the founders of modern science had to storm before they could continue their conquest of nature.

CHAPTER XIV

[1] Werner Jaeger, *Aristotle*, second ed. (*op. cit.*, ch. XIII), p. 368.

[2] *Phys.*, Bk. I, 188a32-188b3. (Hardie and Gaye trans.)

[3] *Ibid.*, I, 188b34-36.

[4] *Ibid.*, I, 191a24-32.

[5] *Met., Lambda*, 1069b32-34. (Ross trans.)

[6] *Ibid.*, 1069a3-8. Brackets and italics in the original.

[7] *Phys.*, I, 192a31-33. Parenthesis omitted.

[8] *Ibid.*, I, 191b10-18.

[9] *Met., Lambda*, 1069b35.

[10] *De Gen. et Corr.*, II, 329a30-32. (Platt trans.)

[11] Cf. *ibid.*, 330a25-331a21.

[12] Cf. *Phys.*, Bk. II, ch. 2; *De Gen. et Corr.*, Bk. II, ch. 9; *De Gen. An.; Met., Alpha*, chs. 1-4, and *Delta*, ch. 2.

[13] *Phys.*, II, 192b18-21.

[14] *De Gen. An.*, I, 715a2-6.

[15] *Phys.*, II, 194b23-24; also *Met., Delta*, 1013a24-26.

[16] *De Gen. et Corr.*, I, 319a19-23.

[17] *Ibid.*, 27-29.

[18] *Cat.*, 2a11-19. (Edghill trans.)

[19] *Phys.*, II, 192b29-32; also *Met., Delta*, 1013a29-32.

[20] *De Gen. et Corr.*, II, 338a20-338b6.

[21] *Ibid.*, 336b37-337a7.

[22] *Phys.*, II, 192b33-66.

[23] Cf. Werner Heisenberg, *Physics and Philosophy* (New York: Harper & Brothers Pub., 1958), p. 41 and John Herman Randall, *Aristotle* (*op. cit.*, ch. XIII), p. 167.

[24] Jaeger, *op. cit.*, p. 389.

[25] *Phys.*, II, 198b17-32.

[26] *Ibid.*, 198b33-199a8.

[27] *Ibid.*, 199a26-33.

[28] *Ibid.*, 198a14-26.

[29] *Met., Lambda*, 1071a13-17.

[30] Randall, *op. cit.*, p. 145.

[31] *Met., Kappa*, ch. 7, 1064a15.

[32] *De Caelo*, Bk. IV, ch. 1, 308a28-31.

[33] *Phys.*, Bk. VIII, ch. 4, 255b14-18.

[34] *Ibid.*, ch. 10, 267a7-9.

[35] *Ibid.*, ch. 7, 261b31-33; ch. 9, 264b8-265a25-28.

[36] *De Caelo*, Bk. I, ch. 2, 269a19-20. Brackets added.

[37] *Ibid.*, ch. 5, 272a5-8.

[38] *Ibid.*, ch. 3, 270b9-25.

[39] Both essays occur in Stillman Drake, *Discoveries and Opinions of Galileo* (Garden City: Doubleday Anchor Books, 1957).

[40] Galileo Galilei, *Dialogue Concerning the Two Chief World Systems* (*op. cit.*, Introduction), p. 50.

[41] Cf. W. D. Ross, *Aristotle*, fifth ed., rev. (*op. cit.*, ch. XIII), p. 95.

[42] Cf. Galileo, *Dialogues Concerning Two New Sciences* (*op. cit.*, ch. XIII), p. 61.

43 *De Caelo*, Bk. II, ch. 13, 25-28. Brackets added. (Stocks trans.)
44 *Ibid.*, 295b20-23.
45 *Ibid.*, Bk. IV, ch. 2, 310b1-5.
46 *Ibid.*, Bk. II, ch. 14, 22-26.
47 Galileo, *Dialogue Concerning the Two Chief World Systems, op. cit.*, p. 139.
48 *Ibid.*, pp. 144-145.
49 *De Caelo*, Bk. II, ch. 13, 294a5-7. Brackets added.
50 Randall, *op. cit.*, p. 160, f.n. 21.
51 Sir Thomas Heath, *Aristarchus of Samos* (*op. cit.*, ch. XI), p. 193.
52 *Ibid.*, p. 218.
53 Cf. *Met., Lambda*, ch. 8, 1074a.
54 *Ibid.*, 1074a12-14. Brackets added.
55 *Ibid.*, 1072a21-27.
56 *Ibid.*, 1072b7-12.
57 *Met., Kappa*, ch. 12, 1068b3-5.
58 *Phys.*, Bk. VIII, ch. 4, 256b7-14.
59 *Ibid.*, ch. 6, 259a7-14.
60 *Met., Lambda*, ch. 12, 1073a29-38.
61 Cf. D. R. Dicks, *Early Greek Astronomy to Aristotle* (*op. cit.*, ch. XI), pp. 213-214.
62 *Met., Gamma*, ch. 1, 1003a21-32.
63 *Ibid., Epsilon*, ch. 1, 1026a10-33.
64 *Ibid.*, ch. 9, 1074b32-34.
65 Cf. *De Anima*, Bk. II, ch. 4, 415b8-12.
66 *Ibid.*, Bk. III, ch. 10, 433a18-20. Brackets added. (J. A. Smith trans.)
67 *Ibid.*, Bk. III, ch. 4, 430a2-5.
68 *Met., Lambda*, ch. 9, 1075a1-4.
69 *Ibid.*, ch. 7, 1072b23-29.
70 Galileo, *Dialogue Concerning the Two Chief World Systems, op. cit.*, Index.
71 Sir Isaac Newton, *Opticks*, Bk. III, Sec. V, fourth ed., corrected (London: 1730), Query 31.

CHAPTER XV

HELLENISTIC SCIENCE

The center of 'ancient science' lies in the 'Hellenistic' period [. . .] .[1]
Neugebauer

The premature death at age 33 of Alexander in Babylon in 323 B.C. brought an end to his conquest of the Near East, Egypt, and Western Asia, but not to the political and cultural domination of these areas by the Macedonians and the Greeks. Aristotle died a year later but, as Sarton states, the "gigantic personalities of Alexander the Great and Aristotle stood at the threshold of a new age." Called the Hellenistic period, this "new age" lasted roughly two centuries, beginning with the death of Alexander and extending to the 1st century B.C. — although in a real sense its legacy includes such culminating Greco-Roman figures as Hero, Galen, and Ptolemy (1st and 2nd centuries A.D.). Greek remained the dominant intellectual and cultural language and Athens the educational and cultural mecca. Although few of the mathematical and scientific giants of this period were Athenians, many were educated in the Academy or the Lyceum, and nearly all visited Athens at some time during their lifetime. This was true even as the centers of learning shifted to other Greek cities such as Alexandria, Pergamon, Rhodes, and Antioch. According to Sarton,

> Athens had been reduced to the status of an impoverished provincial town, yet its spiritual prestige was as great as ever; its schools were still the leading schools of the ancient world and it was still the main center of pilgrimage for every lover of wisdom. [2]

Leaving no living heir (a son was born posthumously who was put to death at the age of 12), Alexander's vast, heterogeneous empire was soon divided among rival military factions. Before he died Alexander gave his signet ring to his Macedonian general Perdiccas, but the intense rivalry among the other generals finally resulted in the empire being divided into three main kingdoms under separate dynasties: (1) Macedonia and Greece ruled by the Antigonids, (2) Western Asia ruled by the Seleucids, and (3) Egypt ruled by the Ptolemies. Although the term 'Hellenistic culture' extends to each of these kingdoms, our focus will be on the Egypt of the Ptolemies because the city of Alexandria was the hub of the mathematical, scientific, and humanistic research during the Hellenistic period. That Alexandria achieved such a position of prominence was due to the illustrious administration of the first Ptolemies who sought to make it a city rivaling Athens as a center of research, learning, and culture. In fact, although the city was originally founded by Alexander, it was Ptolemy I Soter (the savior) and his son Ptolemy II Philadelphus (brother-loving) who were primarily responsible for transforming the city into a great metropolis, the capital of Egypt, replacing the former capital of Memphis. The best known of all the Ptolemaic rulers, and the last, was Queen Cleopatra.

Apart from the renowned scientists and mathematicians associated with the city of Alexandria, two institutions in particular epitomize its ascendance as the new center of research, the Museum and the Library. As regards the first, although the term 'museum' was not new since there are references to a museum in Aristotle's Lyceum, it is due to the Museum of Alexandria that the word and the type of cultural institution it designates are now commonplace. While the Museum was founded and funded by the first two Ptolemaic kings, its development and organization was due to Demetrius of Phaleron and Strato of Lampsacos who served successively under the first and second kings. Both men (some twenty years apart) having been students of Theophrastus who had succeeded Aristotle as head of the Lyceum, it was this school that served as a model for the Museum. The exchange between the institutions was even more direct since Demetrius had been governor of Athens a few years before his appointment by Ptolemy Soter as director of the Museum, and Strato, after serving as tutor to Ptolemy Philadelphus who subsequently appointed him director of the Museum, later returned to Athens upon the death of Theophrastus to become the latter's successor as head of the Lyceum for eighteen years.

In contrast to the Academy and the Lyceum where organized instruction was offered in addition to the research activities of the scholars, the Museum seems to have been strictly a research institute for mathematicians, physicists, anatomists, geographers, etc., where no formal teaching took place. It was particularly Strato who, emulating the empirical orientation of Aristotle and Theophrastus, was responsible for the predominantly scientific research of the Museum which had such a marked effect upon the progress of science during the Hellenistic period. For the first time professional scholars appeared, men who were financially supported by the state in order to devote their lives to the pursuit of knowledge.

> It was because of its creation and because of the enlightened patronage which enabled it to function without hindrance that the third century witnessed such an astounding renaissance. The fellows of the Museum were permitted to undertake and to continue their investigations in complete freedom. As far as can be known, collective research was now organized for the first time, and it was organized without political or religious directives, without purpose other than the search for truth. [3]

While some may not have heard of the Museum of Alexandria, almost everyone has heard of its Library, the most famous in antiquity. Like the Museum, it was founded and supported by Ptolemy Soter and his son Philadelphus who appointed Demetrius its first director. Eratosthenes, the famous geographer and mathematician (the originator of the "sieve," a method for determining all prime numbers less than a given number) was one of his distinguished successors. The prestigious status of the Library and the competition for papyri were such that the Ptolemies resorted to devious means for insuring its prominence, as Sarton relates:

> The kings of Egypt were so eager to enrich their library that they employed highhanded methods for the purpose. Ptolemaios III Evergetes (ruled 247-222) ordered that all travelers reaching Alexandria from abroad should surrender their books. If these books were not in the Library, they were kept, while copies on cheap papyrus were given to the owners. He asked the librarian of Athens to lend him the state copies of Aischylos, Sophocles, and Euripides, in order to have transcripts made of them, paying as a guarantee of return the sum of fifteen talents; then he decided to keep them, considering that they were worth more than the money he had deposited and he returned copies instead of the originals. [4]

Just as the Museum was the center of research for the sciences, the Library was the center of research for the humanities — as indicated by the catalogue compiled by Callimachus, according to which at least six of the eight categories of classification are related to humanistic studies, mainly literature, while the various sciences were lumped under one or two categories. Moreover, the Alexandrian Librarians were not merely the collectors, cataloguers, and guardians of the rolls of papyri, they also had the difficult task of identifying, revising, and editing them. Thus as Sarton claims, "the Alexandria Library was the nursery of philologists and humanists, even as the Museum was a nursery of anatomists and astronomers."

It was under one of the later Librarians, Aristophanes of Byzantium, that a method of punctuation was either invented or systematized, and both he and his successor, Aristarchus of Samothrace (not to be confused with the astronomer from Samos), created the system of grammatical analyses based upon a differentiation of the various parts of speech. It is hard to believe that the magnificent epic dramas and tragedies of Greek literature could have been written without an explicit knowledge of grammar, but this was the case:

> It is marvellous that all of the intricate beauties of the Greek language, a very complex grammar as well as a rich and well-integrated vocabulary, were created to a large extent unconsciously. The main creators of Greek literature did not know grammar, but the Alexandrian philologists extracted grammar from their writings [. . .]. [5]

In one of the major disasters of history, much of the contents of the Library (according to tradition) was burnt during Caesar's siege of Alexandria in 48 B.C. Although thereafter the Library's reputation and importance declined, it continued in existence until either the 5th century A.D. when it was allegedly destroyed by order of Theophilus, bishop of Alexandria, or perhaps until the 7th century when the Muslems took Alexandria.

THE MATHEMATICAL AND SCIENTIFIC ACTIVITIES OF ALEXANDRIA

Within the limits of this study it is not possible to undertake anything like a thorough reconstruction of the thought processes and systematic evaluation of the individual contributions of the mathematicians and scientists who made the Hellenistic Age one of the most important in the history of science. This itself is a specialized field requiring specific training in the mathematics of antiquity, the development of which, though related to modern mathematics, is often presented in a way that makes it seem strange even to modern mathematicians. Moreover, since intuition plays such a crucial role in the solution of mathematical problems, the discovery of theorems and proofs, and the development of mathematics in general, about which we know practically nothing even today, there would be little that we could say in any case. This is particularly true of Greek mathematics (although it applies to Newton as well) since the mathematicians did not indicate the procedures by which they arrived at their mathematical discoveries and solutions, with the important exception of Archimedes' *Method*. As Heath says in his introductory notes to the *Method*:

> Nothing is more characteristic of the classical works of the great geometers of Greece, or more tantalising, than the absence of any indication of the steps by which they worked their way to the discovery of their great theorems. As they have come down to us, these theorems are finished masterpieces which leave no traces of any rough-hewn stage, no hint of the method by which they were evolved. [6]

Even the delineation of a mathematical problem, in contrast to an empirical one, does not have the same degree of conceptual intelligibility, usually appearing esoteric and mysteriously abstract and complex to the uninitiated. Every man can observe that the planets change position, that smoke rises and unsupported objects fall, and wonder about the causes, but investigations of conic sections and of the volume and area of a sphere, or of the trisection of an angle, the duplication of the cube, and the quadrature (squaring) of the circle (the three famous mathematical problems of antiquity), or of mathematical paradoxes such as incommensurate magnitudes, require a unique capacity and specialized training. As Euclid is reputed to have replied to Ptolemy Soter, who asked if there were a shorter way to learning geometry than the *Elements*, "there is no royal road to geometry."

Therefore, rather than attempting to reconstruct the thought processes of the individual investigators, this concluding chapter will present a summary account of the more outstanding Hellenistic achievements in order to understand and evaluate their significance for the development of scientific thought — with the intent of correcting some common misconceptions in the process. As regards the latter, for example, it is usually maintained that what differentiates Greek science (including the Hellenistic period) from modern classical science is that the former was qualitative and descriptive, while the latter is quantitative and experimental. Although this difference is usually expressed as if it were absolute, when one examines closely the nature of Hellenistic science one finds that the distinction is merely one of degree, not reflecting any intrinsic difference but the fact that the science of the Hellenistic period represents an earlier stage in which later, more sophisticated techniques for dealing with scientific problems were beginning to appear.

Indeed, it is not difficult to find clear instances in Hellenistic science of the application of mathematics to empirical problems, both astronomical and terrestrial, as well as examples of simple experiments to test hypotheses. The systematic study of mathematical and astronomical problems in the Academy along with the pursuit of more empirical investigations and cosmological inquiries in the Lyceum bore fruit in the succeeding two centuries. Hellenistic astronomers improved upon the mathematical techniques for measuring the relative distances and sizes of the sun and the moon and for describing more accurately the motions of the planets, while empirical investigations in physics and medicine progressed to the point where the importance of experimentation was clearly recognized. As we emphasized in the last chapter, the realization of the significance of experimentation and of the discovery of mathematical correlations in scientific inquiry could not precede, but depended upon a prior stage of knowledge, that of objective description and classification. It is only when one has come to recognize the limitations of a purely qualitative, descriptive account of nature that one will begin to look at phenomena as manifestations of more fundamental, underlying elements, structures, and processes that require experimentation both for their discovery and for the more precise mathematical designation of their magnitudes and correlations.

For example, once the movements of the planets were cast in the form of the Eudoxean concentric spheres, succeeding astronomers such as Heraclides, Apollonius, and Hipparchus could introduce refinements in and modifications of the theory, or even, as in the case of Aristarchus of Samos, a revolutionary revision, precisely because they had a framework for examining the astronomical data more closely. That is, once a simplifying, unifying framework has been imposed on previously complex, disconnected observations, the phenomena can be examined more closely from the perspective of the framework and predictions made to see how accurately the observations and the interpretation match. This in turn stimulates the development of more refined instruments of observation and/or experimentation to acquire as precise data or measurements as possible which then results in adjustments in the framework to accommodate the new information.

Similarly, as the desire for a better understanding of anatomical structures and physiological processes developed, the greater the interest in dissection, a form of experimentation for acquiring more detailed biological information, as exemplified in the research of Herophilus and of Erasistratus in the Museum. In addition, the preliminary investigations of Aristotle into statics and hydrostatics influenced Strato and Archimedes in their researches in these areas, illustrating how science develops: the differentiation of problems, the interrelation of observations and theoretical interpretations, the refinement of concepts, and the elaboration of methodologies occurring as a natural development of continuous, though at times revolutionary, scientific inquiry into broader and deeper domains of natural phenomena.

Accordingly, the thesis of this chapter is that the Hellenistic period, unlike the 4th century of Plato and Aristotle, was not an age of great scientific-philosophical syntheses, but of brilliant mathematical, astronomical, physical, and medical investigations and discoveries that foreshadow the classical era of modern science. The precursors of mathematical astronomy and physics, of modern trigonometry and infinitesimal calculus, and of anatomy and physiology can be traced particularly to this period: Aristarchus was "the Copernicus of antiquity," Apollonius' research into conic sections provided the mathematical basis for Kepler's first two astronomical laws, Eudoxus' method of exhaustion and Archimedes' method of integration preceded the discovery of differential and integral calculus by Newton and Leibniz, Erasistratus was a forerunner of Harvey in rejecting the humoral theory and surmising the connection of arteries and veins, and Galileo was the heir apparent of Strato and Archimedes. Had not the Hellenistic period "leveled off" (Marshall Clagett's term) and finally ended as a result of the ascendency of Rome and Christendom, modern science might have begun at the beginning of the Christian era or after the achievements of Hero, Galen, and Ptolemy. But if that had occurred, there would have been no "beginning" of modern science in the Renaissance — just a continuous development of Hellenistic science.

It hardly needs to be added that there was, of course, a continuation of mathematical, astronomical, physical, and medical investigations during the Middle Ages, particularly by Arabic scholars. Yet the resurgence of scientific inquiry during the later Middle Ages (13th-15th centuries) and Renaissance (16th-17th centuries) was largely due to the reintroduction of the writings and past achievements of the ancient Greeks to the West. Copernicus, for instance, expressly states that because the 5th century Pythagorean, Philolaus, and the Hellenistic astronomer, Heraclides, had declared that the earth moved he "too began to think of the mobility of the earth." Fabricius and Harvey have numerous references to Aristotle although they were "bred in the Galenic tradition," while Galileo constantly refers to Aristotle even while praising Archimedes whom he had "read and studied with infinite astonishment." The impact of Pythagoreanism and Neoplatonism on Kepler's thought is well documented and even Newton emulated Euclid in the title and form of his major work, "Mathematical Principles of Natural Philosophy." Thus while Western science took root during the Hellenic period, its essential nourishment and cultivation occurred during the Hellenistic Age — which brings us to the most famous of the Alexandrian scholars.

EUCLID

The Greeks having neglected to create a muse presiding over mathematics, a discipline as subject to divine inspiration or cosmic madness as poetry and art, there was no patron goddess to preserve accounts of the personal lives of the greatest mathematicians of antiquity. Thus practically nothing is known of the lives of Pythagoras, Theaetetus, Archytas, Eudoxus, and Euclid. Every schoolboy is familiar with the name "Euclid," but beyond his being the author of the *Elements*, the little we know about him is conveyed mainly in a statement by Proclus, a mathematician and philosopher of the second half of the 5th century A.D.:

> [...] Euclid [...] put together the Elements, collecting many of Eudoxus's theorems, perfecting many of Theaetetus's, and also bringing to irrefragable demonstration the things which were only somewhat loosely proved by his predecessors. This man lived in the time of the first Ptolemy. For Archimedes, who came immediately after the first (Ptolemy), makes mention of Euclid [...]. He is then younger than the pupils of Plato, but older than Eratosthenes and Archimedes, the latter having been contemporaries, as Eratosthenes somewhere says. [7]

This tells us nothing of the dates or places of Euclid's birth and death, but only that he lived at the time of Ptolemy Soter, and since he knew the king personally, that he must have spent some time in Alexandria. The latter is corroborated by a statement from Pappus (a mathematician of the second half of the 3rd century B.C.) that Apollonius 'spent a very long time with the pupils of Euclid at Alexandria,' which implies that Euclid established a school in Alexandria sometime during the first quarter of the 3rd century. Proclus' reference to the theorems of Eudoxus and Theaetetus, who were associates of Plato at the Academy, has led to the inference that Euclid studied there which seems likely since he would have sought out the most prominent school of the time and he would have found the mathematical research of the Academy more attractive than the empirical orientation of the Lyceum.

Although we know very little about Euclid the man, the significant fact is that we do possess his most important work, the *Elements*, one of the most influential books of all time. As Struik asserts:

> The *Elements* form, next to the Bible, probably the most reproduced and studied book in the history of the Western World. More than a thousand editions have appeared since the invention of printing, and before that time manuscript copies dominated much of the teaching of geometry. Most of our school geometry is taken, often literally, from eight or nine of the thirteen books; and the Euclidean tradition still weighs heavily on our elementary instruction. For the professional mathematician these books have always had an inescapable fascination (even though their pupils often sighed) and their logical structure has influenced scientific thinking perhaps more than any other text in the world. [8]

Confirmation of the enduring influence of the *Elements* is the effect it had on two of the greatest scientists, Newton and Einstein; as mentioned earlier, Newton organized the *Principia* after the *Elements*, and Einstein in his "Autobiographical Notes" cites the effect that a book on Euclidean geometry had on him as a boy of twelve: "Here were assertions [...] which — though by no means evident — could nevertheless be proved with such certainty that any doubt appeared to be out of the question. This lucidity and certainty made an indescribable impression upon me." [9]

The thirteen books of Euclid can be divided roughly into three parts: Books I-VI, plane geometry, Books VII-X, arithmetic or the theory of numbers, and Books XI-XIII, solid geometry. [10] Although Euclid is usually called "the father of geometry," several "Elements" had been written before him by Hippocrates of Chios (5th century), by Leon (first half of the 4th century), and by Theudios of Magnesia (second half of the 4th century). According to Sarton, "Theudios' treatise, with which Euclid was certainly familiar, had been prepared for the Academy, and it is probable that a similar one was in use in the Lyceum."[11] Moreover, as Euclid himself acknowledges, many theorems of the *Elements* can be ascribed to earlier mathematicians while sections of the *Elements* are expanded versions of the work of his predecessors. In particular, Book V is based on Eudoxus' theory of proportions applied to incommensurable quantities, Book XII expands the latter's theory of exhaustion, and Book XIII deals with the so-called "Platonic Bodies" or regular solids that so fascinated the Greeks (and later, Kepler). While not especially original, the importance of the *Elements* consists in the fact that Euclid's systematic treatment of the various topics was so much superior to that of earlier works that it immediately replaced all previous books on geometry (just as Ptolemy's *Almagest* was to eclipse all earlier books on astronomy).

Apart from the extended inquiries, new proofs, and improved techniques for dealing with specific problems, the primary significance of the *Elements* lies in its axiomatic method as illustrated in Book I. Beginning with definitions, postulates, and "common notions" or axioms, Euclid illustrates the importance of these for deducing theorems, although even here he was not original, as both Heath and Sarton have pointed out.

> The most amazing part [. . .] is Euclid's choice of postulates. Aristotle was, of course, Euclid's teacher in such matters; he had devoted much attention to mathematical principles, had shown the unavoidability of postulates and the need of reducing them to a minimum; yet the choice of postulates was Euclid's. [12]

It was in fact Euclid's own selection and formulation of the definitions, postulates, and "common notions" (Aristotle's term) or axioms that permanently established the deductive, axiomatic method of geometry (and mathematics) as we know it today. His "fifth postulate" is particularly significant because it was due to investigations and modifications of this postulate that the non-Euclidean geometries of Gauss, Lobachevsky, Bolyai, and Riemann were developed, that of the latter utilized by Einstein in the general theory of relativity to describe the geometric structure of space-time.

In addition to the *Elements*, Euclid wrote a book on plane geometry called *Data* which still survives, several other works on geometry which have been lost, a book on "Conics' and another on "Surface-Loci" which have been lost, and several works on applied mathematics: *The Phaenomena* intended for use in astronomy, the *Optics* which deals with visual perception and optical perspective, and probably a work on music, now lost, which may have been called "Elements of Music." Works on mechanics have been attributed to Euclid on the basis of fragments, but Heath discounts this.[13] Thus Euclid's reputation is based on the *Elements*, "the greatest mathematical text-book of all time."

MATHEMATICAL AND ASTRONOMICAL INQUIRIES OF ANCIENT BABYLONIA AND EGYPT

This would seem to be the moment for a slight digression into the history of mathematics and astronomy to remedy one of the gaps in our discussion of the origins of scientific rationalism. We have described these origins as if they had begun solely with the Ionians, without having any antecedents; however, particularly as a result of the research of James H. Breasted and Otto Neugebauer, we know that mathematical and astronomical investigations flourished in both ancient Babylonia and Egypt. According to Neugebauer, one can pretty much discount any influence of Egyptian mathematics or astronomy on Greek thought, since these disciplines remained at "a relatively primitive level." As he tersely concludes his discussion of Egyptian mathematics and science: "Ancient science was the product of a very few men; and these few happened not to be Egyptians."[14] The one exception to this was the Egyptian calendar which was used by Hellenistic astronomers and which, according to Neugebauer, "is, indeed, the only intelligent calendar which ever existed in human history."[15] In contrast to the strictly lunar calendar of the Babylonians and the numerous, inexact Greek calendars (almost every Greek city-state had its own calendar), the Egyptian calendar divided the year precisely into 12 months of 30 days each, with 5 additional days at the end of the year. In addition, the Egyptians divided the day into 24 hours, though these hours were not of uniform length.

As for the Babylonians, their contributions both in mathematics and in astronomy were much richer. Unlike the Hellenistic Greeks, their mathematical achievements did not progress to the stage where they began looking for demonstrations or proofs of various theorems, and they did not organize their mathematics into an axiomatic system. Moreover, the discovery of incommensurable magnitudes or irrational quantities was uniquely Greek, as were the important theories of exhaustion and integration. But the Babylonians did develop the sexagesimal system (units of measurement based on multiples of 60 still used in geometry and astronomy) and invented the "place value notation" of numbers which Neugebauer justly describes as "one of the most fertile inventions of humanity." On the basis of this place value notation each position of a digit in a group of numbers indicates the power of which it is a coefficient; thus in our decimal system the number 2746 is read as two thousand seven hundred and forty-six.

In addition to these contributions, archeologists have discovered tablets on which numerical problems and tables were worked out to a very high degree, reflecting the application of arithmetic to difficult concrete problems of computation, such as the determination of the coefficients of the quantities of materials to be used in different types of construction, along with the solution of such theoretical problems as the precise predictions of the appearance of the moon based on complex periodic variations or parameters. Besides their arithmetical complexity, these problems often demanded numerical operations involving algebraic functions without the benefit of algebraic notation. According to Neugebauer, even in the Old Babylonian period from 1800-1600 B.C. we "find tables of squares and square roots, of cubes and cube roots, of the sums of squares and cubes needed for the numerical solution of special types of cubic equations, of exponential functions, which were used for the computation of compound interest, etc." [16]

Because of the importance of these numerical and algebraic computations, geometry occupied a lesser role in Babylonian mathematics than in that of the Greeks. And though the Babylonians exceeded the Greeks in these branches of mathematics, it is because their mathematical interests remained tied predominantly to the solution of specific practical problems that they did not advance as far in the direction of mathematical theory or pure mathematics. Greek investigations in geometry led to inquiries into geometrical problems for their own sake, apart from any practical consequences (in fact, both Plato and Archimedes disparaged the latter), as well as the application of geometry to purely theoretical astronomical and physical problems. Moreover, while the Babylonians had discovered individual geometrical theorems, such as the Pythagorean Theorem (a thousand years before Pythagoras), they did not develop a method of demonstration for proving such theorems and consequently did not organize their geometrical discoveries into an axiomatic system. Neugebauer therefore summarizes the achievement of 1800 years of Babylonian mathematics as follows:

> However incomplete our present knowledge of Babylonian mathematics may be, so much is established beyond any doubt: we are dealing with a level of mathematical development which can

in many aspects be compared with the mathematics, say, of the early Renaissance. Yet one must not overestimate these achievements. In spite of the numerical and algebraic skill and in spite of the abstract interest which is conspicuous in so many examples, the contents of Babylonian mathematics remained profoundly elementary [. . .]. Babylonian mathematics never transgressed the threshold of pre-scientific thought. It is only in the last three centuries of Babylonian history [i.e., during the Hellenistic period from 300 B.C. to the beginning of the Christian era] and in the field of mathematical astronomy that the Babylonian mathematicians or astronomers reached parity with their Greek contemporaries.[17] (Brackets added)

Neugebauer's last statement raises the question as to the comparative significance of Babylonian and Greek mathematical astronomy: did the Babylonians reach parity with the Greeks as he asserts? It certainly is not insignificant that progress in mathematical astronomy at the time depended mainly upon the kind of mathematics developed by the Greeks, rather than that of the Babylonians. Ptolemy's geometrical model incorporating circles, epicycles, and eccentrics was based on Eudoxus' original theory of concentric spheres as emended by Heraclides, Apollonius, and Hipparchus. Kepler's first two laws utilized the mathematical investigations into conic sections of Apollonius and the discovery of his third law was indirectly influenced by the Pythagorean and Platonic conception of regular solids. In addition, the theory of trigonometric functions which played such an important role in astronomical measurements was developed by Hipparchus, and Newton's more accurate explanation and prediction of the orbital motions of the planets depended upon the development of integral calculus which, as already mentioned, was initiated by Eudoxus and Archimedes.[18]

These historical facts represent eloquent testimony to the uniqueness of the Greek genius. The Babylonians made very accurate astronomical observations many of which probably came down to the Greeks (e.g., Hipparchus' discovery of the precession of the equinoxes seems to have been based partially on Babylonian data), but there is no evidence that they attempted to organize these observations into a coherent theory of the universe. The relative unimportance of geometry in their mathematical investigations (as well as their primary interest in lunar ephemerides) also adversely affected their astronomical achievements since, as just indicated, the earliest models of the universe were based on geometrical spheres. On the other hand, the Babylonian development of arithmetical progressions and algebraic functions (instead of the "geometrical algebra" of the Greeks), which were later improved upon by the Hindus and the Arabs during the Middle Ages (particularly the famous "Algebra" of al-Khwarizmi in the 9th century), also played their part in the creation of the mathematical analysis of the 17th and 18th centuries.

ARISTARCHUS

Although so much of Hellenistic science reached out to the modern period, nothing approached it more closely than Aristarchus' hypothesis of a heliocentric universe. Roughly eighteen centuries separated Aristarchus from Copernicus, yet his conception of the universe was essentially identical to that of the Renaissance astronomer. Though fruitless, it is intriguing to wonder what the consequences on intellectual history might have been had Hipparchus and Ptolemy not rejected Aristarchus' hypothesis, but attempted to "save the phenomena" on the basis of the heliocentric model — however, as one grandly proclaims, "the times were not ripe." The Pythagoreans, on purely a priori grounds, had been the first to attribute motion to the earth which, along with the counter-earth, the sun, the moon and the other planets, was held to revolve around the central fire or hearth. If by the statement, 'the earth winding round the axis that stretches right through,'[19] Plato meant to attribute a rotary motion to the earth around its axis, then he too would have been one of the first to attribute motion to the earth. But if not Plato, then Heraclides, a student of Plato's successor Speusippus, definitely affirmed the daily rotation of the earth about its axis, thereby accounting for the apparent diurnal rotation of the heavens. Also, in a theory approximating that of Tycho Brahe, Heraclides maintained that the inferior planets, Mercury and Venus, revolve around the sun as their center, while the sun, the moon, and the superior planets, Mars, Jupiter, and Saturn, revolve around the central earth.[20]

But the really imaginative step was taken by Aristarchus who had studied with Strato either in Alexandria or in Athens sometime in the first quarter of the 3rd century. Having previously computed the diameter of the sun to be nearly seven times that of the earth (the actual ratio is 109), he apparently concluded that it is more reasonable to assign a central, stationary position to the larger body. Thus he proposed the heliocentric hypothesis, as described by Archimedes in *The Sand-Reckoner.*

> Now you [king Gelon] are aware that 'universe' [cosmos] is the name given by most astronomers to the sphere whose centre is the centre of the earth [. . .]. But Aristarchus of Samos brought out a book consisting of some hypotheses, in which the premisses lead to the result that the universe is many times greater than that now so called. His hypotheses are that the fixed stars and the sun remain unmoved, that the earth revolves about the sun in the circumference of a circle, the sun lying in the middle of the orbit, and that the sphere of the fixed stars, situated about the same centre as the sun, is so great that the circle in which he supposes the earth to revolve bears such a proportion to the distance of the fixed stars as the centre of the sphere bears to its surface [21] (Brackets added)

For the times this was an incredible hypothesis, requiring astronomers to enlarge their conception of the universe almost, but not quite, to an infinite extent. For the real novelty of the hypothesis was not so much that Aristarchus made all the planets revolve around the central sun, but that he realized that for this to be true the size of the universe must be such that *the dimension of the earth's orbit is as a point to the enormity of the distance and the circumference of the sphere of the fixed stars.* Even eighteen centuries later the magnitude of this supposition constituted one of the major obstacles (particularly for Tycho Brahe) to accepting the Copernican theory. Aristarchus apparently realized, as did post-Copernican astronomers, that in order to explain the absence of stellar parallax (the displacement of the stars owing to the revolution of the earth in its orbit) one must assume that the stars are vastly further away than commonly assumed.

Yet this hypothesis could "save the phenomena." The daily axial rotation of the earth posited by Heraclides would account for the apparent diurnal rotation of the heavens, while the annual revolution of the earth in its orbit would account for the apparent solar year. Copernicus' theory was essentially no different from that of Aristarchus. Since Aristarchus' contemporaries must have raised terrestrial objections to the earth's motion similar to those taken into account by Copernicus and Galileo (e.g., why it is that we do not feel the motion), it would be intriguing to know how Aristarchus might have tried to answer them; unfortunately, however, if there were such arguments, none has come down to us. The danger of introducing revolutionary theories was illustrated in the case of Aristarchus as in the controversy over the Copernican hypothesis, since Cleanthes, a contemporary Stoic, claimed that a charge of impiety ought to be brought against Aristarchus for "moving the hearth of the universe," though nothing came of it.[22]

As prescient as we now know this hypothesis to have been, it was rejected by Hipparchus, one of the greatest astronomers of antiquity. Although this is conjecture, it may have been the more empirical (or more "positivistic") orientation of Hipparchus, renowned for his development of astronomical instruments and for his accurate observations, that contributed to the rejection of Aristarchus' revolutionary hypothesis. We have seen how Heraclides modified the homocentric model of Eudoxus by asserting that the earth revolves on its axis and that Mercury and Venus revolve around the sun. Apollonius, a generation or two after Aristarchus, emended the theory further by introducing epicycles to account for such peculiarities of planetary motion as their apparent retrogression. For if one visualizes a planet moving on an epicycle whose center is on an equant with the earth at its center, as the planet revolves it will appear from the earth to be moving forward in half of its orbit and to retrogress in the other half.

Having thereby simplified Eudoxus' account of retrograde motion with his theory of epicycles, Apollonius then introduced eccentrics to account for the perceptible variations in size and brightness as the planets revolve in their orbits, variations which were known to be inexplicable in terms of the concentric orbits of Eudoxus. If a planet describes a circular orbit whose center does not coincide with the earth but is slightly eccentric (off center) to it, then at different times it will be closer to or further away from the earth, thus accounting for its apparent differences in size and brightness. From the standpoint of the progressive success in explaining specific astronomical phenomena by the extended use of circles, in the form of epicycles and eccentrics from a *geocentric* perspective, it could be argued that Hipparchus was more than justified in rejecting the more conjectural heliocentric approach of Aristarchus, which at that early time undoubtedly could not have offered as specific explanations of astronomical phenomena. At any rate, Hipparchus chose to adopt a modified version of the Apollonian system which became the foundation for Ptolemy's culminating geocentric theory. Actually, as we now know, no system of circles, however ingeniously adapted, could really "save the phenomena," thus they were finally rejected by Kepler along with the assumption of uniform motion.

The second major contribution of Aristarchus was an earlier treatise, "On the Sizes and Distances of the Sun and Moon," which we possess in its entirety. Beginning in the manner of Euclid with six hypotheses describing the angular positions of the sun and the moon when the latter is exactly at half moon, Aristarchus deduces the sizes and distances of the sun and the moon *relative to* the earth or to each other. That is, the final measurements determined by geometrical means (anticipating the method of trigonometry) express ratios between greater and smaller values, so typical of Greek computations; e.g., "The distance of the sun from the earth is greater than eighteen times, but less than twenty times, the distance of the moon from the earth."[23]

Although the numerical values were inaccurate, this is much less important than the attempt itself to calculate the relative distances and sizes of the sun and the moon, illustrating the value of mathematics for empirical inquiry: i.e., given certain quantities and certain unknowns, mathematics provides the computational methods for determining the latter by means of the former. As Sarton states:

> Aristarchus' numerical results were very poor [. . .] but the very fact of "measuring" those celestial bodies was in his age astounding [. . .]. It is as if puny man had reached those two luminaries of the day and the night.[24] [Sarton died before "puny man" did reach one of those luminaries.]

Following his example, later investigators such as Archimedes, Hipparchus, Ptolemy, as well as lesser known astronomers, attempted to improve on his calculations.

ARCHIMEDES

A younger contemporary of Aristarchus, Archimedes was, according to Heath, "perhaps the greatest mathematical genius the world has ever seen." From what Archimedes tells us in *The Sand-Reckoner*, his father was an astronomer who, like Eudoxus and Aristarchus, estimated the size of the sun relative to the moon. We also know that he was from Syracuse, the capital of Sicily, the city that had brought about the disastrous defeat of the Athenian expedition in 413. By the time of Archimedes, Syracuse, though still prosperous and powerful, had passed its golden age and finally was taken by the Roman general Marcellus in 212 B.C. Everyone has heard about the fabulous war machines created by Archimedes for the defense of Syracuse; although the stuff of which legends are made, most of the reports seem to have been true. As described by Plutarch in his *Life of Marcellus*:

> When, therefore, the Romans assaulted them by sea and land, the Syracusans were stricken dumb with terror; they thought that nothing could withstand so furious an onset by such forces. But Archimedes began to ply his engines, and shot against the land forces of the assailants all sorts of missiles and immense masses of stones, which came down with incredible din and speed; nothing whatever could ward off their weight, but they knocked down in heaps those who stood in their way, and threw their ranks into confusion. At the same time huge beams were suddenly projected over the ships from the walls, which sank some of them with great weights plunging down from on high; others were seized at the prow by iron claws, or beaks like the beaks of cranes, drawn straight up into the air, and then plunged stern foremost into the depths, or were turned round and round by means of enginery within the city, and dashed upon the steep cliffs that jutted out beneath the wall of the city, with great destruction of the fighting men on board, who perished in the wrecks. Frequently, too, a ship would be lifted out of the water into mid-air, whirled hither and thither as it hung there, a dreadful spectacle, until its crew had been thrown out and hurled in all directions, when it would fall empty upon the wall, or slip away from the clutch that had held it.[25]

His devices were so effective and launched such terror in the Roman soldiers that Marcellus was forced to give up the attack, capturing the city only after a long siege. It was at the very end of the siege when the victorious Romans entered the city that Archimedes was killed, at the age of seventy-five. There are several versions as to how his death came about, but most converge on the fact that he was so engrossed in a mathematical diagram that he was unaware that the city had been taken and when approached by a Roman soldier he refused to be diverted from his inquiries, whereupon the soldier drew his sword and killed him. It is reported that Marcellus was dismayed at learning of his death.

In addition to his marvellous war machines, he invented compound pulleys, an endless screw, a water-screw for the purpose of irrigating fields and pumping water out of mines, and an orrery for representing the motions of the planets. The latter was seen by Cicero, who said that it depicted the motions of the Moon and the Sun so accurately that eclipses could be demonstrated. As regards his pulleys, he is said to have remarked to King Hieron that "given a place to stand on, I could move the earth," whereupon the king asked for some confirmation. Archimedes then devised a compound pulley which he attached to a three-masted ship loaded with passengers and cargo, such that he could easily move the ship merely by pulling one of the ropes of the pulley.

Although his legendary reputation was based on his mechanical inventions, in true Platonic style he scorned such accomplishments, placing value only on his mathematical research. Again, as Plutarch asserts,

Archimedes possessed such a lofty spirit, so profound a soul, and such a wealth of scientific theory, that although his inventions had won for him a name and fame for superhuman sagacity, he would not consent to leave behind him any treatise on this subject, but regarding the work of an engineer and every art that ministers to the needs of life as ignoble and vulgar, he devoted his earnest efforts only to those studies the subtlety and charm of which are not affected by the claims of necessity. These studies, he thought, are not to be compared with any others; in them the subject matter vies with the demonstration, the former supplying grandeur and beauty, the latter precision and surpassing power. For it is not possible to find in geometry more profound and difficult questions treated in simpler and purer terms.[26]

According to his wishes, after his death there was placed on his tomb the representation of a sphere circumscribed in a cylinder together with an inscription giving the ratio of the cylinder to the sphere, apparently his greatest mathematical accomplishment in his own eyes. When Cicero was quaestor of Sicily in 75 B.C. he discovered the tomb in a ruined state and had it restored.[27]

Since he usually began each treatise with a greeting to the friend or colleague to whom he sent the papyrus (e.g., "Archimedes to Eratosthenes greeting"), most of whom were in Alexandria, it is believed that he spent a considerable time at Alexandria studying with the successors of Euclid and conferring with the celebrated mathematicians of the Museum. In fact, it may have been during his stay in Egypt that he invented the hydraulic screw for the irrigation of fields. While he probably studied with the successors of Euclid, his own mathematical orientation and accomplishments were quite different from those of Euclid. Though the latter extended and perfected the investigations of his predecessors, the *Elements* still was based on previous mathematical inquiries. Quite different was the uncanny originality of Archimedes, who did not follow up the research of others, but who continually opened up new problems or new lines of investigation, while discovering new deductive methods or proofs. As Heath says:

Though his range of subjects was almost encyclopaedic, embracing geometry (plane and solid), arithmetic, mechanics, hydrostatics and astronomy, he was no compiler, no writer of textbooks [. . .] his objective is always some new thing, some definite addition to the sum of knowledge, and his complete originality cannot fail to strike any one who reads his works intelligently [. . .].[28]

The titles of his works, included in the edition by Heath, indicate the nature and the range of his mathematical and mechanical investigations: "On The Sphere And Cylinder," 2 books, wherein he deduces (among other things) that the surface of a sphere is four times its circumference $(4\pi r^2)$ and that a cylinder with its base equal to the circumference of a sphere and its height equal to the diameter will be 1½ times in volume; "Measurement Of a Circle" wherein he deduces by the method of approximation or one similar to integration that the value of π or the ratio of the circumference of any circle to its diameter is less than $3^1/_7$ [3.142] but greater than $3^{10}/_{71}$ [3.141] (prop. 3); "On Conoids And Spheroids" in which he again uses a method similar to integration for determining the volumes and areas of segments of various geometrical figures; "On Spirals" in which he demonstrates the various geometrical properties of spirals; "On the Equilibrium of Planes," 2 books, in which he describes the principles of levers and of equilibrium and deduces other propositions pertaining to statics; *The Sand-Reckoner*, containing the description of Aristarchus' heliocentric view and his own geometrical proof that he could express numbers exceeding in magnitude the amount of sand filling up a universe "as great as Aristarchus supposes the sphere of the fixed stars to be [. . .] ;" "The Quadrature of The Parabola;" "On Floating Bodies," 2 books on hydrostatics in which he demonstrates the "Archimedean Principle" that a body immersed in water displaces its own weight of fluid (it was on the basis of this principle that he solved the problem as to whether the crown made for King Hieron was of pure gold or of gold adulterated with silver); "Book of Lemmas;" "The Cattle-Problem," according to which it is required to find the number of bulls and cows of each of four colors or to solve a problem with 8 unknown quantities; and finally *The Method*, a palimpsest discovered in 1906 by Heiberg, of particular importance because in it Archimedes distinguishes between the method of discovery and the method of deductive proof of theorems. Unlike the other treatises in Greek mathematics where the mathematician seems to go to great pains to hide the process by which he arrived at the discovery of his theory, in *The Method* "we have a sort of lifting of the veil, a glimpse of the interior of Archimedes' workshop [or mind] as it were."[29]

Upon reading Archimedes' treatises one cannot fail to appreciate his intense fascination with mathematical problems and geometrical forms, each work containing a multitude of intricate geometrical diagrams illustrating his proofs. For without algebraic notation, algebraic functions had to be demonstrated geometrically — as they were by Galileo and Newton also. His method of demonstration combined the deductive technique of *reductio ad absurdum* with the method of integration. The former consists of assuming the negation or contradictory of what one intends to prove, expecting to derive inconsistent consequences from it and thereby prove the original proposition. The latter, the method of integration or approximation, consists in approximating areas or volumes by increasing or decreasing figures inscribed within or without the unknown figure until one has approached as closely as possible the desired quantity. That he was able with these methods to attack and solve the kinds of problems indicated displays a mathematical genius of the highest order. He was, as Pappus says, "a wonderful man, a man so richly endowed that his name will be celebrated forever by all mankind [. . .]."

EVIDENCE OF EXPERIMENTAL INQUIRY DURING THE HELLENISTIC AGE

Before concluding this chapter, as well as this phase of the study into the origins and development of scientific rationalism, it is necessary to support our contention that Hellenistic science incorporated experimental inquiry. Experimental inquiry comprises two aspects which, though complementary and usually interdependent, can be distinguished: the *first* consists of exploratory investigations of the antecedent, background conditions, usually by modifying them, to determine which of the elements, structures, or processes among these conditions brought about the occurrence of the phenomena to be explained; the *second* consists of setting up precise experiments to determine which hypothesis among several represents the truest explanation of the phenomena. Though the intent of the two procedures is different, the first concerned with acquiring additional information while the second consists of testing various hypotheses, they are complementary in that even exploratory investigations are made in terms of some theoretical expectations, while the experimental testing of hypotheses often results in the discovery of new information. The difference is that the first aspect emphasizes the experimental conditions and the latter the role of hypotheses.

While the following passages, quoted from the *Pneumatics* of Hero of Alexandria, describe very simple examples of experimentation as compared to modern science, nonetheless, they illustrate the interdependence of theoretical hypotheses and experimental tests. Although Hero lived after the Hellenistic period, his life bridging the 1st and 2nd centuries A.D., the passages quoted are thought to be based on the work of Strato. The purpose of the discussion is to demonstrate the falsity of Aristotle's denial of a vacuum, and in so doing utilizes the atomic theory along with simple experiments and direct observations.

> Some assert that there is absolutely no vacuum: others that, while no continuous vacuum is exhibited in nature, it is to be found distributed in minute portions through air, water, fire, and all other substances: and this latter opinion, which we will presently *demonstrate* to be true from sensible phenomena, we adopt. Vessels which seem to most men empty are not empty, as they suppose, but full of air. Now the air, as those who have treated of physics are agreed, *is composed of particles minute and light, and for the most part invisible.* If, then, we pour water into an apparently empty vessel, air will leave the vessel proportioned in quantity to the water which enters it. This may be seen from the following *experiment*. Let the vessel which seems to be empty be inverted, and, being carefully kept upright, pressed down into water; the water will not enter it even though it be entirely immersed: so that it is manifest that the air, being matter, and having itself filled all the space in the vessel, does not allow the water to enter.[30] (Italics added)

Two factors are involved in the discussion: (1) the experimental evidence that supposedly 'empty' vessels really contain air which, though unseen, is the 'demonstrated' reason the vessels cannot be filled as long as they contain this 'matter;' and (2) the theoretical explanation that the reason the air can prevent the water from entering the vessels is that it is composed of 'minute particles.' Following this initial description and explanation, the experimental conditions are changed by boring a hole in the bottom of a vessel to further 'demonstrate' that it is the presence of air in the vessel that prevents the water from entering.

> Now, if we bore the bottom of the vessel, the water will enter through the mouth, but the air will escape through the hole. Again, if, before perforating the bottom, we raise the vessel vertically, and turn it up, we shall find the inner surface of the vessel entirely free from moisture, exactly as it was before immersion. Hence it must be assumed that the air is matter.[31]

The hypothesis that it is the air in the vessels that prevents the entrance of the water receives additional confirmation from the fact that when the air is allowed to escape the water enters. Furthermore, although the air cannot be seen, its existence can be confirmed because if "we place the hand over the hole, we shall feel the wind escaping from the vessel [...]."[32] Finally, since it is postulated that the resistance and pressure of the air depend upon minute particles which presuppose a 'void' in which to exist, the original hypothesis that "there exists in nature a distinct and continuous vacuum, but that it is distributed in small measures through air and liquid in all bodies"[33] has been indirectly confirmed.

In true scientific fashion, the theory that air is composed of minute, invisible particles is now extended to explain additional phenomena such as the compression and expansion of air, horn shavings, and sponges. The comparison of the behavior of the invisible particles in the air with grains of sand also illustrates an excellent use of analogical reasoning.

> The particles of the air are in contact with each other, yet they do not fit closely in every part, but void spaces are left between them, as in the sands on the seashore: the grains of sand must be imagined to correspond to the particles of air, and the air between the grains of sand to the void spaces between the particles of air. Hence, when any force is applied to it, the air is compressed, and, contrary to its nature, falls into the vacant spaces from the pressure exerted on its particles: but when the force is withdrawn, the air returns again to its former position from the elasticity of its particles, as is the case with horn shavings and sponge, which, when compressed and set free again, return to the same position and exhibit the same bulk.[34]

The conclusion of the argument is then maintained, along with the assertion that the "appeal to sensible phenomena" is superior to arguments which "offer no tangible proof."

> They, then, who assert that there is absolutely no vacuum may invent many arguments on this subject, and perhaps seem to discourse most plausibly though they offer no tangible proof. If, however, it be shewn by an appeal to sensible phenomena that there is such a thing as a continuous vacuum, but artificially produced; that a vacuum exists also naturally, but scattered in minute portions; and that by compression bodies fill up these scattered vacua, those who bring forward such plausible arguments in this matter will no longer be able to make good their ground. [35]

As is true of most theoretical explanations, the "appeal to sensible phenomena" is not a *direct* confirmation of the original theory, but *indirect* evidence in support of it. In the above example, the 'feel of the escaping air' serves as a "bridge principle" or as indirect empirical evidence for the two "internal principles" or theoretical hypotheses, namely, that the vessel contains air which in turn consists of minute particles which explains why the air prevents the water from entering the vessel. The latter hypothesis in turn is taken as evidence of the fact that any substance consists of 'scattered vacua,' since the particles comprising the substance require vacua in which to exist. Thus even a relatively simple example such as the above involves a rather complex interrelation between theoretical hypotheses, experimental conditions, and empirical evidence.

The examples could be multiplied, although most of them would be taken from the centuries immediately following the Hellenistic period because most of the physical writings of the Hellenistic scientists have been lost. There is, however, a clear illustration of quantitative experimentation in physiology attributed to Erasistratus.

> Clearly, then there are emanations from beasts. Erasistratus tries to show this also as follows. If one should take an animal, for example a bird, and keep it in a vessel without food for a given period of time, and then weigh the animal along with the visible excreta, the weight will be found to be much less because of the invisible passage of considerable effluvia. [36]

This is a much simpler example but it still illustrates the test of a hypothesis in relation to varying conditions, and implies that the experiment had been performed.

A final passage, although not referring to an experiment per se, describes scientific inquiry in such a way as to indicate again a precise understanding of the interdependence of theoretical conjectures, experimental tests, and observations. The reference to the theory of 'the four elements' is analogous to the use of biomolecular explanations in medical research today.

> They, then, who profess a reasoned theory of medicine propound as requisites, first, a knowledge of hidden causes involving diseases, next, of evident causes [. . .]. They term hidden the causes concerning which inquiry is made into the principles composing our bodies, what makes for and what against health. For they believe it impossible for one who is ignorant of the origin of diseases to learn how to treat them suitably. They say that it does not admit of doubt that there is need for differences in treatment, if, as certain of the professors of philosophy have stated, some excess, or some deficiency, among the four elements, creates adverse health; or, if all the fault is in the humours, as was the view of Herophilus; or in the breath, according to Hippocrates; or if blood is transfused into those blood vessels which are fitted for pneuma, and excites inflammation [. . .] and that inflammation effects such a disturbance as there is in fever, which was taught by Erasistratus; or if little bodies by being brought to a standstill in passing through invisible pores block the passage, as Asclepiades contended — his will be the right way of treatment, who has not failed to see the primary origin of the cause. They do not deny that experience is also necessary; but they say it is impossible to arrive at what should be done unless through some course of reasoning. For the older men, they say [. . .] reasoned out what might be especially suitable, and then put to the test of experience what conjecture of a sort had previously led up to. [37]

The first part of the quotation refers to the importance of "hidden causes" in medical explanations and then presents from Aristotle (the four elements), Herophilus, Hippocrates, Erasistratus, and Asclepiades examples of such conjectured causes. The last part of the statement indicates the importance of 'testing by experience' the various conjectures. Brief as this account is of the evidence of experimental science during the Hellenistic period, it should be sufficient to dispel the erroneous belief that Greek science was merely qualitative and descriptive.

The above quotation also illustrates how much progress had been made in refining the concept of scientific method since Aristotle. In fact, the Hellenistic period in general represents the natural progression of scientific inquiry within the leading paradigms established in the 4th century. Now, however, the awareness of the interdependence of theoretical hypotheses, experimental tests, and observable evidence is much more explicit and precise. The stage is set for the continued exploration and explanation of natural phenomena similar to the remarkable progress made in science in the 16th and 17th centuries, but the performance was not to take place. The histrionics of the Romans were to be displayed in other areas than mathematical and scientific achievements, while the predominately theological interests of the Middle Ages detracted from any extensive investigation of nature. One would have to await the modern scientific revolution a millenium and a half later before the promise of the Hellenistic Period, as regards the continued development of scientific rationalism, would be fulfilled.

NOTES

CHAPTER XV

[1] O. Neugebauer, *The Exact Sciences in Antiquity*, 2nd ed. (*op. cit.*, ch. XI), p. 1.

[2] George Sarton, *A History of Science*, Vol. II (Cambridge: Harvard Univ. Press, 1959), p. 25.

[3] *Ibid.*, p. 34.

[4] *Ibid.*, pp. 143-144.

[5] *Ibid.*, p. 155.

[6] Sir Thomas L. Heath, *The Works of Archimedes With The Method of Archimedes* (New York: Dover Publications, Inc., 1953), p. 6.

[7] Proclus on Eucl. I, p. 68, 6-20. Quoted from Sir Thomas L. Heath, *A History of Greek Mathematics*, Vol. I (*op. cit.*, ch. XI), p. 354.

[8] Dirk J. Struik, *A Concise History of Mathematics*, 3rd rev. ed. (New York: Dover Publications, Inc., [1948] 1967), pp. 50-51.

[9] Albert Einstein, "Autobiographical Notes," in Paul A. Schilpp, ed., *Albert Einstein: Philosopher-Scientist* (Evanston: The Library of Living Philosophers, Inc., 1949), p. 9.

[10] For a complete annotated account of the *Elements*, the reader is referred to Sir Thomas L. Heath's translation, *The Thirteen Books of Euclid's Elements*, 3 Vols. (New York: Dover Publications, Inc., 1956).

[11] Sarton, *op. cit.*, p. 39.

[12] *Ibid.*, p. 39. Also, Sir Thomas L. Heath, *Mathematics in Aristotle* (Oxford: At the Clarendon Press, 1949).

[13] Cf. Heath, *A History of Greek Mathematics*, Vol. I, *op. cit.*, p. 446.

[14] Neugebauer, *op. cit.*, p. 91.

[15] *Ibid.*, p. 81.

[16] *Ibid.*, p. 34.

[17] *Ibid.*, p. 48.

[18] Cf. Carl B. Boyer, *The History of the Calculus and Its Development* (New York: Dover Pub., Inc., 1959), ch. II.

[19] Cf. Ch. XII.

[20] It was not the same view as that of Tycho Brahe, because on Brahe's view all the planets (except the moon) revolve around the sun which in turn revolves around the stationary earth. For Heraclides, only the inferior planets revolve around the sun and the earth is not stationary but rotates on its axis.

[21] Heath, *The Works of Archimedes With The Method of Archimedes, op. cit.*, pp. 221-222.

[22] Cf. Sarton, *op. cit.*, p. 59, f.n. 10.

[23] Cf. Sir Thomas L. Heath, *Aristarchus of Samos* (*op. cit.*, ch. XI), p. 377.

[24] Sarton, *op. cit.*, p. 56.

[25] Plutarch, *Life of Marcellus*, quoted from Morris Cohen and E. Drabkin, *A Source Book in Greek Science* (Cambridge: Harvard Univ. Press, 1948), p. 316.

[26] *Ibid.*, p. 317.

[27] Cf. Sarton, *op. cit.*, p. 71.

[28] Heath, *The Works of Archimedes With The Method of Archimedes, op. cit.*, p. xxxix.

[29] *Ibid.*, p. 7 of the Introductory Notes to "The Method of Archimedes."

[30] Cohen and Drabkin, *op. cit.*, p. 249.

[31] *Ibid.*, p. 249.

[32] *Ibid.*, pp. 249-250.

[33] *Ibid.*

[34] *Ibid.*, p. 250.
[35] *Ibid.*, p. 251.
[36] *Ibid.*, p. 480.
[37] *Ibid.*, p. 470. Brackets added.

EPILOGUE

Now that an end to this phase of the study has been reached and we are positioned to glance back over the singular achievement of the ancient Greeks in casting off the primitive framework of mythopoeticism and pre-scientific modes of thought to attain a more realistic understanding of the universe, a number of questions emerge that previously were outside our focus of attention. Why, of all the peoples of the world, was it given to a handful of Greeks to seek a glimpse of the features of nature by lifting the veil of myth? In all the aeons of past time, why did this begin in the 6th century B.C. rather than five centuries earlier or three centuries later? Was the awakening predestined or was it merely accidental? Can it be explained as a natural byproduct of a certain intellectual or cultural development so that if that level had been attained in any other locale the birth of objective thought would have occurred there, or were the conditions in Ionia so unique that it could have happened only at that place and time? That is, was the genetic pool in that little area of the world so unusual that it alone was capable of producing a Thales followed by a succession of "pupils and associates" that began the tradition of scientific rationalism, a tradition that has done more to transform the world and man's thinking about it than any other?

Since other peoples did not generate a parallel tradition, could this one have been merely fortuitous or was there something inherent in the composition of the universe itself that made it inevitable? Given the particular world in which we have evolved and the peculiar beings that we are, was the mode of scientific thought that developed the only possible one or could there have been other forms? Was there a kind of conspiracy between nature and nurture that made the type of scientific inquiry as we now know it necessary? If the primary function of our sensory neural system is to process the data supplied by our sense organs to facilitate man's adaptation to the external world, why should subjective fantasy have superseded objective evidence in the early stages of man's conscious awakening? As we look back over this early development it appears as if certain transitions and stages in the cognitive functioning of mankind were inevitable: e.g., the replacement of animistic, anthropomorphic concepts with more naturalistic ones; the progression from a syncretic, egocentric form of thought to a more conceptual, decentered one; the grudging recognition of the significance of empirical observations for testing theories; the gradual realization of the necessity of supplementing qualitative descriptions with experimentation and inferred constructs, etc. — but is the apparent necessity of this development merely a retrospective illusion? Are there in distant reaches of the universe other creatures who, because of their own particular environment and evolutionary history, have undergone vastly different cognitive developments producing radically diverse conceptions of the world which appear as inevitable to them as ours does to us?

What would have been the course of human history had the type of intellectual development described in this book never occurred? Would it have been better for man to have left his destiny in the hands of the gods rather than attempt to wrest some control over this himself? Has the drive toward rational understanding and technological control, particularly in the West, led to a denial or atrophy of other kinds of experiences, such as the parapsychological, intuitive, or mystical, that are equally valid approaches to reality?

Is there a current imbalance in the psyche of Western man owing to the predominance of the rationalistic, scientific tradition?

There are, of course, no answers to most of these questions and that is why they were left in the periphery of this investigation, but acknowledging them now makes one aware how little is ever settled in such inquiries — for every question one faces many more remain unanswered and these usually are the more fundamental. Yet if one were to be put off by this fact no questions would be posed and no progress made in answering them. Thales and Anaximander must have been aware how slight their own solutions were in contrast to the enormous problems they were raising as to what was the first principle or state of things, how did everything arise from this earlier condition, and what controlled the process of coming-to-be and change. These basic inquiries still confront us today and their resolution is as elusive as ever. Contrary to what Wittgenstein asserted, the riddle *does* exist and in all likelihood will remain unsolved.

Though the questions raised by the Milesians seem to be perfectly natural, they are not necessarily what an unbiased query of nature would suggest, but appear to have been inadvertently abstracted from the earlier mythopoetic or theogonic traditions. Once posed, however, we have seen how the Presocratics successively differentiated the problems and refined and extended their answers to them. In Anaximander, particularly, we saw the ancient theogony replaced by a cosmogony in which the first rough outline of the structure of our (galactic) universe appeared, utilizing terrestrial models to explain celestial and meteorological processes. Notwithstanding the legacy of animism and the continued concreteness and diffuseness of thought, one can identify in Anaximander a type of reasoning characteristic of all scientific thinking.

Lacking as yet any one settled tradition of inquiry, various other paradigms or schools of thought arose utilizing radically different abstractions in their attempts to comprehend and explain the universe: Heraclitus' metaphysics of fire and flux; the arithmogeometric cosmology of the Pythagoreans with their farsighted astronomical conjectures; the static, logico-ontological system of the Parmenideans; the efforts of the Pluralists — Empedocles, Anaxagoras, and Democritus — to explain the complexity of phenomena in terms of underlying, irreducible elements, etc. Supporting each of these systems was a basic theoretical discovery, such as the numerical ratios underlying the musical scale, the law of contradiction, or the attempt to avoid violating that law by attributing becoming and change to the rearrangement of the indestructible elements. In place of the hegemony of the gods over the universe, one can discern in the primitive concepts of Logos, Harmony, Love and Strife, Nous, and Necessity a gradual approximation to the conception of a natural law 'governing' nature. Anaxagoras even carried the process of naturalizing phenomena so far as to surmise that the moon had mountains and ravines similar to the earth, concluding that it was either a fragment of the earth or made of the same material.

But the demythologizing process was carried furthest by the Atomists who completely divested their theorizing of all animistic, mythopoetic elements. Not a trace of the archaic, primitive forms of thought can be found in their atomic theory and steady state conception of the universe, proving that in all cultures there are people capable of the highest form of objective, rational thought. For it was they who created the conceptual foundation of the most successful scientific theory of all time, the atomic theory.

Unlike the Atomists, Plato exemplified the Popperean-like notion that all thinking is merely conjecture which led him to offer the boldest speculations, rich in content and insight, but low in truth value. He was the poet laureat of that era of cosmological speculation, creating his highly imaginative — albeit animistic and anthropomorphic — cosmology in the form of "a likely story." Yet Plato has earned everlasting distinction because he was the first to conceive of an independent realm of abstractions as the heuristic basis of mathematical discoveries, absolute values, and objective truth. While a continuation of the primitive tendency to reify concepts, his conception of a transcendent realm of forms represents the first clear distinction between *a priori* and *a posteriori* truths, a distinction that has played such an important role in all subsequent theories of knowledge. Moreover, Plato had a marvellous gift of prophecy as illustrated in his inspired comparison of the Good with the Sun as the source of the illumination, value, and being of Truth, and particularly in the "Allegory of the Cave" which, in contrasting the illusory beliefs of the senses with a more enlightened form of knowledge, forecast the Copernican Revolution. It was because of this extraordinary foresight that Neoplatonism, along with Neo-Pythagoreanism, had such an important effect on the thinking of Copernicus and Kepler.

But it was the towering figure of Aristotle that cast the longest shadow over centuries of intellectual history. While Plato and the Pythagoreans stood for the essential function of mathematics in scientific inquiry and the Atomists for the necessity of utilizing theoretical constructs in explaining phenomena, Aristotle represented the crucial component of sensory observation: i.e., the indispensable role of empirical observations and facts along with mathematical extrapolations and theoretical interpretations in empirical investigations. Even though interpreted within a highly animistic and egocentric framework, Aristotle's respect for observed evidence led him to make the extensive empirical inquiries that served as a first approximation for later scientific investigations. In addition, his extraordinary capacity for analyzing phenomena, clarifying problems, and providing conceptual distinctions and classifications was a further reason his systematic inquiries were often the basis of later scientific research. Finally, the all inclusiveness and coherence of his cosmological system in which the manifold aspects of nature seemed to have their natural places and proper explanations was such a colossal achievement that it overshadowed all other traditions for five centuries. Even now, when Aristotle's scientific concepts have been superceded, one cannot but stand in awe at the power and range of his thought as well as the majesty of his cosmology.

An important reason contributing to the renown of Plato and Aristotle was the fact that they left not just their writings, but a physical legacy in their respective schools for carrying on their research activities. The successful achievements of the schools, mainly in mathematics and astronomy in the Academy and in empirical research in the Lyceum, served both as an inspiration and a model for the famous Museum and Library in Alexandria. It was owing in no small measure to these latter institutions that the speculative, metaphysical, and cosmological inquiries of the 4th century were followed by more specific inquiries and research projects. The Museum and Library not only supported independent research among the resident scholars, they also promoted the exchange of discoveries so important for the crossfertilization of ideas. It was during the two centuries of the Hellenistic period that such luminaries in mathematics and science as Euclid, Apollonius, Heraclides, Hipparchus, Aristarchus, Herophilus, Erasistratus, Erathosthenes, and Archimedes appeared, the latter considered one of the greatest geniuses of all time. These were the scholars who, pursuing narrower topics of research than their 4th century predecessors, made new discoveries and advanced areas of research that provided the background for the later inquiries in the Renaissance.

As a result, the forerunners of each of the major figures, e.g., Copernicus, Fabricius, Kepler, Harvey, and Galileo, in the later scientific revolution can be traced back to this period. It was in the Hellenistic Age primarily that the transition from descriptive to mathematical astronomy and from observational to experimental mechanics and physiology took place. Furthermore, the search for the proof of individual theorems and

Euclid's organization of geometry into a deductive axiomatic system based on prior definitions, postulates, and axioms had a lasting effect on mathematics and theoretical physics. Similarly, Archimedes' brilliant investigations of the areas and volumes of various geometrical figures by the method of exhaustion, as well as his solution of specific problems in mechanics, established the highest precedent for mathematical research in these fields. And, although he minimized his more empirical experiments in statics and hydrostatics, in spite of discovering the first precise physical law in the history of science, and disparaged his marvellous engineering feats, Archimedes demonstrated once and for all the immense practical consequences that could be derived from purely theoretical research in mathematics and physics.

All of which demonstrates that it is primarily to the Hellenic and Hellenistic heritage that we owe the origins, development, and early achievements of scientific rationalism. If, as Einstein claimed, "the most incomprehensible fact about the world is that it is comprehensible," then we are indebted to the ancient Greeks for bringing this to light. Their undying legacy is that the lure of the unknown, although precarious and perilous, is also irresistible and indomitable.

Enuma Elish, *39-41*, 55, 66
Epicurus, 147, 150, 155, 156
Erasistratus, 254, 265, 266, 271
Erastus, 208
Eratosthenes, 92, 251, 255, 262, 271
Eros, 65, 66, 73, 110, 124
Euclid, 253, *254-256*; Elements, *255-256*, 267; 258, 271
Eudemus, 182
Eudoxus, 170, 181, 184; concentric spheres, 239, 253; method of exhaustion, 254; 240,
    255, 256, 258, 260, 261
Euripides, 134
Fabricius, 222, 246, 254, 271
Feigl, H., 23
felting or condensation, 79, 80
Feuerbach, L., 162
Feyerabend, P., 12, 16, 23, 29
FitzGerald, G. F., 68
four elements in Empedocles, 123-124; in Plato, 197-199; in Aristotle, 229
Frankfort, H. A. and H., 35, 36, 43, 46, 47, 49, 52, 53, 55, 59
Frazer, G., 31, 43
Freeman, K., 82, 103, 120, 121, 126, 133, 146, 167
Frege, G., 174
French School of anthropology, 31, 62
Freud, 54, 58, 180
Friedländer, P., 186
Galen, 250, 254
Galilean-Newtonian transformations, 13
Galileo, 11, 16, 19, 20, 24, 27, 29, 42, 52, 58, 74, 93, 100, 101, 102, 103, 134, 142, 156,
    160, 164, 167, 194, 220, 221, 222, 224, 226, 237, 238, 246, 247, 248, 249, 254,
    259, 263, 271
Gamow, G., 166
Gassendi, 24, 147, 160, 203
Gauss, K. F., 256
Gödel, K., 34
Graves, J. C., 11, 120
gravity or gravitational fields, 21, 24, 27; in Aristotle, 235
Grünbaum, A., 112, 114, 115, 120
Guthrie, W. K. C., 82
Hanson, N., 12, 23, 29
Harvey, 246, 254, 271
Heath, T., 75, 168, 170, 184, 186, 187, 240, 249, 252, 256, 261, 262, 267
Hegel, 40, 87, 107, 244
Heidegger, M., 105
Heisenberg, W., 14, 27, 92, 232, 248
Hellenic period, 23, 207, 254, 271, 272
Hellenistic period, 11, 15, 150, 252, 253, 254, 266, 270, 271, 272
Hempel, C., 11
Heraclides of Pontus, 68, 100, 170, 181, 253, 254, 258, 259, 271
Heraclitus, 18, 26, 27, 40, 66, *85-90*, 95, 107, 122, 124, 125, 130, 175, 176, 177, 190,
    198, 235, 270

physis, 23, 67, 68, 69, 70, 78, 80, 85, 94, 106, 163

Piaget, J., *11-14*, 16, 19, 25, 29; directed thought, 36; autistic thought, 36; 41, 44, 47, 51, 55, 59, 60, 133, 181

Planck, M., quantum of action, 20, 57

Planets, 13, 17, 18, 19, 25, 52, 57, 192; in Plato, 194-197; in Aristotle, 238-243; in Aristarchus, 259

Platonic solids, 201-202, 256, 258

Plato, 14, 15; Demiurge or Divine Craftsman, 18, 47, *190-192*, 193, 195, 196, 197, 200, 201; 20, 47; theory of Forms, 55, *171-177*, 198, 207, 226; 56, 72, 73, 78, 79, 80, 84, 86, 90, 91, 92, 94, 104, 111, 120, 134, 147, 148, 150, 153, 154, 163, *168-187;* *anamnesis*, 171; *epistemē*, 173; *doxa, 173; noesis*, 174; dialectic, 176, 192; method of collection and division, 176, 179, 192, *188-206*, Battle of the Gods and the Giants, 189-190; World-Soul, 192-197, 200; Receptacle, *197-204*, 200; geometric description of the four solids and their transformations, 201-204; 207, 208, 210, 212, 213, 214, 220, 221, 222, 226, 228, 231, 246, 247, 254, 257, 259, 270, 271; *Epinomis*, 170, 181, 182, 184, 187; *Laws*, 181, 182, 187, 190, 192, 205, *Meno*, 171, 184, 186; *Parmenides*, 172, 186, 187; *Phaedo*, 170, 171, 172, 186, 189, 205; *Phaedrus*, 179, 187, 192; *Philebus*, 190, 205; *Republic*, 169, 170, 173, 174, 175, 176, 181, 182, 186, 187, 190; *Theaetetus*, 200, 205; *Timaeus*, 47, 169, 175, 184, 186, 188, 189, *190-204*, 205, 212

Plotinus, 173

Pluralists, 197, 169, 270

Plutarch, 100, 131, 184, 161, 267

Poincaré, H., 37, 68

Polanyi, M., 12

Popper, K., 11, 168; doctrine of 3 Worlds, *179-181*; 187, 271

Positivists, 23, 219

Presocratic philosophers, 11, 13, 14, 15, 26, 30, 270

Price, H. H., 22

primitive mentality, 11-12, 30-31, 34-41, 45, 49

principle of sufficient reason, 76, 109

Proclus, 255, 267

Psyche, 80, 84

Pseudo-Plutarch, 73, 74, 76

Ptolemies: Ptolemy I Soter, 250, 251, 253, 255; Ptolemy II Philadelphus, 250, 251, Ptolemy III Evergetes, 251; Queen Cleopatra, 250

Ptolemy, 24, 25, 27, 99, 194, 246, 250, 254, 258, 260

Pythagoras or Pythagoreans, 15, 26, 40, 80, *91-103*, 104, 107, 108, 110, 113, 117, 122, 128, 135, 151, 172, 174, 175, 177, 192, 201, 203, 211, 212, 220, 226, 237, 246, 247, 254, 255, 257, 258, 270, 271

quantum mechanics, 13, 19, 22, 24, 92, 234

Quine, W. V., 24, 25, 29, 120, 140, 146, 178, 187, 208

Quintilian, 208

Radcliffe-Brown, A., 34

Radin, P., 30, 31, 33, 34, 35, 37, 41, 43, 46, 51, 59

Randall, J. H., 211, 215, 220, 223, 228, 232, 234, 239, 248, 249

Reichenbach, H., 11, 16

Riemannian geometry, 92, 256

Ross, W. D., 209, 223, 248